PARKLAND COLLEGE LIBRARY
3 3036 00240 7513

D1769878

HEATING AND VENTILATION

RIETSCHEL-BRABBÉE

HEATING

AND

VENTILATION

A HANDBOOK FOR ARCHITECTS AND ENGINEERS

BY

C. W. BRABBÉE

Translated for American use from the Seventh German Edition of Rietschel-Brabbée "Heizungs-und Lüftungstechnik"

FIRST EDITION

McGRAW-HILL BOOK COMPANY, INC.
NEW YORK: 370 SEVENTH AVENUE
LONDON: 6 & 8 BOUVERIE ST., E. C. 4
1927

PRINTED IN THE UNITED STATES OF AMERICA

THE MAPLE PRESS COMPANY, YORK, PA.

PREFACE TO THE AMERICAN EDITION

Modern Europe has recognized the fact that the scientific development of heating and ventilation practice was inaugurated by the late Dr H Rietschel, Professor of the Technical University of Berlin—Charlottenburg His life work, which is the basis of this book, was printed in five editions French translations have been published.

After the death of Dr Rietschel the work of revision was carried out by the undersigned, who succeeded him in the Chair of Heating and Ventilation at the University of Berlin—Charlottenburg Much development work was undertaken at this University's Research Laboratory of Heating and Ventilation, of which the writer was director for fifteen years The results of these studies, together with simplified formulas for the design of heating and ventilating systems, were included by the author in the sixth and seventh German editions which he prepared The latter have been used extensively in a great many practical installations, both large and small, with uniformly good results

European engineers are in the habit of approaching the subject of heating and ventilation from both the physical and the physiological viewpoints and use mathematical methods extensively to insure economy in design and in operation.

The industry in America has reached a high standard of perfection in the development of its systems. American literature on the subject shows uninterrupted progress In order to contribute to this progress by making available for American heating science and practice the European viewpoints, it was thought desirable to have the seventh German edition translated into English

On this occasion the author has taken certain liberties in the revision. Those portions of the German text which do not apply to American practice have been abridged, others adaptable to conditions in the United States have been emphasized for the convenience of American engineers

The original charts have been improved for rapid computation and the two volumes of the latest German text consolidated into a single book However, the spirit of the "old standard work' has been retained

The author desires to thank Mr Alphonse A Adler for the valuable assistance and painstaking attention he has given to this American edition

C. W. Brabbée

New York, N Y.
October, 1927

PREFACE TO THE GERMAN EDITION

When, at the suggestion of His Excellency the Secretary of Public Works, I undertook the writing of this text on the design of heating and ventilating systems, I did so because of the apparent need of a definite treatment which was not too comprehensive.

The books available in the field of heating and ventilation, while undoubtedly planned to assist the engineer, fail in the matter of designing practical installations The broad discussion of the theory and design often obscures the general perspective and so lacks the terseness required for ready application

This text is to serve practice, only those theoretical discussions are embodied which are required for the proper application of materials to the desired objective

From my experience as consultant, I find a lack of proper relationship between the work involved in the preliminary design for bidding purposes and the final completion of the project In the former, much energy is required of the engineers due to the common insistence on an unnecessarily large number of drawings, specifications, estimates, etc On the other hand, for final installation too little is required of the engineer in the way of technical and hygienic knowledge with faulty systems as the result

In the field of heating and ventilation many phases are still beyond the realm of scientific analysis, as far as possible these topics should be studied for the benefits that might accrue Scientific methods alone can give us the assurance that we are headed in the right direction, particularly in those departures from standard practice occasionally necessary in everyday application

In compiling this text I desired not only to inform owners and architects of the true setting of the problems but also to equip the engineer with rapid means of computation for his designs The treatment of the text material and the Tables contained in Part II will confirm this In numerous examples, the principles discussed are applied to problems arising in practice.

The drawings supplied with this text offer an insight into numerous applications, some of them being designs of the more important installations To limit the scope of the book only the more important descriptions are included It therefore presupposes a certain acquaintance with the subject on the reader's part RIETSCHEL

BERLIN, GERMANY
April, 1893

CONTENTS

PAGE

PREFACE TO THE AMERICAN EDITION v

PREFACE TO THE GERMAN EDITION . . vii

PART I

HEATING

SECTION

I Introduction 3

(A) General, (B) Inside and Outside Temperatures

II Methods of Heating 5

(A) Fireplaces, (B) Conduit Wiring Systems, (C) Iron Stoves, (D) Tile Stoves, (E) Oil Stoves, (F) Gas Heating, (G) Electric Heating

III Central Heating 20

(A) General, (B) Medium-pressure Hot-water Heating, (C) High-temperature Hot-water Heating, (D) Steam Heating, (E) Vacuum Heating Systems, (F) Combination Heating Systems, (G) Warm-air Heating, (H) District Heating, (I) Exhaust Heat Ventilation.

PART II

VENTILATION

I Necessity of Ventilation 121

(A) Introduction, (B) Heat and Moisture Emission from Persons and Illumination, (C) Emission of Carbon Dioxide from Individuals and Animals, (D) Poisonous Effluvia (Ammoniae, Anthropotoxin), (E) Dust

II Required Air Changes 127

(A) General, (B) Temperature Requirements, (C) Humidity Requirements, (D) Carbon Dioxide Requirements, (E) Heat Content of Air, (F) Pressure Requirement, (G) Air Changes Determined by Experience.

III Means of Insuring Air Changes 131

(A) Pressure Distribution in a Closed Room, (B) Non-mechanical Room Ventilation, (C) Ventilation by Gravity Circulation, (D) Ventilation Systems with Forced Circulation

IV Cooling of Rooms 154

(A) General, (B) Cooling Methods

PART III

THE DESIGN OF HEATING SYSTEMS

I Heat Transmission 159

(A) Heat Transmission through Materials for the Steady State, (B) Heat Loss Prior to Attaining Steady State, (C) Detailed Computation of Heat Loss According to Rietschel, (D) Determination of Heat Loss from Cubical Contents

SECTION PAGE
II Computation of Heating Surface 178
(A) General Theory, (B) Countercurrent and Parallel-current Apparatus, (C) Application of Equations
III Piping Equations 184
(A) Introduction, (B) General Theory, (C) Summary of Formulas
IV Design of Various Types of Heating Systems 188
(A) Local Heating Plants, (B) Hot-water Heating, (C) High-pressure Steam Heating, (D) Low-pressure Steam Heating Systems, (E) Vacuum Heating Systems, (F) Combination Heating Systems, (G) Warm-air Heating Systems.

PART IV

VENTILATION SYSTEMS

I. Determination of the Required Air Changes 257
(A) Heat Standard, (B) Humidity Standard, (C) Carbon Dioxide Standard, (D) Pressure Standard
II Formulas for the Design of Details of Ventilating Systems 262
(A) Filters, (B) Apparatus for Humidifying, Washing, and Drying Air, (C) Heating Surface, (D) Duct Systems, (E) Fans
III Cooling & Drying Air 281
IV. Approximation 283

LIST OF TABLES

TABLE
I Diameter, Weight, External Surface and Value of $\frac{k}{l}$, 287
II Latent Heat of Evaporation of Water for Temperature to 212° F 287
III Pressure, Temperature, etc, of Water Vapor 288
IVa Coefficient of Heat Conductivity 289
IVb Coefficients of Heat Conductivity of Insulating Materials 290
V Radiation Constant C of Building Materials 291
VI Heat Transmission Coefficients 292
VI Heat Transmission Coefficients (Continued) 293
VI Heat Transmission Coefficients (Continued) 294
VI Heat Transmission Coefficients (Continued) 295
VI Heat Transmission Coefficients (Continued) 296
VII Allowances for Special Conditions 297
VIII Heat Transmission Coefficients K for Water Heated Surfaces 298
VIII Heat Transmission Coefficients K for Water Heated Surfaces (Continued) 299
IX Heat Transmission Coefficients K for Steam Heating Surfaces 300
IX Heat Transmission Coefficients K for Steam Heating Surfaces (Continued) 301
Xa Vento Hot-blast Heaters 302
Xb Final Temperatures and Condensations 303
Xc Friction of Air through Vento Hot-blast Heaters 304
XI Steam at 227° F and 5 lbs Press 306
XII Coefficient K of Heat Transmission of Air or Flue Gases through Thin Sheet Iron Surfaces to Air 306
XIII Density of Water at Temperatures from 100–212° F 307
XIV. Additional Pressure Heads and Increase of Heating Surface for a Hot-water Heating System with Overhead Distribution, Taking the Heat Loss of the Piping into Consideration 308

TABLE		PAGE
XIV	Additional Pressure Heads and Increase of Heating Surface for a Hot-water Heating System with Overhead Distribution, Taking the Heat Loss of the Piping into Consideration (Continued)	309
XIV.	Additional Pressure Heads and Increase of Heating Surface for a Hot-water Heating System with Overhead Distribution, Taking the Heat Loss of the Piping into Consideration (Continued)	310
XV	Loft Heating Systems Table for Preliminary Design	311
XVI	Proportion of the Resistance of the Fittings to the Total Resistance of a Piping System	312
XVII	Diameter of Condensation Pipe Lines for Steam Heating Systems	312
XVIII		313
XVIII		314
XVIII		315
XIX	Volume of Dry Air Necessary to Supply or Remove 1,000 B t u per Hour	316
XIX	Volume of Dry Air Necessary to Supply or Remove 1,000 B t u per Hour (Continued)	318
XX	Hourly Air Change in Cu Ft at Room Temperature for Fully Occupied Rooms Based on the Humidity Standard and the Following Conditions	320
XXI	Moisture in Pounds to be Supplied to 1,000 Cu Ft of (Room) Air, Taken from Outside to Raise the Humidity to 50 Per Cent after Heating	321
XXII		322
XXIII*a*	Stove Heating Surface in Sq Ft Required for Rooms with Windows Computed by Means of Length of the Outer Wall Single	323
XXIII*b*	Stove Heating Surface in Sq Ft Required for Rooms with Double Windows Computed by Means of the Length of the Outer Wall	324

List of Charts[1]

CHART

I Gravity Hot Water Heating System Temperature Drop = 30° F

II Forced Hot Water Heating Systems

III High Pressure Systems

IV Vacuum Steam

V. Low Pressure Steam

VI Air Heating and Ventilation Systems

VII Equivalent Diameters d_e and Crosssectional Areas for Rectangular Ducts

[1] See inside back cover for these charts

PART I
HEATING

HEATING AND VENTILATION

SECTION I

INTRODUCTION

A. GENERAL

At the present time solid, liquid, and gaseous fuels are used for heating Their characteristics and chemical reactions in the process of combustion are outside the scope of this text

The computations required in detailed design of installations are treated in the second part of this work The subdivisions of that part are as follows:

Calculation of heat loss from buildings

Computing sizes of boilers and radiation

Piping layouts for heating and ventilating systems

Determination of air duct and chimney sizes

The first part considers:

Inside and outside temperatures

Forms of heating systems

In what follows it is assumed that architects and engineers in general are not required to build boilers, radiators, ventilating equipment, pumps, motors, etc. But it is important that they be in a position to criticise and appraise the value of such equipment in the practice of their profession It is not the function of this text to give methods of construction and manufacture of heating apparatus Those phases of design of apparatus must be considered, however, which apply to heating installations.

B. INSIDE AND OUTSIDE TEMPERATURES

Sick rooms	68–72° F
Living rooms (servants' rooms)	65–68° F
Halls, lecture rooms, prisons, and business offices	65° F
Public meeting halls, and exhibition rooms, entrances, stairways, according to use	50–65° F
Churches	50° F
Jails, overnight use only	50° F
Bathrooms	72° F
Dressing rooms, kitchen, and stalls	59° F
Toilets ..	At the temperature of the anteroom
Greenhouses	According to need, 50–86° F

The temperatures given above should be measured in the center of the room or along the warm inner walls at eye level (5 ft from the floor) [1] It should be noted that the required room temperature depends upon weather conditions For example, 65° F is sufficient for lecture halls during winter On the other hand, during the summer months the same temperature would be decidedly too cool The required room temperature is also dependent upon the relative humidity and the air movement With higher relative humidity the room temperature may be relatively lower, and *vice versa*

The effect of the air change is of considerable importance As a rule, with increased ventilation, room temperatures should be kept higher and *vice versa*[2] (see also Required Air Changes under the section on Ventilation) Furthermore, the condition of the surrounding walls should be considered. With well-distributed radiation throughout the room, the occupants will be comfortable at comparatively lower temperatures, whereas concentration of the radiation requires a greater supply of heat for the same comfort

Experiments have shown that it is permissible to reduce room temperatures without discomfort when the heating surface emits a strong and hygienically correct amount of radiant heat Contrary to the general method of measuring the room temperature at eye level (5 ft above the floor) the following may be substituted [3]

Undoubtedly the ideal conditions for room heating are expressed in the words "warm feet and cool head" In recognition of this it is preferable to change the measurement of the important room temperature to the temperature at knee height (1 5 ft above floor level)

The existing low outside temperature varies with the geographical location [4]

[1] See also "Measurement of Air Temperature in Closed Rooms," from HANSEN "Zur Messung von Luft Temperaturen in Geschlossenen Raumen," *Gesundh -Ing*, July, 1921 For data on temperatures near ceiling, see part III, p 173

[2] See investigations reported by "A S H & V E Guide," Chap XVIII, and reports in *Trans* A S H & V E, 1924–1927

[3] BRABBEÉ, "The Heating Effect of Radiators," *Jour* A S H & V E, 1925–1927

[4] "Prevailing Temperatures in Germany," from MARX, "Die Temperaturverhältnisse Deutschlands," *Gesundh -Ing*, 1902 See also "A S H & V E Guide," Chap I, pp 19–20, 1925

SECTION II

METHODS OF HEATING

The following classification may be noted:

Local Heating Systems	Central Heating Systems
Fireplaces	Hot-water heating
Conduit heating	Steam heating
Iron stoves	Combination systems
Tile stoves	Warm-air heating
Oil stoves	District heating
Gas heating	Exhaust steam heating
Electric heating	

A. FIREPLACES

The open fireplace is the oldest form of local heating. A typical design is illustrated in Fig. 1. Attempts to imitate the open fireplace

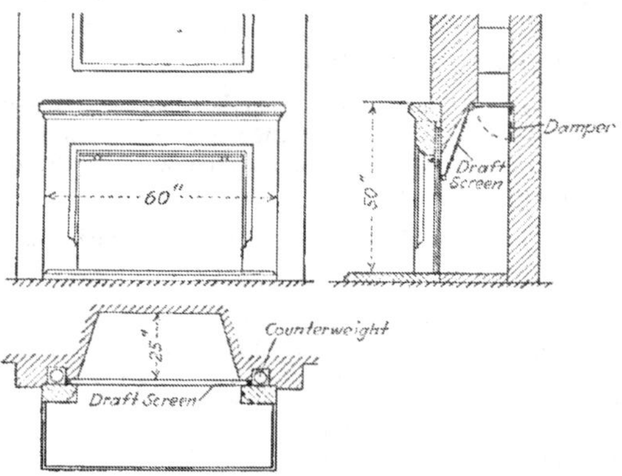

Fig. 1.—Open fireplace.

by means of other forms of heaters may seem artificial. Certain special cases, however, have been successful. Lighted more for the pleasant effect it creates then for its efficiency, the open fireplace affords heat mostly by radiation and only a negligible amount by convection.

Room ventilation and the attraction of the open fireplace are the advantages of this form of heating. Among the disadvantages may be cited the large consumption of fuel, sometimes discomfort due to intensity of radiation, and the rapid cooling of the room when the fire is extinguished. The efficiency of the fire place as a heater is estimated at from 5 to 10 per cent.

B. CONDUIT HEATING SYSTEMS

Conduit systems are still used for heating greenhouses. For churches the warm-air or steam systems have supplanted them.

In conduit systems one or more ducts are used to convey the flue gas. The combustion takes place in a chamber which is located at a lower level and may operate continuously or otherwise. The ducts are of brick or cast iron placed either in a horizontal or slightly inclined position. They may be exposed in the room or depressed in channels located below the floor level and covered with suitable gratings. The temperature of the combustion gases in the conduits is estimated at about 1500° F. This temperature must not be reduced below that temperature which will impair the draft in the chimney.

Some of the disadvantages of conduit heating systems are: escape of flue gas due to leaky ducts, accumulation of dust on the heating surfaces, roasting of the dust which settles on the ducts due to the high temperature of the conduits, discomfort due to overheated air, odors, and low efficiency.

C. IRON STOVES

Application of Iron Stoves

Recent statistics show that although direct-steam and hot-water systems are used extensively, the iron stove still plays an important part in house heating. The major application of the stove is in the smaller homes of rural communities and less frequently in urban centers and in cities.

To save fuel, heating is often restricted to certain rooms. In the case of settlements, workmen's homes, the smaller apartment blocks and certain other buildings where a variable demand for heating exists, the need is satisfied by the introduction of individual stoves. Groups of buildings can use central heating to advantage when exhaust steam in sufficient quantity is available from nearby steam plants. Where peak loads exist on the steam plants, making the exhaust-steam supply insufficient, the following illustration will suggest a solution.

Assume that in a group of houses supplied by central heating it is desired to maintain an inside temperature of 60° F. when the prevailing outside temperatures are around 5° F. In this case peaks are taken care of by individual stoves located where needed. By peak loads are meant such loads as occur during exceptionally low prevailing temperatures, windstorms, etc.

TECHNICAL CONSIDERATIONS

1. Advantages.

Iron stoves permit rapid heating of rooms and easy regulation during operation, occupy small space, and are simple to install and low in first cost.

2. Air Circulation in Rooms.

Stoves are as a general rule erected near the rear walls of a room. The heated air ascends at the stove (Fig. 2), is then cooled at the windows, where it descends, and returns along the floor to the stove to be reheated. This results in drafts near the windows. The cool currents of air are augmented by the infiltration of outside air through the crevices in the lower parts of the windows (see Pressure Distribution in Closed Rooms, Part II, p. 131). The presence of drafts is a disadvantage with stoves as compared with radiators which may be placed directly under windows (see Fig. 17, p. 20).

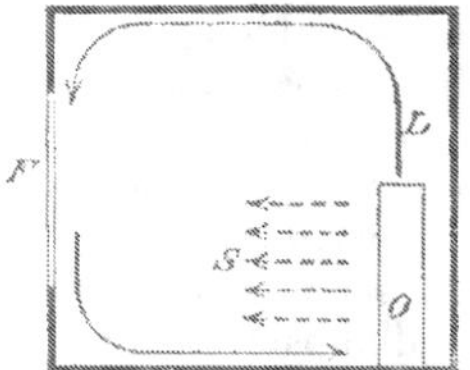

FIG. 2.—Air currents in a room heated by a stove. F = window; O = stove; L = air current; S = heat rays.

For stove heating, therefore, the draft should be minimized by good window construction, making them tight by means of weather strips or other means of securing a closer seal.

3. Space Requirements.

Whereas radiators may frequently be placed in relatively unimportant locations, stoves have the disadvantage of taking up the more valuable space. This is of particular importance in small houses where room is at a premium. Because of the space required for stoves, small rooms adjacent to heated rooms are sometimes not equipped with stoves.

4. Fuel and Ash Handling. Firing Periods.

The fuel- and ash-handling problem for each stove is an unpleasant duty particularly in the larger houses where the servant problem is of importance. This is of lesser consideration in the small home. The interval between firing periods is an important factor. For good modern installations once a day should suffice in mild weather, while twice a day may be necessary in extremely cold weather.

5. Room Ventilation by Means of Stoves.

While the fuel is burning in the stove, ventilation (sometimes overestimated) of the room takes place due to the air requirements of combustion. This ventilation ceases when the furnace doors are shut.

6. Installation Costs and Operating Expense.

Installation costs for stove heating systems are considerably lower than for steam or hot-water heating systems. The same is true for the operating expense. With stoves the occupant may adjust the operating expense to suit the needs by control of the operation or the number of stoves installed The variable needs of a dwelling have an important bearing upon the possible economy of operation It must be remembered, however, that installation costs are increased when numerous stoves are installed due to the separate flues required

Stove Design

The following factors should be embodied in the design of a good stove:

1. Avoidance of Overheating.

Red-hot stove surfaces should be avoided For instance the grates should be so designed that the adjacent surface shall not be overheated A suitable refractory substance should be used where overheating is likely to take place

2. Rapid Heat Emission.

The refractory lining should not unduly retard the heat transmission by being unnecessarily thick

3. Prolonged Heating without Undue Attention

Though heat energy cannot be stored conveniently as heat, it may be stored in potential form as fuel The design of magazine stoves is based on this principle

4. Regulation.

This should be easily effected by control of the air supply under the grate as well as by damper control in the chimney

5. Cleaning of Exterior Surfaces.

Every unnecessary form of decoration should be eliminated, and in general smooth surfaces should be used

6. Stove Jackets.

Increase of convection is attempted by jacketing the heating surface of the stove, forming a channel for the passage of air (ventilation stoves)

7. Stove Radiation.

This is an important element to include in the design

Stove Testing

Up to the present the stove efficiency was determined by flue-gas analyses According to this method the combustion efficiency η_F is given by the equation

$$\eta_F = \frac{pH - V}{pH}$$

where p = total amount of fuel fired

H = heat value of fuel in B t u per pound

V = heat loss as determined from flue gas (generally by means of Orsat gas-analysis apparatus)

The value of η_F in the trade literature was given as high as 95 per cent

Fudikar,[1] however, showed that these methods of test were open to objection and that the figures so derived were too high

In various publications[2] it was indicated that the determination of the combustion efficiency η_F is not sufficient to estimate the heating effect of a stove but that a comparative test[3] would be necessary under identical conditions, *i e*, rooms of similar construction and equivalent heat loss

In general it is practicable to place the stove on a scale, thus checking up the fuel charges

Stove Sizes

The size of the stove to be used for a room is determined from the heat loss W of the room in question (see p 189) From tests on stoves under similar conditions of load and location within the room, the heat emission K is determined The required surface of the stove F will then be

$$F = \frac{W}{K}$$

Usually the size of the stove is estimated by the volume of the space to be heated

A more exact method based on the heat emission from the stove surface was published by Barlach and Brabbée[4] For a simplified method of determining the required stove heating surface by Wierz,[5] see page 189 The usual estimates of the size of stove required, based on the cubical contents of a room, are unreliable.

Typical Stove Designs

1. "Cannon" Stoves.

The iron stove was developed from the "cannon" stove illustrated in Fig 3 It consists of a single iron cylinder with a fire chamber beneath as shown Owing to its inefficiency this type of stove is obsolete In the "cannon" stove the heat in the flue gas has little opportunity to be

[1] "Research on Tile Stoves," from FUDIKAR, "Untersuchung an Kachelofen," *Mitteilung der Berliner Versuchanstalt* (*Bull* 24 of the Research Laboratory, Charlottenburg), R Oldenbourg, Munich and Berlin, 1917

[2] "Fuel Economy in the Home," from BRABBÉE, "Beitrag zur Brennstoffwirtschaft im Haushalt," Vereins Deutscher Ingenieure, Berlin, 1920

"Tests of Tile Stoves," from (*Bull* 32 of the Research Laboratory, Charlottenburg), R Oldenbourg, Munich and Berlin, 1921 "Untersuchung von Kachelofen," *Mitteilung der Berliner Versuchanstalt*

[3] "Research in Testing Heating Apparatus," from BRABBÉE, "Relatives Forschen oder wissenschaftlich praktisches Versuchsverfahren in der Heizungstechnik,' *Gesundh -Ing*, p 157, 1923.

[4] Berlin, 1921 World Trade and Technique, "Weltwirtshaft und Technik"

[5] WIERZ-BRANDSTATTER, "The Iron Stove," issued by the German Society of Iron Stove Manufacturers, R Oldenbourg, Munich and Berlin, 1923.

transmitted into the room and so passes up the chimney at a very high temperature, with a consequent waste of the fuel. Various attempts were made to remedy these conditions. Meidinger used ribbed sections surrounded by an air casing. Keidel used smooth surfaces inside the air casing, provided flues, and later lined the iron body with firebrick where it would otherwise be in contact with the burning fuel. Numerous other forms were also developed along similar lines, and some were constructed with means for regulating the air supply to the fire.

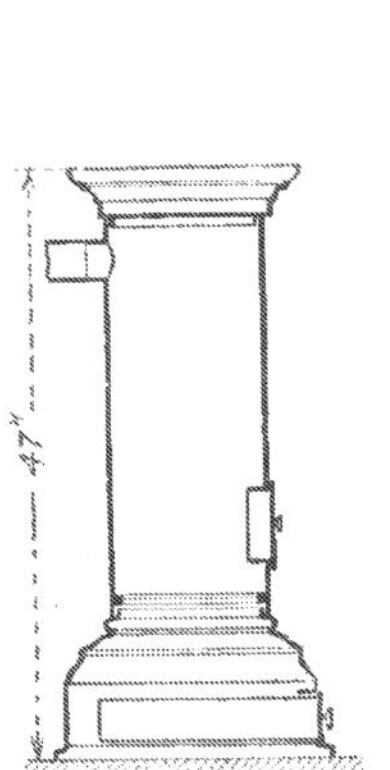

Fig. 3.—Cannon stove.

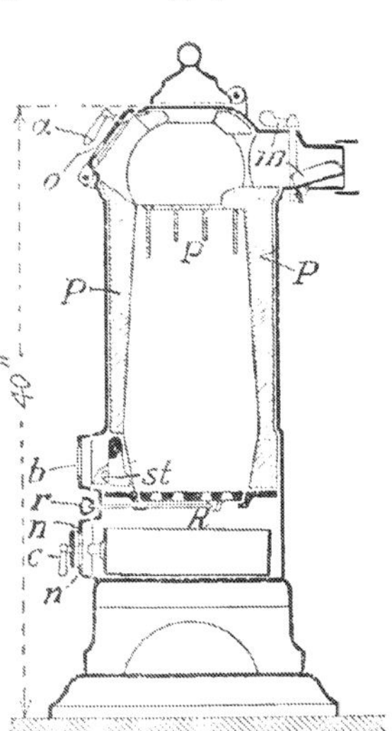

Fig. 4.—Irish stove. (*Vosswerke, Hannover-Sarstedt.*)

2. Irish Stoves.

A new line of development ensued when the Musgrave stove appeared on the market. It came out as a magazine stove when introduced about the middle of the last century. This was the precursor of the so-called "Irish" stove of which a particular design is shown in Fig. 4.

Special features of this design are: the large combustion chamber, a grate *R*, frequently of the shaking type, the vertical grate *St*, the regulation of the air supply *n*, the control of the flue gas (sometimes with cold-air check *m* to the chimney), the lining of the magazine with firebrick *P*, the upper feed door *a*, and the elimination of special flues.

Stoves were occasionally designed by artists and were enameled or tiled for hygienic reasons and to improve their appearance. Anthracite, coke, and many other fuels were used.

3. American Stoves and Their Modifications.

In the year 1875 Perry constructed a magazine stove (Crown Jewel) shown in Fig. 5. The stoves were constructed to burn anthracite and had the following main features: horizontal grates *P*, comparatively large magazines *F*, a basket grate *K*, through which the flue gas passed to the lower sections of the stove, heating them and the floor at the same time and finally discharging to the chimney. Preliminary heating to establish the draft was accomplished by opening the bypass shutter *L* so connecting the fire pot with the chimney. This form of stove has been further developed by various German firms.

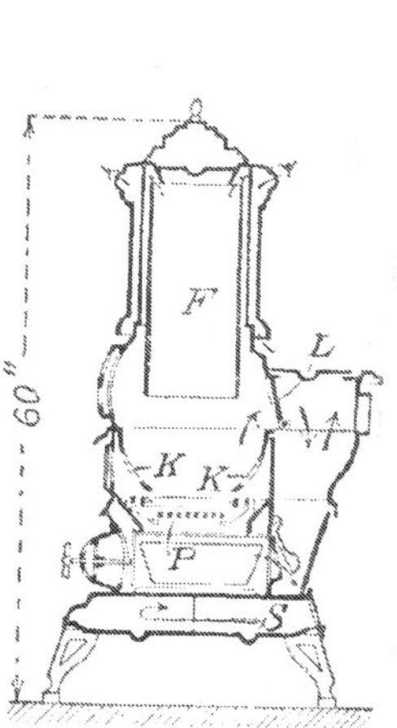

Fig. 5.—Crown Jewel stove.

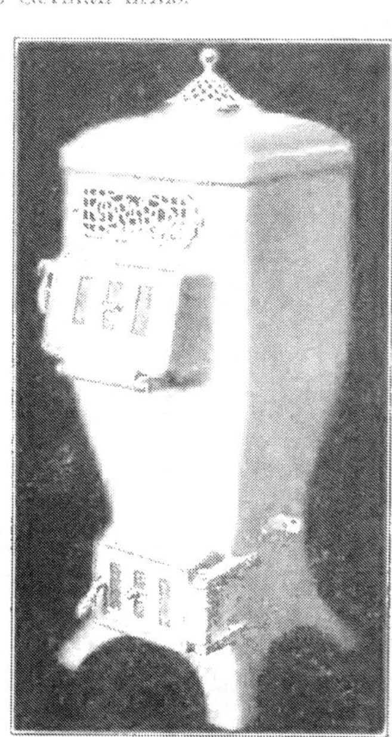

Fig. 6.—Brabo stove. (*American Radiator Company, New York.*)

4. The Brabo Stove.

The author developed the Brabo stove shown in Figs. 6 and 7. It was intended to incorporate the features given on page 8. Corresponding headings in the following are numbered alike:

(1) Incandescent radiating surfaces are avoided by use of firebrick lining, thus isolating the fuel bed from the heating surface.

(2) Thinness of firebrick lining to avoid undue retarding of heat transmission.

(3) Magazine capacity of sufficient size to operate 8 hours in ordinarily cold weather without replenishing fuel charge.

(4) Regulation possible by means of feed-door slide damper, ash-door slide damper, and stove-pipe damper.

(5) Unnecessary decorations are eliminated to promote cleanliness.

(6) No fixed jacket to interfere with the direct radiation to the room.

(7) Special construction of the firepot to insure effective radiation toward the floor and the lower air strata.

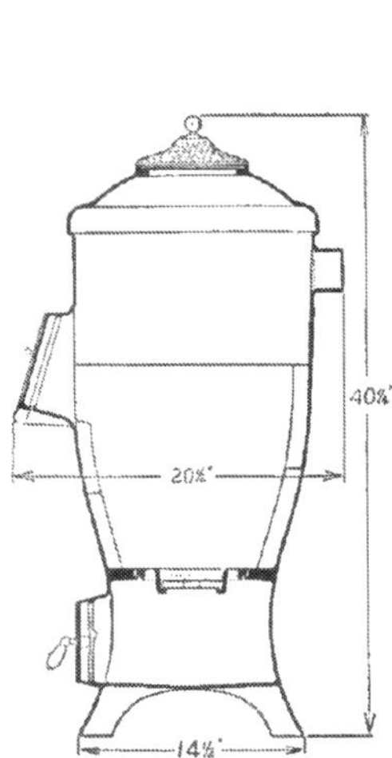

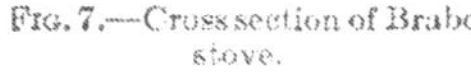

Fig. 7.—Cross section of Brabo stove.

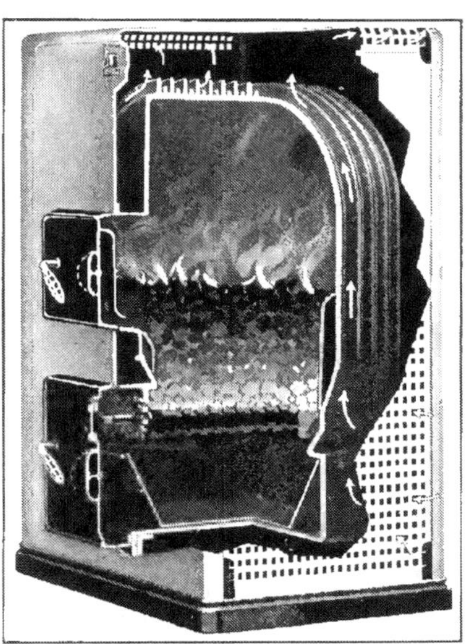

Fig. 8.—Vecto heater. (*American Radiator Company, New York.*)

(8) The use of a secondary air supply through the fire door to make it possible to burn any form of fuel including soft coal. The latter is burned smokelessly. It is economical in operation and in time required for attention. Its low cost brings it within the purchasing power of any prospective owner.

(9) Stoves are enameled for beauty and hygienic purposes. These stoves heat two normal rooms (4,000 cu. ft.) and take the chill out of a third room.

5. Jacketed Stove.

An application of the jacketed stove is shown in Fig. 8. This is a combination of a stove and a warm-air furnace. The stove is located in one of the living rooms where it heats just as a radiator. The other rooms of the same floor and the second floor of such installations are heated by means of the current of warm air which issues through the grilles. The cold-air supply enters the base of the stove. By providing suitable openings to the second floor and inducing a sufficiently rapid current of warm air, the one stove is made to heat a house of four to seven rooms. During recent years a large number of these stoves were distributed in the United States.

D. TILE STOVES

Tile stoves are used extensively throughout Europe, particularly in the small homes and in the rural districts. Their erection requires a

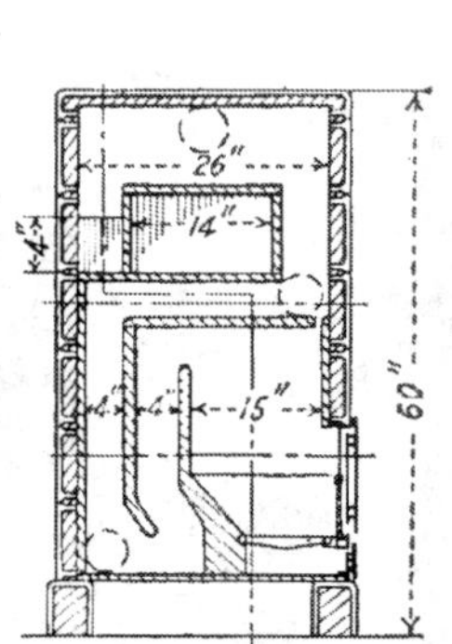

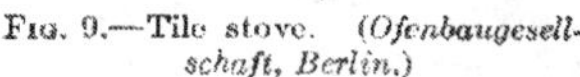

Fig. 9.—Tile stove. (*Ofenbaugesellschaft, Berlin.*)

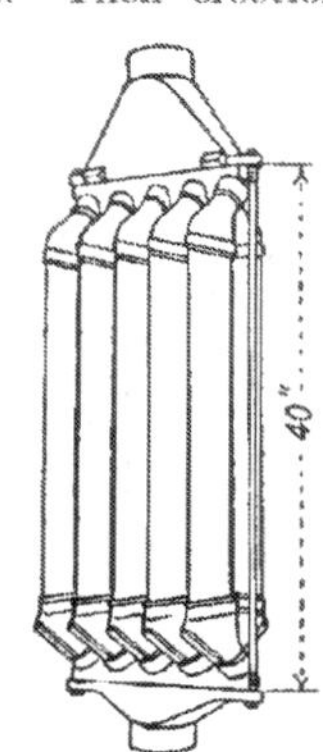

Fig. 10.—Flue gas heater. (*Kleiro Werke, Karlsruhe i. B.*)

particular type of skilled labor which is not available in America. In view of this, tile stoves will not be shown in great detail in the present edition. A single type will suffice as an example.

Figure 9 shows a tile stove designed by the author. In general two or more of these are installed in the house. It will be noted that the stove consists of a small grate, a large combustion chamber, a revertible flue, and three other flues which surround a chamber through which air may pass and which in addition is used for cooking purposes. The flue gas leaves the tile stove through a metal stove-pipe and is then discharged to the chimney.

Well-designed iron and tile stoves have a reasonably low flue-gas temperature. In inferior designs the flue temperatures are excessive, reaching 1000° F. or more. A remedy for excessive flue temperatures lies in the use of heaters which are fitted to the smoke pipe and are so made effective heating surface. Thus this heating surface effects economy, since it utilizes otherwise waste heat which would have passed to the chimney

Flue heaters of the type just described should be made readily accessible for cleaning and should offer a minimum of resistance to the flow of exit gases

A flue-gas heater is shown in Fig 10 In the design shown, the products of combustion are led through the several pipes, giving up a substantial amount of heat to the room The heater permits of cleaning When flue heaters are used, cold-air inlets to the stove should be kept closed, and anthracite or coke used as fuel

Flue-gas heaters may be used to advantage when two rooms are to be heated by a stove either of iron or tile construction In this case the stove is located in one room, while the flue-gas heater is located in the other

E. OIL STOVES

Oil stoves in general can be regarded only as an auxiliary form of heating These stoves are easily transported, erected, and put into operation An occasional filling of the tank is the only attention they require One advantage is the small space occupied by oil stoves The products of combustion usually are permitted to enter the room

From several observations the average CO_2 content in rooms heated with oil stoves was found to be.

After 2 hours operation,	3 parts per 1,000
After 4 hours operation,	6 parts per 1,000
After 6 hours operation,	8 parts per 1,000

Since 1 5 parts per 1,000 is considered a maximum CO_2 content permissible for continuous occupation of rooms, this form of heating should be used with care Oil stoves often create odors which are unpleasant to the occupants, but have been greatly improved in the last years

F. GAS HEATING

Principles and Their Applications

Among the advantages of gas heaters it is noted that the cost of installation is low, they are simple to install, they are clean in operation since there is no fuel or ash handling, they are put into service in a short time, they admit a simple regulation

The disadvantages include surface temperatures often exceeding those considered hygienically satisfactory, the dry distillation and the roasting of the dust in the contacting air, the necessity of erecting them at inner walls to make proper chimney connections with consequent drafts from the windows; the negligible heat-storage capacity

The principal field for gas heaters is in cases where rooms are not used constantly and yet require speedy heating During mild weather when the main heating plant is not in operation, gas heaters are also used as auxiliaries in houses equipped with steam or hot water. They are further used for the occasional need to provide domestic hot-water service.

Typical Gas Radiator Designs

The German Society of Gas and Water Engineers has rendered an important service in publishing the "Guide to Correct Construction, Installation, and Operation of Gas Heating Appliances." A part of the following is taken from this book:

Gas stoves should be connected to effective chimneys. They should be so constructed that in case of failure of the vent neither incomplete combustion nor the extinction of the flame is possible. For this purpose back-draft "diverters" are used to keep the gas jet aflame even though a strong back pressure exists. Gas flames must not come in contact with the heating surfaces. Burners should be set so as to permit observation of the flame. Revertible flues should be avoided in cases where the chimney does not assure complete venting of the products of combustion at times of unfavorable (high) outside temperatures. Heating surfaces should be so dimensioned that the flue gas is not cooled below the dew point; otherwise large quantities of water are condensed. The condensate in combination with the acids contained in the flue gas may cause considerable destruction in the gas passages.

Older forms of gas heaters use either the radiant heat of the flame itself or the heat emitted from the flue gas. The newer forms seek to combine both effects. The customary method of testing, in general use, hardly affords a solution as to the value of the combined effects of radiation and convection. For an analysis of the actual relations, supplementary methods of comparative testing are essential.[1] The commercial forms of gas radiators will be described.

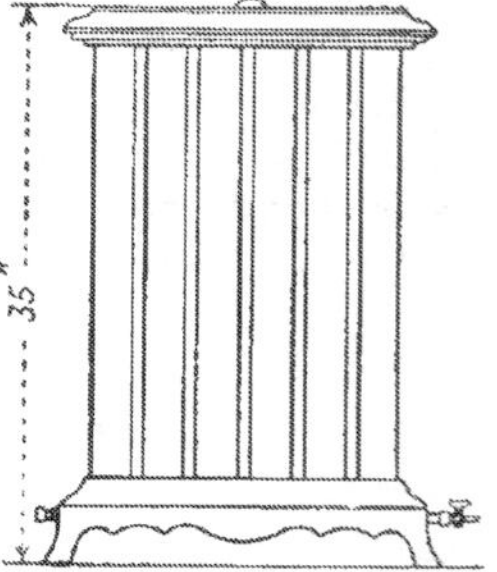

Fig. 11.—Gas heater. (*Vosswerke, Hannover-Sarstedt.*)

Hot-air Gas Radiators

The radiator shown in Fig. 11 is made of steel. In this design the burner is placed at the base, and the products of combustion pass through oval sheet-metal ducts into the dome and thence to the chimney. The ducts are situated over the burner outlets so that a back draft cannot extinguish the flame.

A similar design is also made of cast iron.

[1] "Research in Testing Heating Apparatus," from Brabbée, "Relatives Forschen oder wissenschaftlich-praktische Versuchsverfahren in der Heizungstechnik," *Gesundh.-Ing.*, p. 159, 1923.

Professor Junkers' Gas-fired Warm-air Heater

In Fig. 12 a reflector R emits the radiant heat. The flue gas passes through the oval tubes S in the direction of the single-shafted arrows and thence out to the chimney A. The air to be heated circulates around the tubes S as indicated by the double-shafted arrows. These heaters are equipped with a back-draft diverter to prevent contamination of the air in the room.

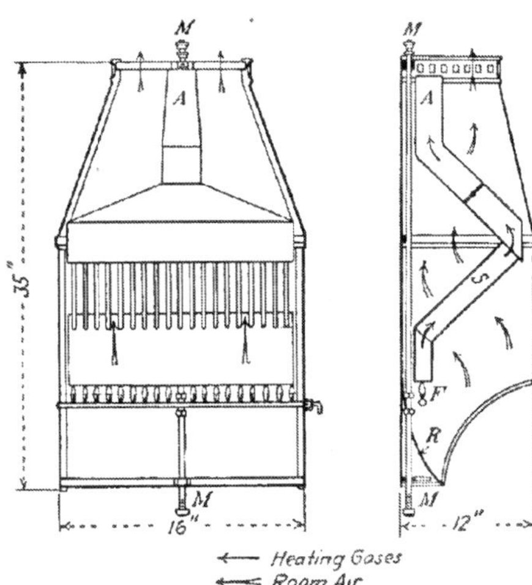

Fig. 12.—Gas heater. (*Junkers & Co., Dessau.*)

Radiant Heaters

In the design shown in Fig. 13, the refractory elements located in the base of the heater are brought to incandescence by means of a gas flame. The reflecting radiating surface S is of firebrick. The flue gas passes through metal radiator tubes on its way to the chimney, and additional heat is thus abstracted from the heater.

Vents and Chimneys

As mentioned previously, all gas stoves in general should be equipped with chimney vents. Exceptions can be made only in cases of very small stoves, located in relatively unimportant places. Chimney vents should be located where certainty of operation is assured even under the adverse condition of high outside temperatures. Therefore they should be placed at the inner walls, if possible, though sacrificing the favorable position of the radiators underneath the windows. Owing to the possibility of the separation of water from the flue gas, the vent pipe should not end in the attic space because of the consequent destructive effect upon the woodwork.

When vents are not provided for, gas heaters are often connected directly to fireplaces. This practice may lead to difficulties. It was shown that all gas stoves should be equipped with back-draft diverters. When the stove is not in use, cold air passes through the diverters into the chimney and so impairs the draft. As a consequence the additional connection of a gas heater to a chimney operated at capacity may lead to failure of the remaining connected stoves. Moreover, it must be noted that carbon monoxide (CO) may be formed by incomplete combustion and may pass through the diverters of connected stoves into the occupied rooms.

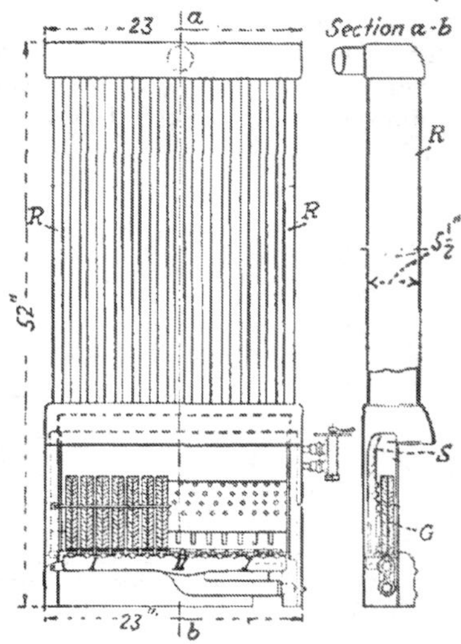

Fig. 13.—Radiant gas heater. *(Gasbetriebs-gesellschaft, Berlin.)*

Because of the dangers, it is essential that gas stoves be supplied with separate and specially constructed vents. The dimensions of the exhaust vents of gas stoves according to the "Guide" previously referred to are.

Hourly gas consumption, cubic feet	Inside diameter of gas pipe, inches	Inside diameter of vent, inches
5	¼	2
20	⅜	2½
50	½	3
75	¾	3½
150	1	4½
250	1¼	6
500	1½	6¾
1,000	2	9

Estimating Sizes of Gas Stoves

(See Part III, p. 190.)

G. ELECTRIC HEATING

Some of the more important advantages of electric heating are: simple construction, low cost of installation, no sensible heat loss, no chimney requirements, speedy heating, excellent regulation, cleanliness, since no coal or ashes are handled, and low surface temperatures where heating surface is properly proportioned.

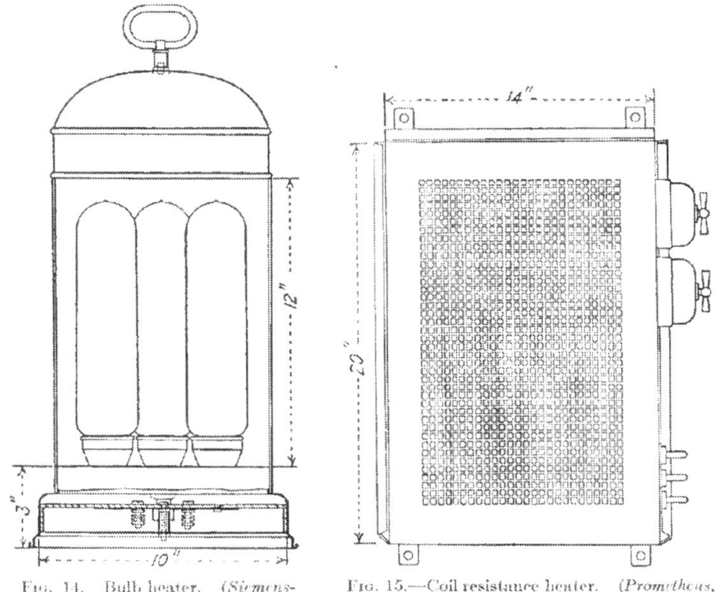

FIG. 14. Bulb heater. (*Siemens-Schuckert-Werke, Berlin.*)

FIG. 15.—Coil resistance heater. (*Prometheus, Frankfurt a. M.*)

Among the disadvantages might be included the expense of operation (higher than either gas, coal, coke, etc.), although the Siemens Company of Dresden maintains that in case of correct distribution of electrical heaters in a room much less heat is needed for sufficient warmth than by any other means.

Electrical heaters are used in Norway and Switzerland where electricity generated by water power is available at low cost. It is also used to heat individual cabins on ships, and trolley cars. Electrical heating has made rapid strides in recent years.

Types of Electric Heaters

1. Bulb Heaters.

Figure 14 is self-explanatory. It is to be noted that bulbs with carbon are used in preference to metal filaments.

2. Resistance Heaters.

Resistance heaters are made either of the coil or plate type. A coil resistance heater is shown in Fig. 15. In this form three coils are used behind a perforated enclosure, the number of coils being varied to suit requirements. The heat emission is restricted by the enclosure. This is not the case with the plate type shown in Fig. 16. Here the radiant heat of the plates issues directly into the room. Resistance heaters of this form are used for diverse purposes—electrically heated cushions, foot warmers, medical work, etc.

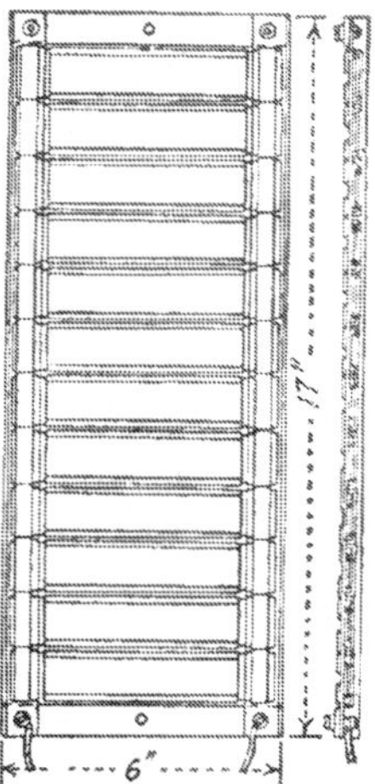

Fig. 16.—Plate resistance heater. (*Prometheus, Frankfurt a. M.*)

Heater Calculations

The heat emission is readily computed when the voltage and the current intensity are known (for these details see p. 190).

Testing of Electrical Heaters

Comparative tests are the only reliable means of determining the serviceability of the various designs. This subject is referred to in the literature.[1]

[1] "Research in Testing Heating Apparatus," from Brabbée, "Relatives Forschen oder wissenschaftlich-praktische Versuchsverfahren in der Heizungstechnik," *Gesundh-Ing.*, p. 159, 1923.

SECTION III

CENTRAL HEATING

GENERAL

Central heating as practiced in Europe includes hot-water, steam, and warm-air heating. In the three cases mentioned, the heat necessary for warming the various rooms is generated in a central place and transmitted by the heating medium (water, steam, or air) to the separate rooms.

The following are advantages common to all three forms of central heating: Since the heat is generated in a suitably designed central location, attention to the fire is limited to that place. In Europe coke is usually used as a fuel, insuring smokeless combustion. This circumstance is noteworthy in so far as it has been established that in the large cities it is the firing of domestic heaters that is the important cause of the smoke and soot nuisance. With central heating the carrying of fuel and ashes in rooms is eliminated. Likewise attention to the heaters is reduced to a minimum. Moreover, the fact that the servant problem is acute led to rapid development in central heating systems.

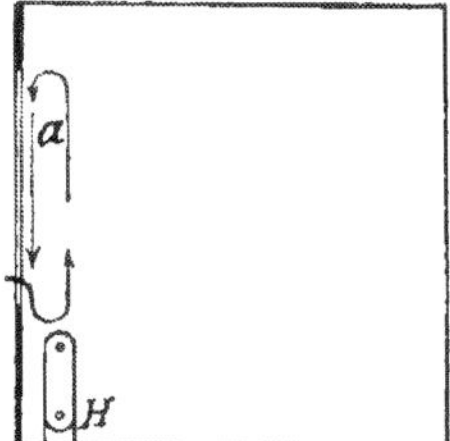

FIG. 17.—Air currents with radiator located under window.

Small radiators in steam and hot-water heating systems may be placed directly beneath windows or near large cooling surfaces. The cold air entering or descending along the window F (Fig. 17) is heated by the radiator H and forms a local air current at a which rises as shown and does not in general reach the occupants. The effect here noted is opposite that shown in Fig. 2 (p. 7) when stoves are used. The fact that it is possible to eliminate the disturbing draft in steam and hot-water heating is an important advantage.

The use of numerous low radiators and the necessarily longer piping distribution system increases the cost of the installation. Moreover, since in general the risers are located near the outer walls, the greater heat loss increases the operating expense. Because of this the radiators are sometimes located at inner walls.

Other advantages lie in the facts that the individual radiators require a smaller space in comparison with stoves, that there is freedom from fire hazard, that it is simple to provide heating in halls, stairways, bathrooms, etc., and that excellent ventilation systems are possible in combination with central heating systems. The foregoing advantages are obtainable at a correspondingly higher cost of installation. The expenditure, however, must be justified by the convenience afforded, including the economy in operation.

Central heating is installed in any one of the following ways:

Low-pressure hot-water heating
Medium-pressure hot-water heating
High-temperature hot-water heating
High-pressure steam heating
Low-pressure steam heating
Combination heating systems
Warm-air heating
District heating.

A. LOW-PRESSURE HOT-WATER HEATING

Low-pressure hot-water heating systems are those which operate at water temperatures below 200° F even in extreme weather. They are characterized by the fact that the water content of the system is open to the atmosphere and hence operate at or near that pressure. The following types are included in this form of heating:

Gravity systems
Single-floor heating
Greenhouse heating
Forced-circulation systems
Pump systems

Gravity Heating

1. General.

a Two-pipe Overhead Distribution System.—The hot water from the boiler K (Fig. 18) flows through the main riser S to the distribution main V which supplies the hot water to the radiators. An expansion tank A (which is open to the atmosphere) is connected to the riser S. Since water boils at 212° F under atmospheric pressure, in ordinary operation of the heating system it is prevented by operating at a lower temperature.[1] From the main V the hot water is distributed to supply lines F which are piped to the radiators. The cooled water from the radiators flows through the return lines R to the return main G and thence back to the boiler. As shown in Fig. 18 all lines pitch upward toward A so that any air in the boiler, distribution mains, risers, or radiators may escape by venting to

[1] Formerly systems were designed to operate at temperatures not exceeding 190° F but lately this limit was extended to 200° F.

the high point. As will be noted, each radiator is connected to a supply and a return riser; hence the arrangement is styled a two-pipe system.

b. Two-pipe Basement Distribution System.—The arrangement shown in Fig. 19 differs from that shown in Fig. 18. In this design the hot water is distributed by a basement main V to supply risers S and is cooled in the

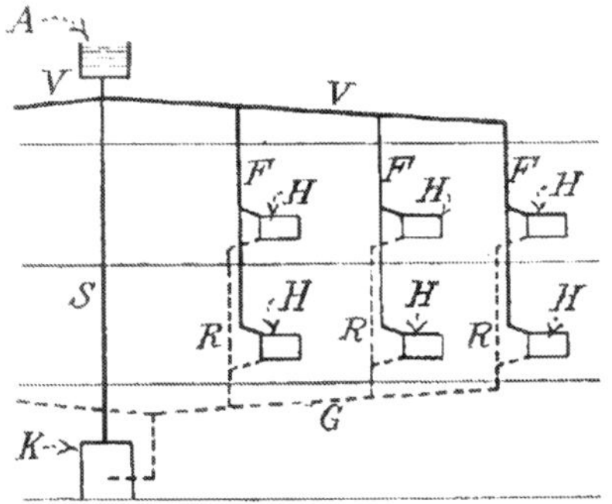

Fig. 18.—Two pipe overhead distribution system.

radiators, from which it flows to the returns F and main G to the boiler. As in the previous case all mains including air vents L are pitched upward towards the expansion tank A. An exception is made in the case of the connecting air vent lines L' which are pitched downward towards A.

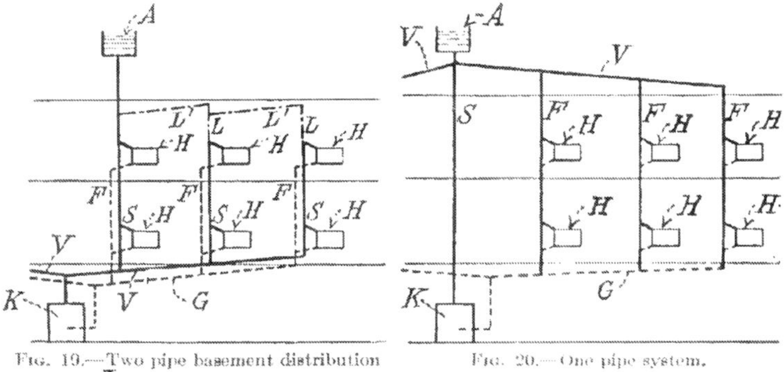

Fig. 19.—Two pipe basement distribution system.

Fig. 20.—One pipe system.

c. One-pipe System.—In the one-pipe system shown in Fig. 20 the hot water from the boiler flows through the pipes S to the supply main V, whence it flows downward through F to the radiators H. The water cooled in the radiators is returned to the same lines F flowing back to the boiler through the common return main G. It should be noted that only one riser is used for both flow and return, from which the system derives the name of one-pipe system

2. Field of Application, Advantages, and Disadvantages.

Systems with gravity circulation operate by the difference in pressure caused by the cooler (denser) return water forcing the warmer (lighter) water up the supply risers Installations of this kind have a positive action which is dependable The water temperature in the supply main is readily changed by regulating the boiler load It is possible as a consequence to adapt the heat delivery in the rooms by regulation from a central place Systems are usually designed so that even with extreme outside temperatures the supply-water temperature does not exceed 200° F. For average winter conditions, the heating surfaces have maximum temperatures of about 150° F., a temperature which is hygienically desirable Corresponding to the low radiator temperatures is the mild and even heating of the rooms, and that is the reason why low-pressure hot-water heating is considered best hygienically

Radiators which for any reason receive a limited supply of hot water will cool to a greater extent, so that in a limited way they automatically receive more water This inherent regulation is a valuable characteristic

If several rooms are located so that their operating conditions differ (*e g*, north or south exposure, etc), each group may be given separate supply lines and sometimes separate return lines By checking the flow in the section desired, a simple control of the system is possible.

Direct regulation of individual radiators is provided by means of the radiator valve The hot water stored in the system continues to emit heat even though the supply at the boiler is curtailed The fuel stored in magazine-type boilers also lends to continuity in operation Hot-water systems are noiseless

The water supplied initially into the system should be brought to a boil The small amounts which evaporate through the expansion tank should be replaced by rain water or other clean water which is previously boiled In this way the water content is free from oxygen and therefore will limit corrosion In fact the carefully operated hot-water systems have a life that is practically unlimited They are subject only to such corrosive influence which results from frequent change of water in which the precautionary measure outlined above has not been taken In other words the water supply should remain unchanged unless necessary for some reason

Hot-water systems have three disadvantages First, there is a possibility of freezing, with the result that the radiators, piping, or boiler may burst and cause a flood This is mostly due to shutting off radiators in extreme weather, and at the same time opening windows near the radiation A protective measure lies in the use of hot-water-type radiator valves which are provided with a small hole in the seat of the valve which prevents the complete shutting off of the water supply A second disadvantage is that hot-water heating is sluggish because of considerable heat stored in the water of the system As a consequence hot-water heating is unsuited to rooms which must be heated rapidly or where the demand for heat is very variable For cases of this kind the steam systems described in the text (p 78) are preferable A third disadvantage lies in the higher installation cost In general the disadvantages are compensated for by the many advantages

Owing to the large water content of a hot-water heating system there is a considerable amount of heat storage. Its economical use in the usual periods of light heating load offers some difficulties as Wierz[1] shows, particularly if there is unequal distribution The improper proportioning of pipe sizes causes improper circulation and consequently poor distribution At present, efforts are being made to combine gravity systems with pump systems In such installations fuel economies have been effected Other efforts are directed toward reducing the water content by suitable design of radiators and boilers In this connection the new Classic radiators and Narag boilers of the National Radiator Company of Europe are noteworthy developments.

The foregoing considerations indicate the field of application of low-pressure hot-water heating If it is a question of comfort and of uniform heating, hot-water systems take first place. Moreover, the extension to larger houses, hospitals, greenhouses, etc is recommended The system lends itself to heating of homes wherein the rooms are kept at a reasonably low temperature to insure health In this way the body develops a resistance to climatic changes, and the ill effects of temperature variations are ameliorated

3 Overhead and Basement Distribution. One-pipe vs. Two-pipe Systems.

From the computation given on page 212 it is seen that more rapid circulation takes place with overhead mains than with basement distribution On the other hand with basement mains the heat loss from the supply piping assists in heating the house, while with overhead distribution it is less effective so far as the lower rooms are concerned but it is effective in heating attic rooms and keeping the cellar cool At times one or the other of these advantages is of importance The cost of installation for basement mains is the lesser Accordingly a number of elements enter into the choice of the distribution system depending upon local circumstances All things being equal, basement distribution is used for the small and average-sized installations, while overhead distribution is used for horizontally extended buildings and such other cases where circulation is likely to be deficient

[1] "Improvements in the Economy of Central Heating and Particularly in Gravity Systems," from Wierz, "Die Verbesserung der Wirtschaftlichkeit der Zentralheizung, insbesondere der Schwerkraftheizung," *Gesundh -Ing*, p 477, 1923

As a general rule the two-pipe system is common The one-pipe system has the following disadvantages a tendency of the radiators connected to the same riser to disturb circulation, an increase in heating surface for the radiators on the lower levels; an overheating of the lower rooms when the upper radiators are shut off, a sluggish heating up of the system The important advantage of the one-pipe system is its simplicity of installation, particularly when the piping is exposed The one-pipe system is also an advantage in forced-circulation systems (pump systems) and for heating numerous floors when the heat requirements of such rooms are the same

The final selection of the system must be based upon local considerations and will be influenced by the computation of the piping lay out

4. Boilers for Hot-water Heating.

a. Steel Boilers (1) Large Boilers —Large hot-water heating systems use steel boilers. Horizontal return tubular boilers are preferred, due to their large water content This is particularly important in installations which make use of exhaust steam heating In such cases the boilers serve as storage tanks for the heat in the steam which cannot be utilized immediately. After the steam supply is cut off, the heating of the building is accomplished by means of the heat stored in the hot-water boilers Installations of this kind have the disadvantage of slow heating up When rapid heating up is desired, water-tube boilers are to be preferred In cases where large heating surface must be installed in a small space, combination boilers offer a solution. For extreme winter weather, continuous operation of the system is arranged for in that the boilers are stoked at night during severe weather and operate with banked fires when weather gets milder.

The design of hot-water boilers is essentially the same as for steam boilers, and the reader is referred to the texts devoted to this special field in mechanical engineering The boiler trimmings for hot-water practice are much simpler Superheaters and steam domes are not used Since the boilers are entirely full of water when in use, such devices as safety valves, gage glasses, try-cocks, water-feeding apparatus, etc are not needed The arrangements for efficient combustion and furnace design, however, are the same as for steam boilers Stokers, forced draft, etc are also used particularly on the larger units For these details the reader is referred to the texts which deal with these subjects. For safety devices used in hot-water practice, see page 35.

(2) Small Boilers —For heating small, medium-sized, and also large houses steel boilers have been used frequently in the past, but the tendency is markedly in favor of cast-iron sectional boilers

b. Cast-iron Sectional Boilers (1) Medium-sized Boilers —The first European sectional boilers were introduced in 1898 by Strebel whose design was patterned after a study of the practice in the United States

His main objects desired were: use of the cheaper cast iron instead of steel so as to lend itself to mass production; to extend the heating surface by adding similar sections; to provide a design to insure a low water line (for steam boilers); to permit continuous operation with little attention; to avoid expensive boiler setting; and to insure smokeless combustion. The boiler design shown in Fig. 21 met these requirements. In the design shown the eight intermediate sections are similar. Each section, moreover, is self-contained and is a complete boiler unit in itself namely, grate A, ash pit B, magazine C, flues D, and flue header E to which the smoke hood is attached for right, left, or downward connection.

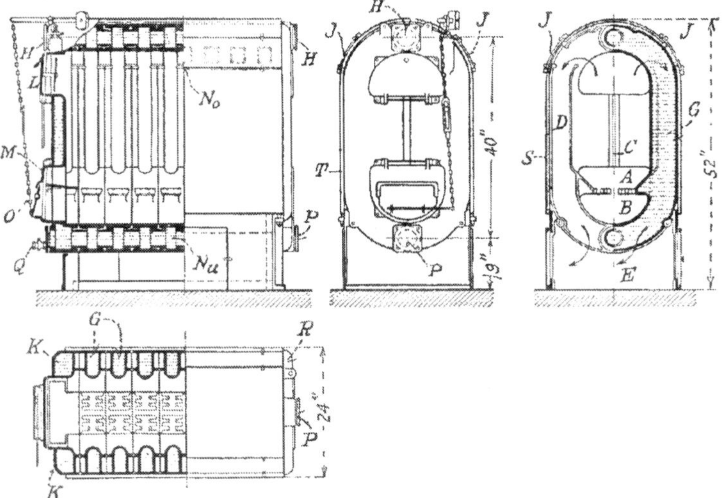

Fig. 21.—Cast iron sectional boiler. (*Strebelwerk, Mannheim.*)

The cold water flows through the lower row of nipples N_u and passes through the hollow boiler sections G in opposite direction to the fluegases. It then passes through D to the upper row of nipples N_o and thence through H at the front or rear section into the supply main of the system. The boiler has two openings J which permit of cleaning even though the boiler is in operation. The intermediate sections have a front section K which has a feed door L for the fuel supply, the clinker and ash door M, air supply O, connections H or P for hot water, and the fill or drain cock Q. The rear section R also has the water connections H or P.

The sections are assembled by means of push nipples (conical) which are forced into place under heavy pressure making the metal-to-metal joint. The boiler requires no brick setting. Radiation loss is minimized by the insulation S. A simple sheet-metal jacket T completes the boiler. By varying the number of sections the boiler ratings are varied from 600 to 3,000 sq. ft. of hot-water heating surface.

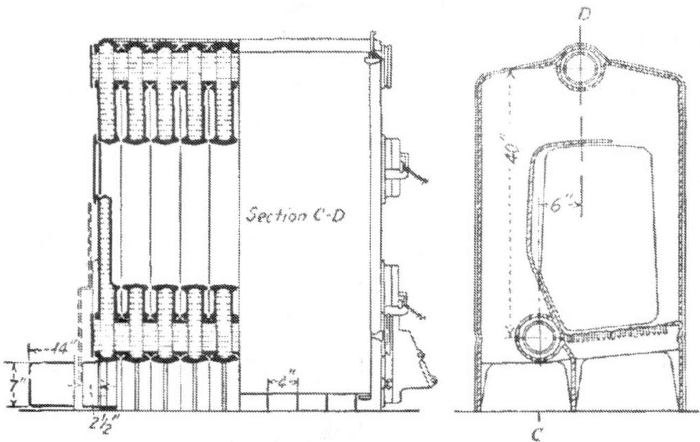

Fig. 22.—Lollar boiler. (*Buderussche Eisenwerke, Wetzlar.*)

The Lollar boiler shown in Fig. 22 is of similar construction. The flue gas, however, is taken off from one side only in contrast with the divided path of the Strebel boiler. The remaining constructional features will be clear from the illustration.

In the types of boilers described the gases of combustion pass through the fuel bed and cause the coke to glow.

(2) Small Boilers.—For the smaller heating installations, hot-water supply systems, small greenhouses, etc., small boilers ranging from 250 to 1,000 sq. ft. of hot-water rating are available. The general design is the same as for the larger units. The sectional construction is also suitable for these conditions. A design of a small boiler is shown in Fig. 23.

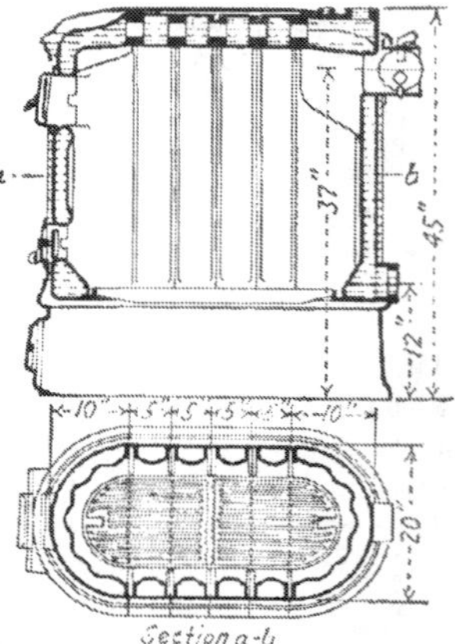

Fig. 23.—Strebel small boiler. (*Strebelwerk, Mannheim.*)

For installations operated from the kitchen range (*e.g.*, single-floor heating systems) the boiler is of a design shown in Fig. 24. A range boiler known as "Cookanheat" is used extensively in small houses subjected to the milder English climate.

(3) Large Boilers.—The large boilers are now made in sizes ranging from 8,000 to 16,000 sq. ft. of hot-water rating. It might be noted that these boilers are smaller than is customary in American practice. As shown in Fig. 25, the general characteristics of the sectional construction are present. One departure, however, is noted in that the flue gases travel sidewise and upwards. The magazine portion remains cold with black fuel, and combustion takes place only in the zone from the grate to the first uptake flue. The boilers usually have an upper feed opening so

FIG. 24.—Cookanheat. (*National Radiator Company, England.*)

that the magazine may be charged by means of small trolleys running above the boilers.

An innovation by the Strebel Works shown in Fig. 26 consists of boiler sections set up from front to back in addition to sections added sidewise. This arrangement is known as the Catena boiler. For safety in operation and easy access it is general not to provide more than two grates assembled in one group. The groups are then separated at suitable intervals.

If the load of an installation is so large as to require several cast-iron boilers of large size, it is possible that the steel boilers previously mentioned, may be preferable.

There are many other boilers which are not described here. In general their serviceability is easily determined from field tests.[1] On page 191 will be found the method of determining the required boiler capacity.

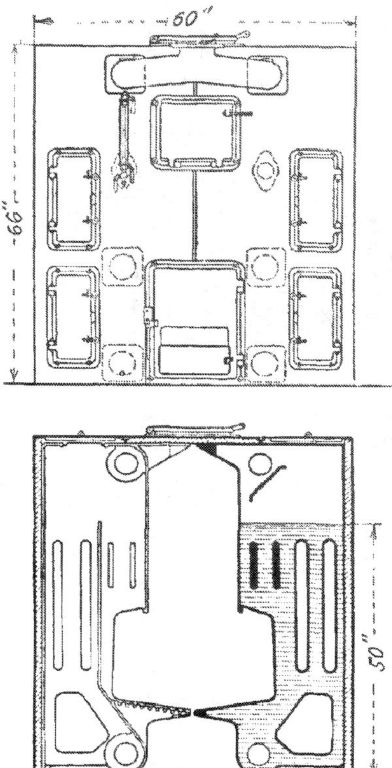

FIG. 25.—Large sectional boiler. (*Nationale Radiatoren Gesellschaft, Berlin.*)

The boilers previously discussed are arranged for the burning of coke, since this is the fuel in general use throughout Germany. Recently, however, there is a tendency toward the use of mixtures of coke and lignite briquettes (brown coal). Lignite is a low-grade fuel quite different

[1] "Tests on a Strebel Boiler," from "Untersuchung eines Strebelkessels," and "Tests on a Lollar Boiler," from "Untersuchung eines Lollarkessels," Bulletins 2 and 17, Research Laboratory, Charlottenburg, published by R. Oldenbourg, Munich and Berlin.

in its combustion characteristics from the bituminous coals used in the United States.

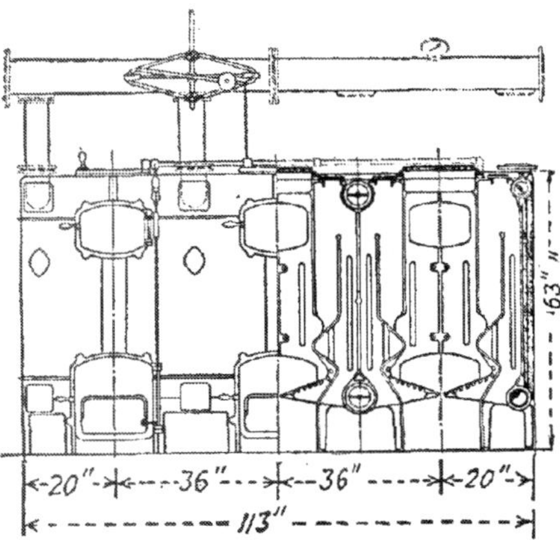

FIG. 26.—Catene boiler. (*Strebelwerk, Mannheim.*)

For the combustion of the low-grade fuels (lignite, peat, etc.) boilers of a somewhat different design are employed. The type illustrated in

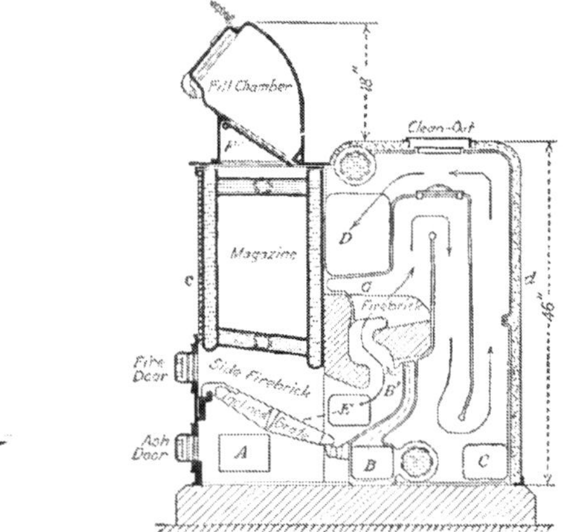

FIG. 27.—Boiler for low grade fuel. (*Gebr. Körting Hannover.*)

Fig. 27 has the following features worthy of note: the inclined grate with air supply *A*, secondary air supply *B* or B′, the incandescent refractory *G*, the magazine with a drop lid and the gas passage ascending, then descending, and finally ascending to the chimney connection. The sections are assembled sidewise rather than one behind the other. With the larger boilers double sections are used.

Oil boilers at the present time are formed by installing specially designed oil burners in boilers of the types generally used for domestic heating. Figure 28 shows a typical combination of the two. In this regard it should be noted that for the majority of oil burners the radiant effect of the flame, so important in the heat transmission from a coal fire, is reduced. Long flues are therefore required in the oil-burning boiler to absorb the heat by connection.

FIG. 28.—Ideal water tube oil-burning boiler. (*American Radiator Company, New York.*)

With this in mind the efficiency of an oil burner depends on the boiler to which it is attached, together with the connected heating system. If oil burner, boiler, and heating system are of good design and properly installed, the actual fuel cost is very reasonable.

In recent years, domestic heating using gas boilers has made rapid strides. Without a doubt, gas house heating represents the highest degree of cleanliness, readiness at all times for use, constant source of heat, and regular performance at high efficiencies. Only electric heat surpasses gas in these respects. Figure 29 represents the latest development in gas-burning boilers. These are suitable for the types of hot-water heating systems already described. Thermostatic regulation should be provided, for reasons of economy.

While special gas heating rates are not found throughout the United States generally, the price trend is downward, whereas for other fuels it has been upward. Increased distribution and the development in utilizing gas heat efficiently are making the rates for gas more favorable.

That the use of gas boilers has increased so rapidly in the last few years is the best sign of the qualities of its service. If the gas companies can solve the problem of distribution of excess coke, a still more rapid development will take place. In this connection, the use of two boilers for good-sized houses, one a small gas boiler, the other for coke, will prove economical. Gas could be used for fall and spring and part of the winter,

FIG. 29.—Gas boiler. (*American Radiator Company, New York.*)

covering the greater portion of the heating season. For the cold period the coke boiler would be run a much shorter time, but would still supply a large part of the seasonal heat demand at a lower cost than gas heating throughout.

c. Water Heaters, Heated by Means of Hot Water or Steam.—For steam-hot-water heating systems water heaters (countercurrent apparatus) are used, which are heated by steam and not by coal or coke (see also p. 89). Similar forms of heaters are used in special cases where the countercurrent apparatus is fed by hot water.

5. Boiler Trimmings.

a. Thermometer.—Every boiler should have a thermometer in the supply line to indicate the outgoing water temperature. For accuracy the mercury bulb should be set in the water streams or it should be inserted in a cup filled with oil which is immersed in the stream.

b. Means for Filling and Draining.—The lowest point of the boiler must be provided with a fill or drain cock, arranged to be accessible for inspection in the event of leakage.

c. Altitude Gage.—The height of water is ordinarily indicated by a mark on the altitude gage mounted on the boiler. The following arrangement is somewhat more positive. A small pipe is connected at the normal water level in the expansion tank (p. 44) which terminates at a drain in a suitable location as, for instance, in the boiler room. A valve is placed in this line, and from time to time it is opened by the attendant. The system is filled when water begins to overflow in the pipe line. The valve is then closed, and the system is in operating condition.

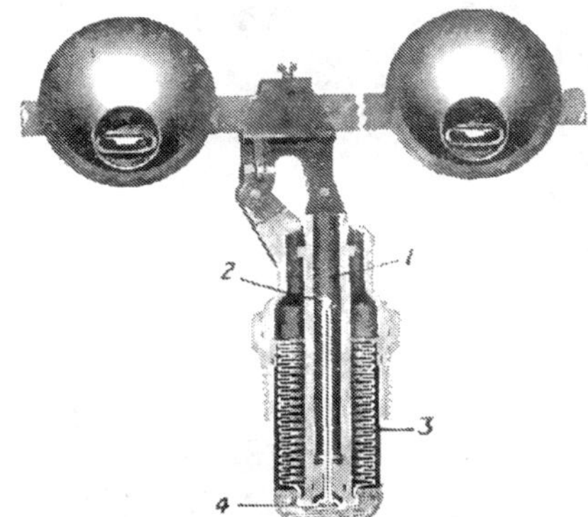

FIG. 30.—Damper regulator. (*American Radiator Company, New York.*)

d. Damper Control.—Every cast-iron sectional boiler is provided with a damper regulator. One form is shown in Fig. 30. It consists of a stem 1 which transmits thrust from the bottom of the bellows distributing the travel equally among the folds, a filling tube 2 provided with a positive method of sealing the liquid chamber, a one-piece bellows 3 providing for extra-long temperature range 100 to 220° F., and a volatile liquid 4 which develops a pressure when exposed to heat. In operation the lever is attached by means of a rod or chain to the damper. If the temperature of the water rises above that desired, the gas pressure generated by the volatile liquid compresses the bellows, forces the thrust rod or stem upward, thereby tilts the lever, and so adjusts the draft. As the water cools, the gas pressure is released and the counterweight opens the draft. Adjustment for temperature is made by changing the position of the weights on the lever.

By the means provided, the regulation is adjusted to the outside weather conditions by controlling the water temperature of the heating system. The following temperatures are suggested:

	Degrees Fahrenheit					
Outside temperature	−10	0	+10	30	50	60
Supply water temperature approximately	200	190	175	150	115	95

The figures given are at best approximate and should be determined for each particular case. The influences affecting the water temperature are: exposure, wall construction, overnight cooling, etc. There are other forms of regulators, both with and without bellows construction, which are suitable for regulating purposes.

c. Dampers.—The correct use of the damper makes efficient operation possible with a saving in fuel. Its incorrect use may be a source of danger.[1] Dampers with fixed openings are objectionable since, if the aperture is too small, a dangerous checking of the flue gas results, while if too large, the control is ineffective. The proper size of opening depends upon the size of boiler, and for equal sizes of boilers it depends upon the operating conditions (*i.e.*, outside temperatures, periods of operation, wind, etc.). Dampers in the main flue are generally undesirable. Dampers should be placed behind each boiler, however, before the gas passes into the main flue or in the case of a single boiler directly into the chimney. The damper should if possible be controlled from the boiler front by the use of chains and pulley mechanisms.

6. Apportioning Load among Boilers.

When a single boiler is used, it may be difficult to control during the periods of extremely light loads. In severe weather it may be necessary to drive the boiler beyond its economical capacity even at the expense of additional fuel.

Where two boilers are used, the former practice was to divide the boiler heating surface into units of one-third and two-thirds of the required boiler capacity. It is then possible to use the smaller in mild weather, the larger in cold weather, and both in extreme weather. Should the larger boiler fail when needed most, the smaller cannot handle the load. For this reason some prefer to use two boilers of equal capacity so that in case one fails, the other may under forced conditions maintain satisfactory operation.

For larger installations where continuity of service is important, three equally large boilers are used so that any two can carry the entire load

[1] MARX, *Gesundh.-Ing.*, 12 and 13, 1917.

without excessive fuel waste. It must be noted that with increase in subdivision of the boiler capacity, the time required for attention is increased.

In boilers with revertible flues and insufficient draft a bypass B should be provided as shown in Fig. 31, which may be closed when not needed. When the bypass is open, the length of flue travel in the boiler is reduced and the flue gas escapes at a very high temperature. While the draft is improved by opening the bypass, the heat absorption by the boiler is less, so that operation is accompanied by an increase in fuel consumption.

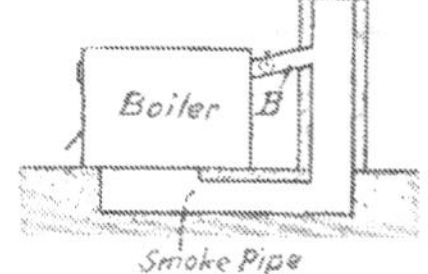

Fig. 31.—Boiler with bypass.

7. Safety Devices.

When several boilers are connected to a line, it should be possible to disconnect any unit. The important reasons are to prevent the heat loss caused by the water flowing through the boilers not under fire and to enable any boiler to be repaired without interruption of the heating plant.

Unsuitable arrangements for this purpose have caused boiler explosions when through oversight a closed-off boiler was put under fire. These accidents have led to the enactment of ordinances in Europe, the more important measures of which will be given.

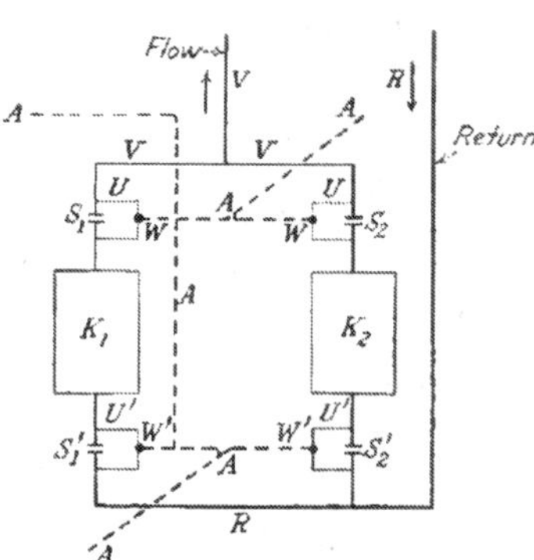

Fig. 32.—Safety valve for hot water boiler installations.

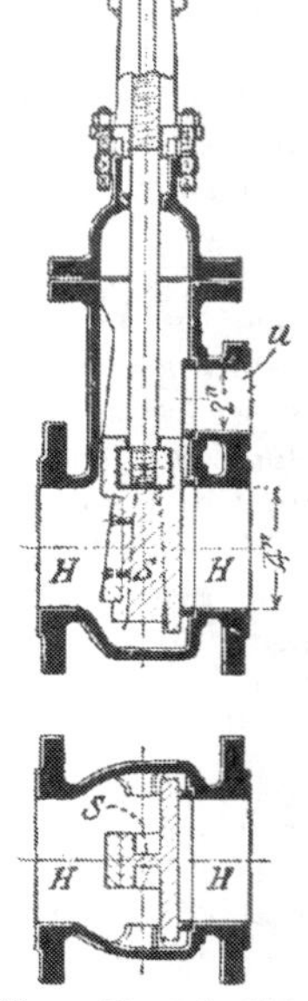

Fig. 33.—Schmidt's bypass valve.

When two boilers (Fig. 32) K_1 and K_2 can be cut out from the supply line V and the return line R by means of the valves S_1 and S_2 as well as S_1' and S_2', these valves must be provided with bypass lines U and U' in which bypass valves W and W' are located. The bypass valves will either open the bypass line or connect the boilers with the atmosphere by means of the vent lines *A* and *A'*. Intermediate settings of the valves cannot be made.

The bypass valve may be made integral with the stop valve as shown in Fig. 33. As shown, the main *H* is closed by the gate *S*. The main *H*, and the bypass *u* are in a definite relation, as will be seen in the illustration. The arrangement makes a simple safety device, allows for a clear pipe connection, and limits the unavoidable losses of water.

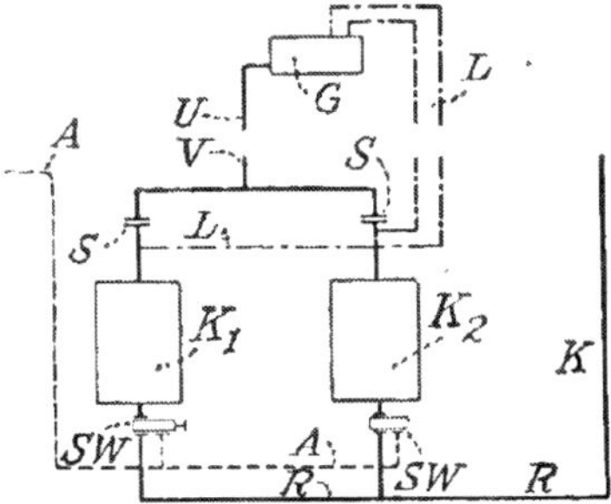

Fig. 34.—Safety lines for hot water boiler installations.

Instead of the bypasses *u* on the supply mains *V*, safety lines may be used which connect the boilers with the atmosphere without the possibility of being shut off. These lines must be carried to the expansion tank, in such a way that they can empty above the level of the highest water line into the tank from above. Figure 34 shows a layout where the bypasses of the return valves are arranged as safety valves *SW* of the kind shown in Fig. 33. The proportions of bypass vents and lines have been determined experimentally.

8. Boiler Testing and Rating.

The industry appreciates that progress in boiler development cannot take place without study of boiler tests. These may be made in the laboratory and in the field. *Bulletins* 2 and 17 of the Research Laboratory of Heating and Ventilation at the University of Berlin-Charlottenburg[1] give a rather complete report on the methods and the results of boiler tests.

Field tests are necessary to determine the operation of boilers under everyday conditions. In these tests it is possible to discover errors common in installation such as pertain to chimneys, flue connections, etc. A number of such tests have been published.[2]

Based on research of this kind, methods of determining the required boiler capacity have been devised (see p. 191). Briefer methods for sizing boilers and boiler rooms were developed by Uber for public buildings[3] (schools, administration buildings, court houses, and prisons). The following table is taken from his book:

[1] Hereafter referred to as the Research Laboratory, Charlottenburg, Professor Dr. Brabbée, Director.

[2] "Economy of Central Heating," from Dr Grahl, "Wirtschaftlichkeit der Zentralheizungen," 2 Aufl., R. Oldenbourg, Munich and Berlin, 1920.

[3] "Construction and Operation of Central Heating Systems," from Uber, "Bau- und Betriebstechnisches für Zentralheizungen in Preussischen Staatsgebäuden," Ernst & Sohn, Berlin, 1915.

Gross enclosed volume, R in cubic feet	Volume H to be heated, in cubic feet	Heat w required for 1 cu. ft. of H in B.t.u. per hour
Up to 150,000	$0.50R$	3.4
150,000– 350,000	$0.60R$	3.0
350,000– 700,000	$0.65R$	2.7
700,000–1,500,000	$0.70R$	2.5
1,500,000 and over	$0.75R$	2.3

Using the table, the effective volume H to be heated is obtained from the gross cubical content of the building R. For every effective cubic foot to be heated, it will require w B.t.u. per hour. Denoting by W the hourly heat requirement, presupposing good construction under ordinary conditions, $W = Hw$. If, as Uber assumes, 1 sq. ft. of hot-water boiler heating surface delivers at least 1,500 B.t.u. per square foot per hour, the required boiler heating surface F in square feet is obtained from the relation

$$F = \frac{W}{1,500}$$

For steel and cast-iron boilers the heating surface is the surface exposed to the fire and hot flue gas. The efficiency[1] of the boiler for specified loads should be guaranteed by the boiler manuacturer. It should not be required that the manufacturer guarantee the fuel consumption of the installation, since this depends upon the building construction, weather conditions, and above all the skill in attendance. In general, test efficiencies may be as high as 80 per cent, while the average annual mean in daily operation may be from 50 to 60 per cent.

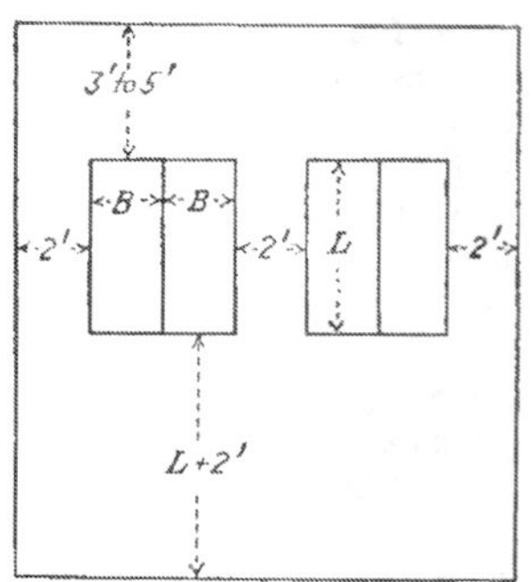

Fig. 35.—Dimensions of a boiler room.

9. Boiler Room.

After estimating the boiler heating surface, the dimensions of the boiler room (Fig. 35) may be fixed from the figures given by Uber. If B is the width of the boiler and L the length, then the following table applies for sectional cast-iron boilers:

[1] By efficiency is meant the relation between the useful heat transmitted to the water and the total heat contained in the fuel. The former is obtained from tests, and the latter is the product of the weight of fuel fired and the heat value per pound.

Hot-water boiler heating surface in square feet	B in feet	L in feet
750– 2,000	$2\frac{1}{2}$	$2\frac{1}{2}$–$4\frac{1}{2}$
2,000– 5,000	3	3 –5
5,000–12,000	5	4 –6

Small boilers up to 750 sq. ft. heating surface require approximately 10 sq. ft. of floor space and an equivalent area for firing. For medium and large boilers only the minimum area required is given in the above table.

For the smaller boilers the fuel is usually stored on the same level with the boilers. In the larger installations elevated storage with downfeed arrangements have been introduced. With inadequate draft, downfeed systems introduce an element of danger. The feed openings should be protected so as to prevent attendants from falling into the feed openings and to prevent as much as possible the escape of furnace gas and the liability to explosion. By means of simple trolley arrangements the conditions might be improved. This is especially so in the larger plants as shown in Fig. 36 where the fuel bunkers are placed above the boiler room and the fuel is charged by means of metal chutes.

The fuel storage space should be designed to take care of about a half years' supply. In larger installations 2 months' needs may be ample. When coke is used, as is the case in Germany and other parts of Europe, certain conditions must be observed. Extra storage space is desirable so that the coke may be bought during the summer months when its price is low and the coke is dry. The space should be divided into pockets of equal size so that the coke may be purchased by volume and not by weight. Gas coke in particular takes up large amounts of moisture. Moreover, the divisions of the space makes it possible to check the amounts delivered and also to tell at a glance the available supply.

Cooling pits should be provided on hot-water systems in which the discharge water is held until cool enough to permit draining to the sewer. These are often required by police regulations. The pit water should be cooled to about 100° F.

Boiler rooms should be provided with supply and exhaust air ducts. The supply ducts which provide the air required for combustion should be approximately double the chimney area. There should also be two outlets, one near the ceiling and another at the floor, joined after rising about 10 ft. The outlet duct should be led through warmed rooms and should have an area of about one-third that of the chimney. The end of the outlet should face opposite to the direction of the prevailing wind so that its operation will be unaffected by it.[1]

[1] UBER, *loc. cit.*

The chimney should be designed for adverse conditions likely to occur. This should not, however, be carried to extreme since, if the chimney is too large, it is difficult to heat, and may cause back drafts. If a number of boilers are used, two or more chimneys may be required. No outside connections to the boiler chimney should be made. Long horizontal flues are not practical, as they tend to create dangerous conditions, especially in mild weather. At the base of the chimney a cleanout door should be provided which also serves in making a preliminary fire for warming the chimney (see Operating Instructions). The architectural

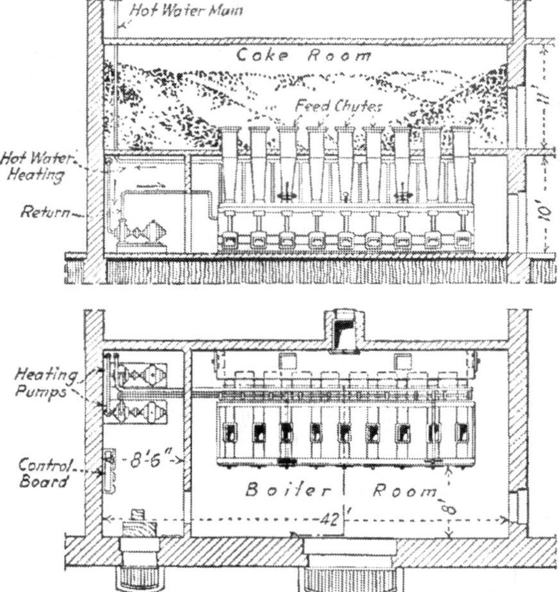

FIG. 36.—Boiler room with coal bunkers above.

features must be subordinated to the safety requirements. The design of chimneys is considered on page 40.

a. Operating Instructions.—Important operating instructions should be pasted in a conspicuous place in the boiler room. Instructions should not only contain general directions but also such instructions adapted to local conditions. In calling attention to possible accidents, the following points should be noted:[1]

[1] "Accidents in Operating Central Heating Systems," from MARX, "Über neuere Unglücksfälle beim Betrieb zentraler Feuerungsanlagen," *Gesundh.-Ing.*, 1917, 12 and 13.

b Firing Boiler.—After interruption of operation and especially in mild weather:

(1) Before firing boiler, open all dampers wide. Test and determine whether slides are tight in the guides; if not, the leaks should be stopped.

(2) Before firing boiler, a preliminary fire should be made at the base of the chimney, to overcome inadequate draft in mild weather.

(3) The chimney fire should be continued sufficiently to warm the chimney walls. After the fire is out, close all openings in the flues and chimney and fire boiler.

(4) Should smoke develop in the boiler room, open all doors and windows immediately. Occupants feeling distressed should leave promptly.

(5) Careful attention to these instructions is more important with the poorer grades of fuel.

The instructions should be amplified by diagrams if they add to clearness.

c Fireman.—The fireman should be trained in the duties he is to perform. The duties consist of attending to the fire, removing ashes, adjusting dampers, and maintaining the proper temperature in the building. He should also strive to burn the fuel economically and to prevent accidents. The latter is dependent to a large extent on the design and the physical condition of the heater.

d Control Board.—For the larger heating systems when a control panel is used, it should be placed conveniently with respect to the boilers. It may have the following equipment mounted: supply and return valves, remote control appliances, thermometers, motor starters and controls for ventilation equipment, and any other equipment necessary for complete operating control (see also District Heating).

10. Chimneys.

In the operation of stoves and boilers, the draft through the chimney is of the utmost importance. It is desirable that this subject be discussed in some detail.

A simple theoretical investigation will assist in solving many problems in the case of house chimneys.[1] In Fig. 37 let

h = vertical height of chimney in feet

l = total length of chimney in feet

t = flue-gas temperature in chimney in degrees Fahrenheit (assumed for simplicity to be the same throughout the chimney)

γ = density of flue gas in pounds per cubic foot

t° = temperature of outside air in degrees Fahrenheit

γ° = density of outside air in pounds per cubic foot

R = frictional resistance of chimney for a length of 1 ft. in inches water column

[1] An approximate determination of chimney sizes is given on p. 283.

Z = local resistances in the chimney in inches water column (including changes of direction, variation of area of cross-section, effect of connecting flues, elbows, tees, etc)

The items R and Z increase rapidly with increase in velocity of the flue gas. Thus at a reduction in flue area at any point there is a substantial increase in resistance or its equivalent, a substantial loss of draft Increasing the number of stove connections may also impair the draft

If H is the draft intensity in inches water column, the following equation must be satisfied

$$H = h(\gamma^\circ - \gamma)0\ 192 = \Sigma(lR + Z)$$

In words, the force exerted by the difference in weight between the column of atmospheric air and the corresponding column of heated flue gas must equal the summation of all the resistances which oppose the flow

From this equation the following conclusions are drawn·

(1) The higher the chimney the more effective is the draft

(2) The chimney is most effective when γ° reaches a maximum, i e , in winter On the contrary, if the outside air is warm, the draft is impaired

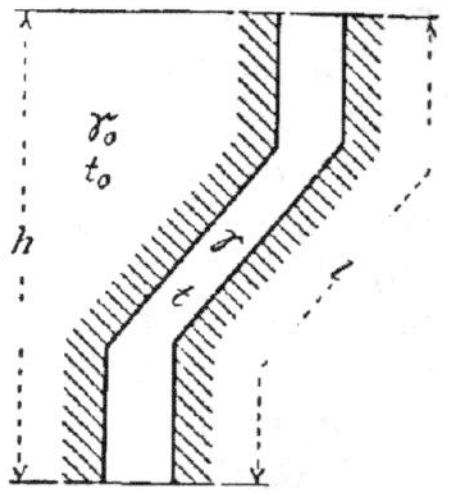

(3) The draft intensity increases as γ decreases, i e , with increase in flue-gas temperature. For this reason chimneys within the building walls are more effective than those placed outside By insulating the chimney where it passes through attic space and by insulating steel-pipe additions, the flue temperature remains higher and the effectiveness of the draft is thereby increased

Fig 37.—Diagram of a chimney

(4) In cases of stoves or boilers having reveitible flues, their added resistance is sometimes responsible for the failure in starting up In such cases the use of a bypass in starting the fire is advantageous.

(5) If γ should be greater than γ°, the chimney action fails This condition may be brought about during a cool night which lowers the temperature of the air within the chimney, the outer air may later be heated by the sun A remedy is to heat the air within the chimney gradually so that it becomes warmer than the outside air

(6) If a wind should blow downward on the chimney, the pressure of the wind may be sufficient to overcome the upward pressure of the chimney action and so cause it to fail in its function The remedy is to extend the height above the roof level or *to mount* a suction head[1] on top of the chimney proper If in an adverse case the chimney top should be in a "pressure angle," the suction head may be inadequate, in which case

[1] See p 139

the remaining remedy is to elevate the chimney above this level (usually a neighboring roof)

(7) The chimney draft decreases when R or Z increase Insufficient flue area, deposits of soot, numerous connecting flues, and the like impair the draft For good operating conditions the chimney must be proportioned properly Chimney extensions should maintain in general the same cross-sectional area. Excessive area may cause countercurrents

(8) An increase in l decreases the draft. Of two equally high chimneys the one with the shorter flue travel is the more effective

(9) An increase in R affects the chimney draft adversely An otherwise correctly proportioned chimney may fail when the inner surface is rough due to projecting stones, etc , in masonry constructions Chimneys therefore should have a smooth inner surface

(10) The chimney draft improves when Z decreases; *i e* , every change in direction impairs the draft, abrupt changes in cross-sectional area are objectionable, improperly designed suction heads offer resistance, excessive penetration of smoke pipes into the chimney decreases the available flue area at that point and so increases the resistance.

(11) Air leakage into the chimney increases γ. It may become sufficiently important to nullify the chimney effect The infiltration of air may occur by leaving cleanout doors open or ajar, by open or partly closed fire doors or ash doors of stoves not in operation but connected to same chimney, or by leakage through partition walls of neighboring chimneys On this account stoves often fail to draw when placed on an upper floor whereas a similarly located stove with an independent short chimney may be effective

(12) In general not more than two stoves should be connected to the same chimney, in exceptional cases three. Gas heaters should be separately connected

11. Piping and Piping Details.

The German standard pipe and fittings are not applicable in American practice. For this reason they will not receive consideration here. These subjects are treated at length in the standard texts[1] to which the reader is referred.

12. Pipe Covering.

For hot-water heating the use of substantial insulating material is important It reduces the heat loss from piping and with it the operating expense Furthermore, these losses may influence the circulation of the system and therefore are of importance in the proper functioning of the system as a whole The use of poor insulation is to be avoided

To determine the insulating qualities of the materials used, the physical standards are set by measuring the coefficient of heat transmission

[1] HARDING and WILLARD, ALLEN and WALKER, "Handbook of the National District Heating Association," etc

Investigations in Germany have for the most part been carried out at the Laboratory for Technical Physics at the University of Munich by Knoblauch[1] and his coworkers (see also p. 70). For a practical comparison of the insulation value in regard to the calculation of heating mains, Rietschel chose the following method with success.

If W represents the heat emission of the bare pipe and p the insulation efficiency of the covering in per cent, then under otherwise similar circumstances the heat loss of a covered pipe is expressed by $\left(1 - \frac{p}{100}\right)W$. In this way p forms a standard of comparison for the insulating value of coverings. These values have been determined by Rietschel in a number of experiments on hot-water piping at the Research Laboratory, Charlottenburg. The more important values are given in the following table:

Number	Type of covering	Values of p in per cent for a thickness of			
		½ in.	¾ in.	1 in.	1¼ in.
1	Infusorial earth (kieselguhr) covered and glued	50	60	67	74
2	Infusorial earth with cork covered and glued	68	73	76	80
3	Infusorial earth, molded	64	69	73	76
4	Cork, molded	50	63	71	78
5	Silk hair	73	78	80	81
6	Felt	79	83	86	87

All these materials can be applied to the pipe without further preparation. When using infusorial (diatomaceous) earth the piping must be heated. This material is a good insulator and is moreover cheap. A disadvantage is that the covering is easily injured. Molded cork is more substantial but also more expensive. Felt is an excellent insulator but is subject to vermin. Asbestos is used wherever coverings must be strong mechanically and operate at high temperatures.

Covering the pipe flanges will decrease the heat loss from the pipe line materially.[2] The covering should be removable so that the flange and bolts are accessible. After completing the pipe covering, it should be wound with a strong material such as canvas and painted a protective coat.

[1] Literature on heat transmission of insulating materials by Knoblauch, Nusselt, Kammerer, Hencky, Raisch, Reiher, Schmidt etc.

Gesundheitsingenieur 1908-1927, also Archiv. für Wärmewirtschaft 1910–1927, and Zeitschrift des Vereins deutscher Ingenieure 1908–1927.

[2] EBERLE, *Zeit. Ver. deut. Ing.*, 1908, 1909.

13. Piping Layouts, Valves, and Expansion Tank.

The piping system of a low-pressure gravity system consists of the supply headers to the boiler, the supply mains (and branches), the supply risers, the radiator connections, the return risers, the return mains, the return header, and return connection to the boiler. The accurate sizing of these elements is of fundamental importance in the balanced operation of the system. These computations are treated in greater detail later on. The correctness of the method of design is confirmed by a very large number of both small and large installations.

In laying out the system, it is important that in the event of damage to any radiator it may be possible to operate all or most of the remaining installation. This is done by providing each radiator with a supply and

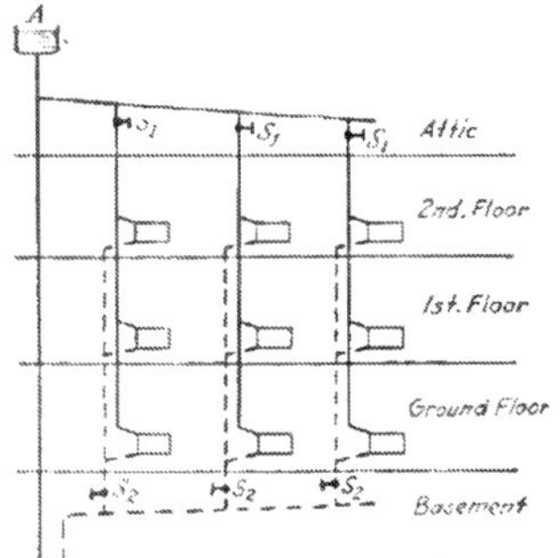

FIG. 38.—Installation allowing risers to be disconnected.

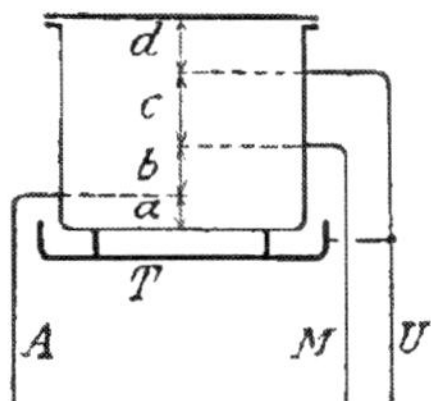

FIG. 39.—Expansion tank.

a return valve, one of which may be used for regulation (see p. 56). Normally this is expensive on account of the large number of valves required. To avoid this no attempt is made to isolate each individual radiator, but means are provided to disconnect the risers. For this purpose each riser is equipped with two valves (Fig. 38) S_1 and S_2 and is also provided with air and drainage vents. If a radiator requires repairs, the particular riser on which it is installed is drained and the remainder of the system continues to function. For shutting off risers, gate valves are recommended, since they offer less resistance to the circulation.

The expansion tank of a low-pressure gravity water system is shown in Fig. 39. It is connected to the system by the pipe A a distance $a = 4$ in. from the bottom. The indicating pipe M is placed a distance $b = 6$ in. above the point a (see p. 33). The distance c is taken so as to provide for about twice the expansion of the total water content of the system. For this allow 1 gal. of content for every 40 sq. ft. of radiator surface. At the elevation $a + b + c$ an overflow connection U is provided, the distance d being from 4 to 8 in. The cover of the expansion tank may be loose or, if tight, provided with a small vent pipe. Below the tank is a drip pan T which leads any water that may boil over to the overflow connection.

If an altitude gage is used in the boiler room, the expansion tank is designed with a smaller area and a greater height.

In accordance with the safety ordinances the expansion tank and the pipe connection A should be protected against frost. This is not always accomplished by locating it near the chimney, since the stackgas at this height may be already too cool for full protection. Placing the tank in the stair well is recommended; in that event the well must be enlarged for its reception.

14. Radiators, Enclosures, and Regulation.

a. Location of Radiators.—The following points should be observed in locating radiators. From physical considerations as to relative densities of the air, determine the air current that will prevail and locate the radiator in that position which will cause the least discomfort to the occupants. For example in Fig. 40, with the arrangement of radiators shown, the alcove will be inadequately heated without regard to the amount of heating surface installed in the adjoining room. In this case the air current will not reach the alcove because of the interference caused by the partition U. If on the other hand all the heating surface is installed in the alcove, it will for similar reasons be overheated, while the adjoining room will remain cold. The evident remedy is to divide the heating surface and install the required amount in each room.

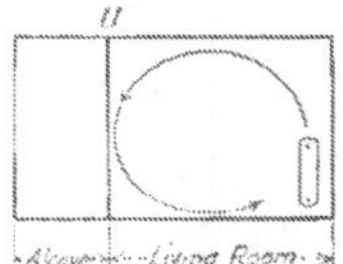

Fig. 40.—Faulty installation of a radiator.

When the air currents are divided in a single room, it is desirable in general to subdivide the radiation so that the proper proportion is located in each cooling area. In rooms with very high ceilings it is desirable that no unpleasant currents be permitted. It is therefore proper to locate radiators beneath large church windows, to heat skylights by individual units, and to place proportionate heating surface beneath glass or sheet-metal roofs, cupolas, etc.

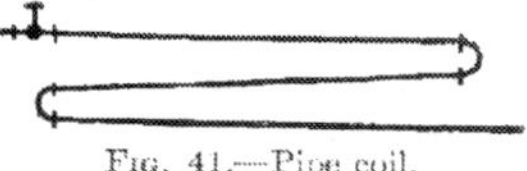
Fig. 41.—Pipe coil.

In rooms where it is probable that partitions may be used at some future time, the general scheme of dividing the radiation at the window locations is proper. For the effectiveness of exposed or enclosed radiators, see page 50.

b. Forms of Heating Surface.—In what follows, only commercial types of radiators will be considered.

(1) Pipe Coils.—Commercial pipe built up in the form of heating surface (Fig. 41) is very efficient. It is made of single or multiple coils, is erected near the floor, and so becomes effective in heating the lower parts

of the room. This form of heating surface is economical to install, since there is a reasonable temperature difference between the coil and the surrounding air, thereby giving a high heat transfer in addition to a substantial heat emission by direct radiation. From a sanitary viewpoint the coil is easily kept clean. A disadvantage of a single-pipe coil is its cost and the limited amount of total heat emitted. When

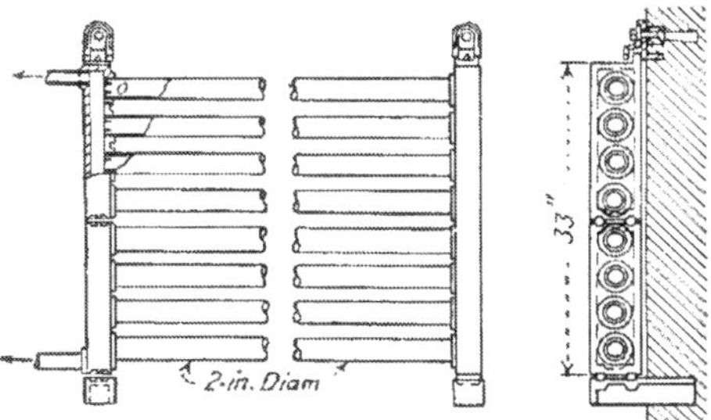

Fig. 42.—Multiple pipe coil.

several pipes are used and connected at the ends by a suitable manifold (Fig. 42), this objection no longer exists. The heat emission from a multiple coil is somewhat less per unit of area than for a single pipe coil, due to the increase in the temperature of the contacting air and consequently lower heat transfer. Multiple coils are themselves easily cleaned and against a smooth wall offer little opportunity for collecting dust.

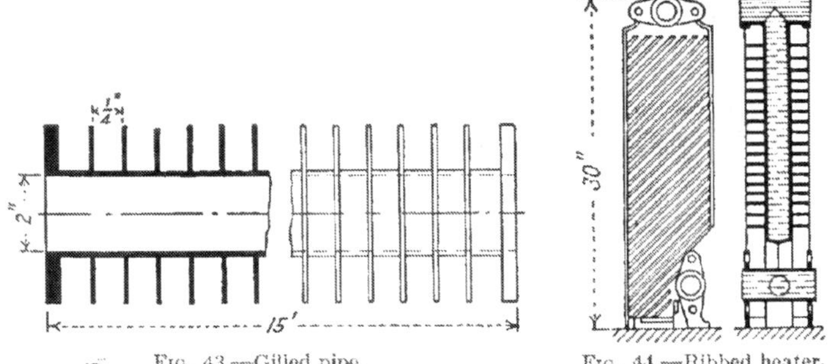

Fig. 43.—Gilled pipe.

Fig. 44.—Ribbed heater.

Uniformity in the heating of individual pipes is obtained by means of a suitable throttling orifice at the inlet side of the pipes.

(2) Fin Surfaces.—Fin surfaces are an attempt to concentrate a large amount of heating surface into a small space. From Fig. 43 it will be seen that this surface is difficult to keep clean and as a result cannot be considered sanitary. The same applies to ribbed design shown in Fig. 44.

(3) Cast-iron Radiators.[1]—The main ideas embodied in the design of cast-iron radiation are: mass production so that by the use of similar sections any desired heating surface is obtainable, small floor-area requirements and ease of cleaning by the use of vertical surfaces, unobtrusive form, good heating effect, and the use of thin cheap gray iron castings. It will be noted that the two-column radiator similar to the one shown in Fig. 45 fulfils these requirements. The surfaces are usually smooth, since the decorative are being abandoned. Radiators are made up with one or more columns (Fig. 46) and in heights from about 12 to 45 in.

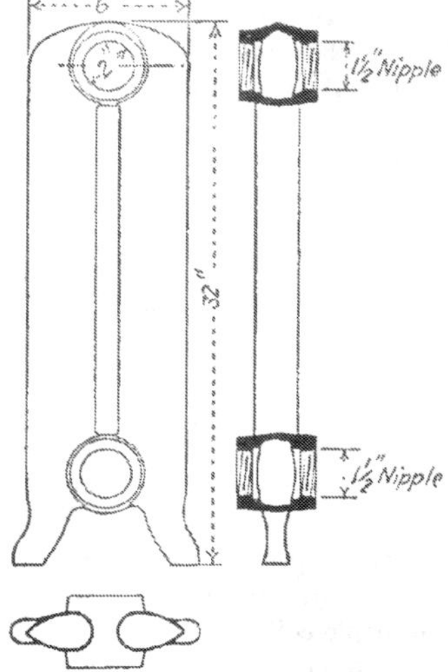

Fig. 45.—Two column radiator.

Although high radiators insure a large amount of heating surface in a small space, it is recognized in scientific analysis that the lower forms are the better. For equal heating surface the lower types are more expensive.

Radiators with four or more columns permit concentration of heating surface in a limited space, but at the same time they are difficult to keep clean. For this reason the low

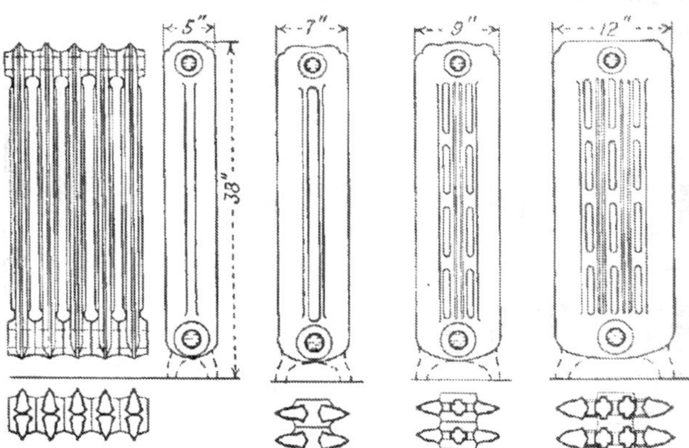

Fig. 46.—One to four column radiators.

[1] The word radiator is a poor choice since the heating effect of a radiator is only partly due to radiation.

radiator with only a few columns receives preference in heating hospitals even though its cost is greater. In cases where the cost of radiation must be limited, high types are used and placed along inner walls instead of under windows; inconvenient drafts may be the result.

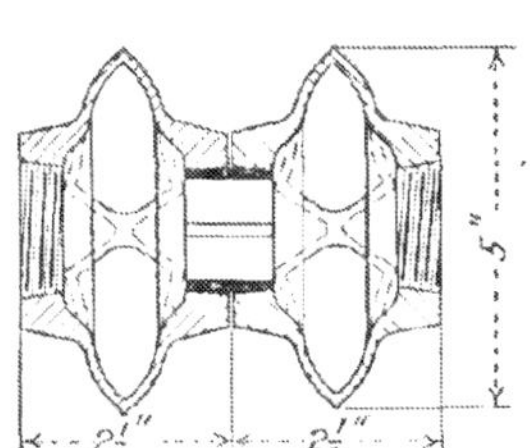

FIG. 47.—Radiator nipple connection.

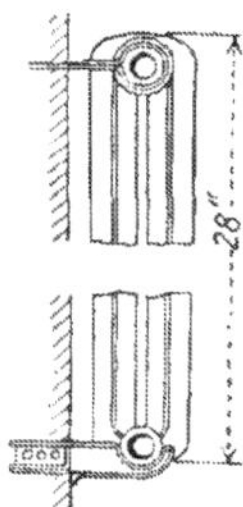

FIG. 48.—Radiator supported on brackets.

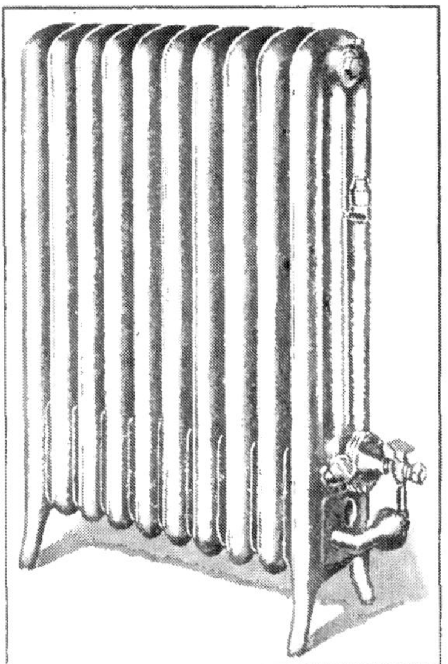

FIG. 49.—Gas steam radiator.

Sectional radiators are joined by means of right and left nipples (Fig. 47) forming a metallically tight joint. Radiators are best when supported on properly designed brackets, as shown in Fig. 48. The use of legs on radiators is often not desirable since it is difficult to clean floors underneath. The sections should be so designed that little material projects above the nipples; otherwise venting becomes difficult and there will be disturbances in the circulation of the water.

Radiators in general are tested to a hydrostatic pressure of 85 lb. per square inch before shipment. Special forms of radiator construction allow for the insertion of gas heating elements, as shown in Fig. 49, for use in mild weather.

On account of the increasing price of cast iron in proportion to its weight, there is a tendency to obtain the greatest heat emission for a given weight of iron. This is expressed in the formula

$$W = \frac{Hk}{g}$$

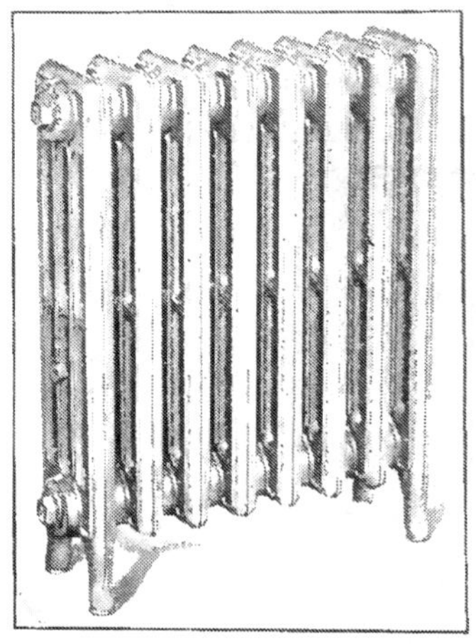

FIG. 50.—American Corto. (*American Radiator Company, New York.*)

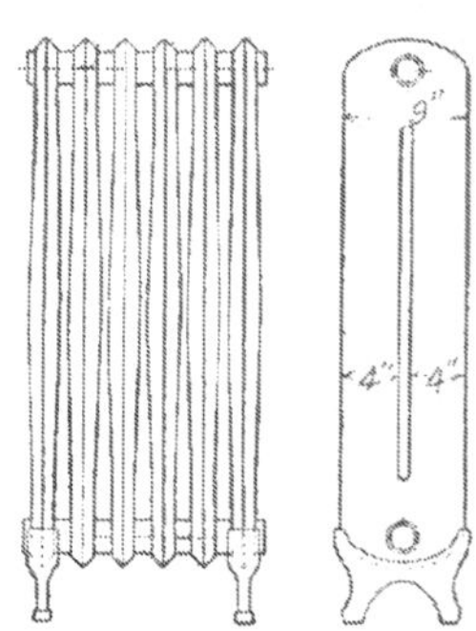

FIG. 51.—Steel radiator.

in which

W = economic factor in B.t.u. per pound of iron
H = heating surface in square feet
K = heat transmission coefficient in B.t.u. per square foot per hour per degree Fahrenheit
g = weight of radiator in pounds.

A new type of radiator giving a high economic factor is shown in Fig. 50, and it is to be seen that this effect is obtained by using a greater number of slender columns. Also it has a graceful appearance.

(4) Steel Radiators.—Recently steel radiator designs of a form shown in Fig. 51 appeared on the market. They are manufactured in sections, each section being welded. It has been stated erroneously that the heat transmission of the thin steel sections is greater than for the thicker cast-iron radiators. The author has shown in numerous tests that the heat transmission is frequently less for steel radiators than for cast-iron designs. In addition steel radiators have important defects in design and will not be durable in low-pressure steam systems.

(5) Ceramic Radiators.—A radiator was also attempted, made from pottery clay and glazed on the outer surfaces. It is sanitary in so far as it permits cleaning. On steam heating systems its heating effect is pleasant. It is not recommended for hot-water heating systems on account of the flooding that may result in case of breakage. From tests the heat emission is lower than for the cast-iron designs.

c. Radiator Enclosures.—The ideal position of a radiator from a heat-transfer viewpoint would be central and in the open so that the air would have free access and it could be cleaned easily. Installation in an

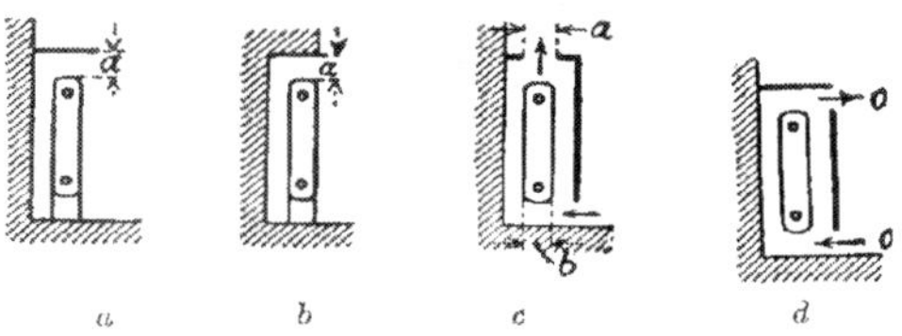

FIGS. 52*a*, *b*, *c*, *d*.—Diagrams of radiator enclosures.

open recess is also to be recommended. It is desired, however, to render the radiator as inconspicuous as possible, and for this reason sacrifices are made. It is important that the wall surface behind the radiator be smooth so that dust will not lodge there. In some cases the radiator and its recess as well as the adjoining walls may be decorated in a way to render the radiator less conspicuous. The heat loss through the outer walls is reduced if an insulating material is placed behind the radiator.

Where radiators are enclosed, in general additional heating surface must be installed. The effect of enclosures on the heat emission has been determined in numerous tests,[1] the more important results of which are given below:

(1) Mantels as in Fig. 52*a* reduce the heat output of radiators as follows:

With space *a* in inches	=	1½	3	4
Reduction of heat emission in per cent	=	5	3	2

[1] *Bull.* 4 of the Research Laboratory, Charlottenburg, published by R. Oldenbourg, Munich and Berlin.

(2) Open recesses of the type shown in Fig. 52*b* reduced the output of the radiator as shown below:

With space *a* in inches	=	$1\frac{1}{2}$	3	4
Reduction of heat emission in per cent	=	11	7	6

(3) Enclosures of the type shown in Fig. 52*c* showed that in a radiator with $b = 9$ in. wide the reduction in the heating effect was as follows:

Depth of register *a* in inches	=	6	7	9	10
Reduction in output in per cent	=	25	19	13	12

(4) Enclosures of the form indicated in Fig. 52*d* with the same radiator and free inlets and outlets *O* showed a reduction in the heating effect of

FIG. 53*a*.—American Recesso Radiator. (*American Radiator Company, New York.*)

25 per cent. When the openings *O* were screened with 50 per cent free air in the screens, the total reduction was 40 per cent.

From the data given it is apparent that adequate provision must be made when using radiator enclosures. A good form of enclosure is shown in Fig. 53*a*, with an American Recesso radiator. The radiators are designed with a flat front and recessed beneath window. Panels are provided at each side removable and providing place for supply and return connections, radiator valve, and trap, concealed but still accessible. The valve handle is located on the window sill, an air inlet is through a grille at the floor, and an air outlet through a grille on the window sill. Proper color treatment will make the radiator unnoticeable; yet its direct radiation is utilized, and only a 10 per cent increase in surface required. The radiator brackets form part of the window construction, making the installation practical and the appearance attractive.

Figure 53*b* shows a construction along similar lines. The radiating effect has been increased making possible a reduction in amount of surface installed, with corresponding economy in operation and fuel cost.

Fig. 53*b*.—Recessed radiator. (*American Radiator Company, New York.*)

An increase in the heat emission results if the enclosure forms a high channel (Fig. 54) by which the output of convection increases. Arnold and Henky[1] made experiments along this line, and the latter has determined that an increase of 15 per cent takes place in the heat transmission.

[1] "Maintaining Specified Temperatures by Regulation," from Arnold, "Über die Einstellung und Einhaltung bestimmter Temperaturen in Räumen durch die Regelung der Heizvorrichtungen, erläutert an Schulheizungen," *Gesundh.-Ing.*, pp. 361, 373, and 381, 1917.

"Recessed Radiators in Art Galleries," from Henky, "Die Nischenheizung in Gemäldegalerien," *Gesundh.-Ing.*, pp. 69 and 81, 1918.

For heaters in which air is circulated under pressure, see heading Warm-air Heating pages 94 and 95.

15. Testing and Rating of Radiators.

A large number of radiators (of different design) have been tested at the Research Laboratory, Charlottenburg.[1] It was found that the heat emitted by a radiator depends upon several circumstances such as location along a thick or thin outer wall, air movement within the room, size of the room, etc. For use in design, the minimum values of the heat transmission coefficient K should be used to provide for unfavorable conditions. In view of this the minimum values have been established.

The heat emission of a radiator is determined from its coefficient of heat transmission K which is the amount of heat given off in the steady state by 1 sq. ft. of radiator surface in 1 hour when the average temperature difference between the radiator and the room temperature is 1° F. at eye level. The heat transmission coefficient K based on condensation measurements is quite variable, and the different factors entering will be discussed later.

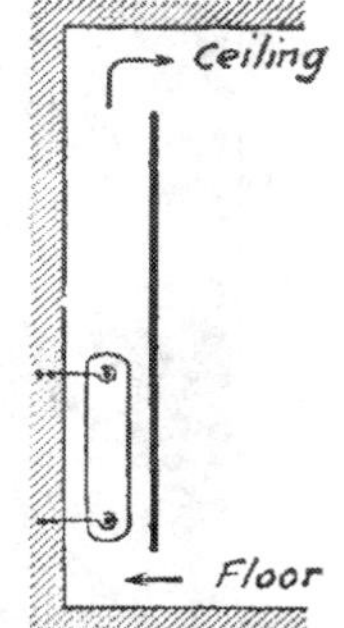

FIG. 54.—Increased convection.

In recent years development along quite different lines has taken place. The effect of radiators in general is not and shall not be considered as merely warming a house or building, but as rendering comfort to the occupants. In other words the basis for judging a radiator must be the heating effect in the lower portions of the room heated as related to human comfort and not the total output as measured by condensation. The goal in heating should be the hygienic ideal, as expressed by the statement, "Warm feet and cool head." The author started investigations along these lines about 1917, and since then has had the opportunity for further studies in a special laboratory devoted to this purpose.[2]

It is the author's belief that results already obtained have been so convincing that a new phase in testing and determining radiator values has begun and will lead in the near future to new conceptions in the heating of buildings for human occupancy.

[1] "Prize Study," RIETSCHEL, "Preisarbeit" *Gesundh.-Ing.*, p. 327 et seq., 1896; also *Bull.* 1 and 4 of the Research Lab. Chlbg. (*Mitteilung der Anstalt*, Heft 1 und 4).

[2] "Testing Methods in Heating Practice," from BRABBÉE, "Relatives Forschen oder wissenschaftlich praktische Versuchsverfahren in der Heizungstechnik," *Gesundh.-Ing.*, p. 159, 1923: BRABBÉE, "The Heating Effect of Radiators," *Jour* A. S. H. & V. E., November, 1925, 1926, 1927; "Heating Effect of Radiators," from BRABBÉE, "Beitrag zur Frage der Heizwirkung von Radiatoren," *Mitteilung der Berliner Versuchanstalt* 36 (*Bull.* of the Research Laboratory, Charlottenburg).

Nevertheless the various factors affecting the heat transmission coefficient K, determined by condensation measurements only, are given below:

a The value of K for a radiator as experimentally determined should represent the minimum value under normal conditions to insure safe application in actual practice This condition is met in the testing method devised by Rietschel and used in this text.

b The values of K become of real importance only when they are determined under the same conditions as in practice

In this connection the Research Laboratory, Charlottenburg, was particularly favored, since the results were gained in making a very large number of tests with practically all types of radiators Comparison of the values of heat transmission K determined from various test installations under widely different conditions is not only of questionable value, but must lead to erroneous results

c The heat emission of radiators is dependent upon the height of the room The higher the room and the larger the cooling effect of the surrounding surfaces, the greater will be the air velocity around the radiator These influences will increase the value of K

d Tests show that a radiator when placed near an inner wall opposite to a window emits more heat than when placed beneath a window, because of greater convection Nevertheless, the latter location is preferred in many cases because of hygienic and other reasons (avoidance of objectionable currents, uniformity of temperature distribution, saving of wall space)

e It is sometimes recognized that the heating effect of radiators from hygienic and economic viewpoints cannot be judged solely by the heat transmission coefficient K

f *Form of Radiator*.—The heat emission from a radiator takes place by radiation, conduction, and convection [1] The radiant effect is materially reduced if the radiator surface is counterradiant For this reason all surface possible should be placed to radiate towards the center of the room When the heating surfaces face each other with narrow spaces between, much of this radiation is ineffective The clear space between two sections of a radiator should not be less than 1½ in on account of the difficulty of cleaning The heat emission by convection is to a large extent dependent upon the air velocity For this reason any hindrance of the air supply at the radiator will render it less effective.

g Increasing the height and the number of sections reduces the heat transmission coefficient Hence for differing radiator heights there will result differing values of K If the radiator has a surface of about 40 sq

[1] "Radiation from Radiators," NUSSELT, "Die Warmestrahlung des Heizkorpers," *Gesundh -Ing*, p 293, et seq, 1919.

ft an increase beyond this figure does not materially affect the heat transfer coefficient The influence of thickness of the metal wall within practical limits may be neglected

h The effect on the heat transmission of ribbed or other forms of fine surface diminishes as these surfaces are extended Experiments of Rietschel showed that a length of 1½ in is economically the best It must not be overlooked, however, that this figure holds only for the radiators tested where the spacing of the parallel ribs was not more than ¾ in

i *Outer Surface of the Radiator*—Rough surfaces increase the radiation effect and therefore increase the heat transmission, while smooth surfaces, on the contrary, reduce the transmission Experiments of Rietschel abroad and the author in this country showed that the final influence of painting the surface is small, with metallic paints such as bronze and aluminum it may reduce the heat transmission 8 per cent

j *Radiator Distance from Wall*—The heat transmission is affected by its distance from the walls It decreases when this is too small as well as when too large The best distance was found to be from 2 to 2½ in

k *Average Water Temperature*—If an average temperature t_m was used in the calculations of the heating surface, it is important that this be maintained in the radiator. Otherwise it will be necessary to correct for the new transmission coefficient K which will prevail at the differing temperature Heating surface becomes less effective if the water is insufficiently mixed. For example, horizontal pipes of large diameter have a lower coefficient than smaller pipes Therefore with flat heating surface baffles should be built into the construction to insure requisite mixing or the radiator must be cross-connected

l *Water Velocity*—An increase in the velocity of the water is accompanied by an increase in the heat transmission. Within the limits used in gravity systems the influence may be neglected

m *Air Velocity.*—Variation of the air velocity over the heating surfaces has considerable influence on the heat transmission Experiments have shown that with certain forms of heating surface a fivefold increase in the heat transmission is obtained with a tenfold increase in the air velocity This is found in the heaters used for ventilation and in the air heating systems considered in greater detail on page 90 *et seq*

n. Influence of Room Dimensions—The room dimensions influence the heat transmission of the radiators in two ways

(1) Plan Dimensions of Rooms—The larger the floor area of a room the less favorable it is for obtaining uniform temperature by mixing of the warm and cool air Also if the heating surface is concentrated in a small space, the average room temperature must be higher for the same degree of comfort.

(2) Height of Room.—The higher the room the more will the heated air rise to the ceiling and consequently the more difficult it will be to maintain the proper temperature at the floor.

16. Computing Radiation.

The amount of heat W lost by the room per hour is determined by means of the heat loss computations considered on page 174. This amount of heat must then be supplied by the radiation in order to maintain the required temperature. The coefficient of transmission of the heating surface F in B.t.u. is found from the experiments previously mentioned and is expressed in terms of the heat transmission coefficient K in B.t.u. per square foot per degree Fahrenheit per hour. These values are given in Tables VIII and IX (p. 298 *et seq.*) for the usual cases arising in practice.

The equations generally applicable for computing the heating surface are also developed on page 178. For hot-water radiators the final equation is expressed in the following simplified form:

$$F = \frac{W}{K(t_m - t)}$$

where

F = radiation required in square feet.
W = heat loss from room in B.t.u. per hour.
K = heat transmission coefficient in B.t.u. per square foot per degree Fahrenheit per hour.
t_m = average water temperature in degrees Fahrenheit.
t = required room temperature in degrees Fahrenheit at eye level.

17. Useful Heat Output of Radiators.

The first experiments in this matter were conducted with steam radiators and, for that reason, it is herewith referred to page 81.

18. Hot-water Radiator Control.

a. Regulating Valves.—Regulating appliances serve a twofold purpose: for use in equalizing the flow through the radiators to secure satisfactory distribution (steam fitters regulation), and for use of the occupant in adjusting the room temperature (hand regulation).

For valves whose combined function is to equalize flow and to regulate, the heat emission of the radiator should be in proportion to the movement of the regulating handle (hand regulation). The unique conditions imposed and the importance of the practical results attained are generally known. For a detailed discussion the reader is referred to *Bulletin* 25 of the Research Laboratory, Charlottenburg.[1]

[1] "Regulation of Steam and Hot-water Heating Systems," AMBROSIUS, "Untersuchungen an Regelvorrichtungen fur Dampf- und Wasserheizungen," *Mitteilung der Anstalt* 25, R. Oldenbourg, Munich and Berlin, 1919.

A valve to fulfil the foregoing requirements should have the characteristics shown in the diagram illustrated by Fig. 55*a*. In this case the regulating lever in moving from the cold to the warm position admits the correct flow from the zero to the wide-open position. Intermediate positions of the valve are found on the line *K*. The valve shown in Fig. 55*b* fulfils this requirement within practical limits. The performance of this valve within the limits of adjustment is shown in Fig. 55*c*. Intermediate adjustments will fall between curves I and III. Hence the requirements of good regulation are incorporated in the design.

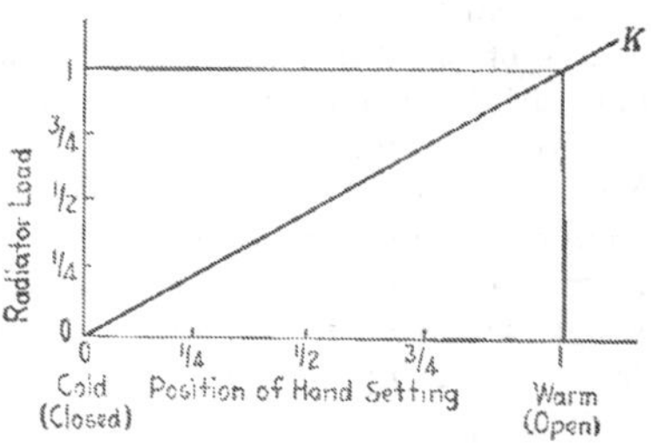

FIG. 55*a*.—Characteristic of an ideal valve.

An extreme case of the valve of poor performance is shown in Fig. 56*a* giving a performance curve shown in Fig. 56*b*. In the figure, *S*

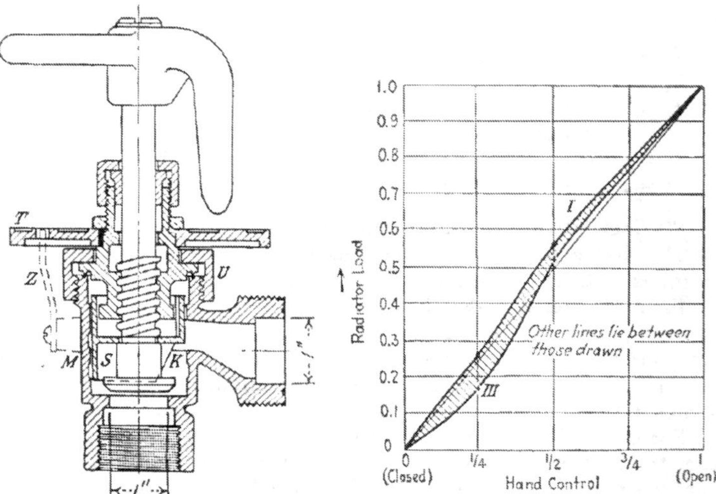

FIG. 55*b*.—Water valve. (*Rietschel & Henneberg, G.m.b.H., Berlin.*)

FIG. 55*c*.—Performance curve of valve shown in Fig. 55*b*.

represents the adjustable cylinder for equalizing the flow. The following may be noted from the diagrams:

(1) The performance curves are different for every adjustment (I to V) and deviate considerably from the ideal performance shown by the

diagonal line. For this reason and because of the fact that the occupant of the room does not know of the adjustment made, good regulation is impossible.

(2) From the performance with adjustment IV it will be noted that the valve may be turned from the warm position to almost two-thirds of its travel without appreciable reduction in the heat emission of the radiator. It corresponds to a radiator load of from 0.7 to 0.8 between these limits. It is only in the last third of the valve movement that regulation takes place. Since a very small movement in the latter interval is accompanied by considerable flow, close regulation is impossible.

(3) The poorest performance is indicated by curve V. If the handle is turned from the position indicating warm to the position 2 with the

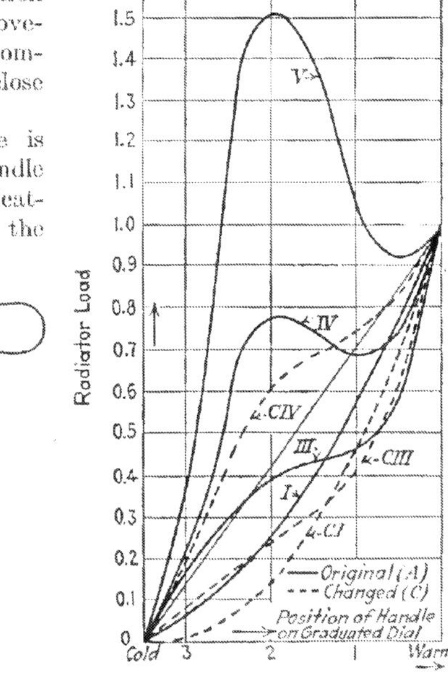

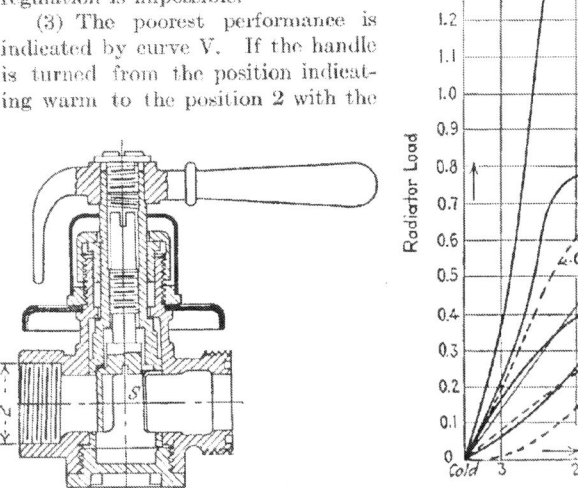

FIG. 56*a*.—Water cock.

FIG. 56*b*.—Performance curve of cock shown in Fig. 56*a*.

idea of shutting down on the flow, the valve increases the flow from 1 to 1.5, *i.e.*, increases the heat output 50 per cent. The effect is therefore the reverse of what should take place. It requires no additional proof that as a means of regulation this valve does not answer the requirements.

b. Automatic Radiator Regulation.—There are two fundamental types of automatic regulators:[1] (1) those in which there is a sudden interruption in the heat supply, and (2) those in which the circulation is affected gradually. The former operate electrically, by air pressure or similar means. The latter use the expansion of liquids as an operating

[1] See also Automatic Regulators for Low-pressure Heating Systems, p. 84.

means. Quick-acting regulators in general, are not desirable for hot-water heating systems. For example, if a room temperature of 70° F. is desired and if the regulator closes at this temperature, overheating will occur because of the hot water stored in the system. In the reverse case if the regulator opens at 70° F., appreciable cooling will take place below this temperature before the system becomes filled with hot water.

The second type gives promise of better results. It should be noted that at present regulators are unable continously and automatically to maintain the desired room temperature for all weather conditions. This, in addition to the fact that regulators require attention, frequent adjustment, and are expensive to install, has had a deterrent influence in extensive applications in Europe.[1]

Fig. 57.—Parlor Arcola. (*American Radiator Company, New York.*)

In the United States, however, thermostatic control has received considerable attention. Its application has resulted in economy and has made a wide appeal to the public in that it has reduced to a minimum the attention required at the boiler.

Single-story (Loft) Heating

The form of boiler shown in Fig. 57 is used for low-pressure hot-water heating systems where the boiler and radiators are on practically the same level. No cellar is required in such systems, though if one is available its use is feasible. As mentioned on page 27 (Fig. 24) combination ranges and heaters with radiators attached have been used in Europe, more especially in England. Small boilers in compact form for use in the living room of a small house are available. By means of an outer jacket they are made attractive in appearance.

The advantages of a hot-water heating system for small homes, independent buildings, etc. are that it is available for heating floors of buildings otherwise unheated and that the regulation is under control of the occupant.

The disadvantages are that if the rooms are to be warm at an early hour, the heater requires early attention; that fuel and ashes are handled

[1] Various forms of regulators were tested at the Research Laboratory, Charlottenburg. See *Bull.* 2 of the Research Laboratory, Charlottenburg, published by R. Oldenbourg, Munich and Berlin, 1910.

in the living room or kitchen and that mains and risers are exposed in the rooms.

For the design of this form of heating system, see page 203.

Greenhouse Heating

Greenhouse heating requires a low-pressure hot-water system using a boiler similar to the one shown in Fig. 57 and pipe coils for heating surface as shown in Fig. 42. The system described above under Single-story Heating is well adapted for heating the smaller greenhouses. The design of such systems is considered on page 203.

Accelerated Circulation Systems

Forced-circulation systems are hot-water systems in which the velocity of the water is increased by flowing steam or air into the supply line. The Reck heating system shown in Fig. 58 is representative of the possible systems[1] and was the first installation of this type. Steam from the boiler A flows through pipe B and enters the counter-current heater D through pipe C, where the hot water for the system is heated. The hot water is then piped through E and passes through a second countercurrent cooler F and thence to the mixer G. Here steam from the pipe B is sprayed into the supply, whereupon the light steam-water mixture passes through the motor line H into the expansion tank J. Here the steam separates to the top, flows to the condenser K where it is condensed to water, and returns through the pipe L to the boiler. The line N of the first countercurrent heater D connects with the common return M. From the lower part of the tank J the supply water flows through the upper supply main O into the heating system and returns from the system through pipe P to the first heater D.

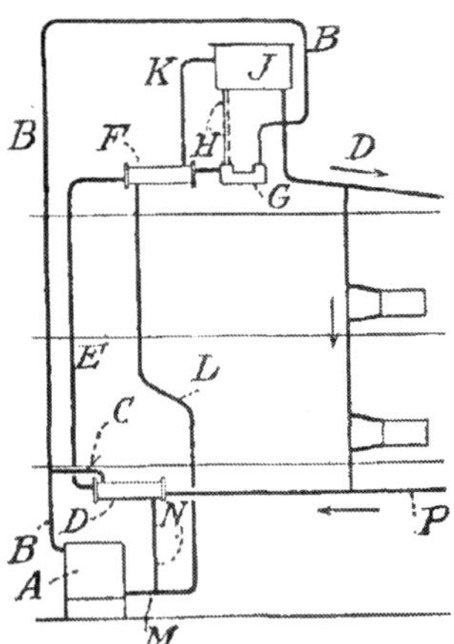

Fig. 58.—Reck heating system.

1. Advantages.

The circulation is accelerated by the force applied in the motor line, so that smaller piping may be used. The increased circulation permits the installation of radiators below the boiler level.

[1] "Accelerated Hot-water Systems," from Meter, "Warmwasser-Schnellumlaufheizungen," *Gesundh.-Ing.*, p. 469, 1907.

2 Disadvantages.

The installation requires a number of specialties which need inspection and attention At times the system is noisy Water enters the supply line at approximately 212° F, which makes it less desirable from a hygienic viewpoint than the gravity hot-water system The system lacks the simplicity of central regulation common to gravity systems

3 Field of Application.

Installations of the kind described were used when a large percentage of the radiation was located below the boiler level In cases of this kind, forced-circulation systems operated in conjunction with pumps are better During a certain period the Reck system practically became the standard, much work and many patents having been issued in its connection The system, however, is now losing its popularity In the case of district heating some of Reck's ideas have been retained (see p 105)

Rooms below the boiler may also be heated without forced circulation by one of the following stoves, one-pipe system, locating radiators on ceiling with special air supply, and pumping systems

If the radiator is placed near the ceiling, an enclosure is necessary through which air circulation is forced to the floor This form of installation is not recommended

The design of accelerated circulation systems is considered on p 203

HOT-WATER SYSTEMS WITH FORCED CIRCULATION

A pump system of forced circulation applied to a hot-water system is shown in Fig 59, where P is the pump installed in the return R

1. Advantages.

Among the advantages of this system are the following heating of large horizontally extended buildings and groups of buildings from a central location, unified control from a central point, heating of rooms and buildings which are below boiler level, reduction of installation costs by use of smaller pipe sizes, freedom in layout of mains and risers, attention to proper venting, single firing center, centralization of fuel and ash handling, use of low-grade fuel in properly designed furnace, simplification of attendance and supervision of installation, possibility of extensive heat utilization

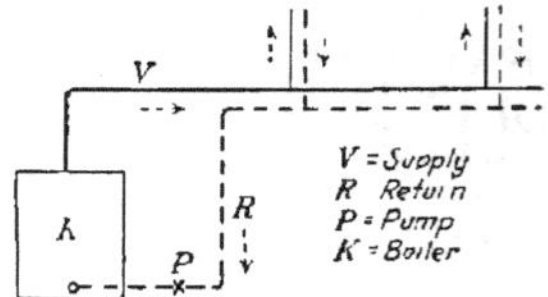

FIG 59 —Diagram of a pump heating system

2. Disadvantages.

Such a system requires a pump, and has an operating expense for electricity, or other motive power.

3. Field of Application.

Pump systems should be used in place of gravity systems when the installation cost and operating expense (including interest, depreciation, etc.) warrant and when proper attention to the system is assured. The applications are for extended buildings, plants,[1] hospitals, communities using exhaust heating, power plants with heating, ventilating, and hot-water supply systems, district heating systems, etc.

4. Installation.

The boiler, piping, and radiators for a pump system are fundamentally the same as for a gravity system with the exception that in the larger installations, steel boilers, automatic stoking, etc. are likely. Due to the greater difficulty of venting on account of the more rapid circulation, proper precautions must be taken. In many cases a layout shown in Fig. 60 has been found satisfactory. In this the water from the boiler flows through the riser S to the open expansion tank A placed at the highest point of the system. The supply lines pitch upward to the last supply riser, where closed venting tanks E are located. Increasing the pipe size at T improves the venting. Soon after starting the system the tanks E should be vented. F indicates supply lines, P pump, and K boiler.

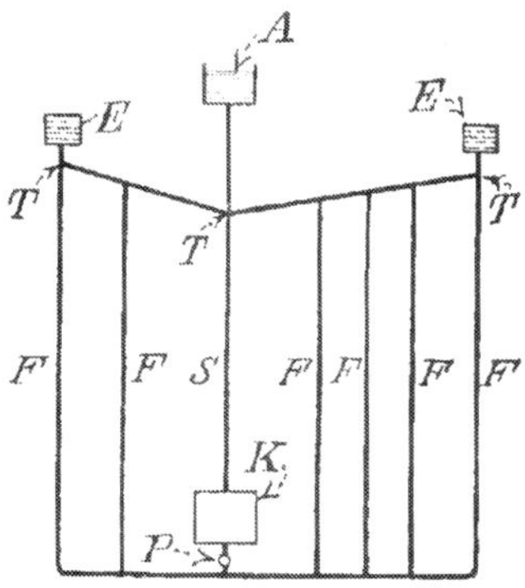

Fig. 60.—Venting of a forced circulation system.

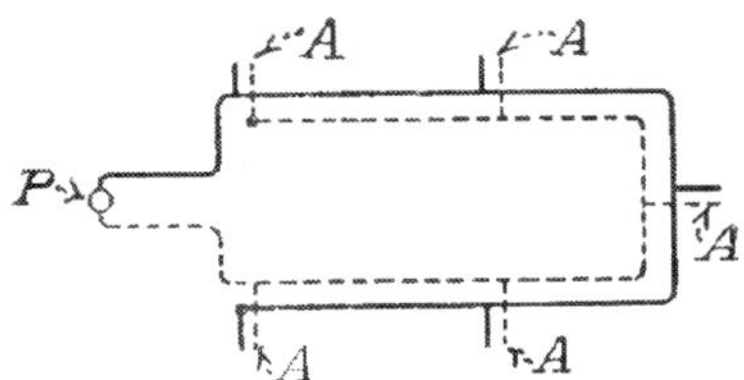

Fig. 61.—Layout of Tichelmann.

It is sometimes necessary in pump systems to check the flow in the risers close to the pump. This may lead to some difficulties and can be avoided by using the layout of Tichelmann (Fig. 61). For details of installing pump and expansion tank connections, see page 102.

At times it may be desirable to supplement a gravity system by the addition of a pump for use in severe weather. In this case the pressure should be such as to exceed that of the gravity system by only a slight amount. In computing the pump system the gravity effect of the system itself must be taken into consideration. The detailed computation and layout for pump systems is given on page 203.

[1] Whittemore, E. H., "Forced Hot Water System Heats Ford's Twin Cities Plant," *Jour. A. S. H. & V. E.*, vol. 31, no. 12, December, 1925.

In installing the pump, care must be exercised so that the noise of the pump is not transmitted to the piping system This in a measure is attained by installing the pump on a suitable foundation, by choosing a quietly operating motor with a low operating speed, by using heavy rubber gaskets between pump and main flanges, etc The pump, if motor driven, should have direct current for wide range of speed regulation or a steam turbine drive Provision should be made to insure continuity of operation by duplication of equipment or by having a supply of spare parts on hand The use of a day and a night pump may be economical. Safety devices[1] should be provided so that even for long periods with pumps out of service boiling over or hammering will be prevented

In very large installations, as for example when several buildings are connected to one pump system, it becomes virtually a "district" system and these are considered in greater detail on page 100

B. MEDIUM-PRESSURE HOT-WATER HEATING

Hot-water heating systems using medium pressures differ from low-pressure systems in that they operate at somewhat higher temperatures In low-pressure systems the maximum temperature of the water in extreme weather is around 200° F, whereas the medium-pressure systems employ a temperature around 250° F under the same conditions The medium-pressure systems have a higher surface temperature on the radiator even for moderately cold weather, which is above that hygienically desirable

Systems of this kind differ from those employing low pressure in that the expansion tank is closed The overflow leads to a compression valve so that a pressure of about 15-lb gage is maintained to insure the necessary water temperature

C HIGH-TEMPERATURE HOT-WATER HEATING

High-temperature hot-water heating systems employ temperatures as high as 400° F For hygienic conditions this temperature is entirely too high for dwellings Its use in certain industrial processes as in baking, drying, varnishing, etc has much to recommend it As shown in Fig 62 it consists of a closed piping system with pipes of 1-in inside and 1½-in outside diameter. The piping is tested to a pressure of about 3,600 lb per square inch The boiler K is formed of one part of the piping system while the other part is placed in the rooms to form the heating surface H connected by the piping R

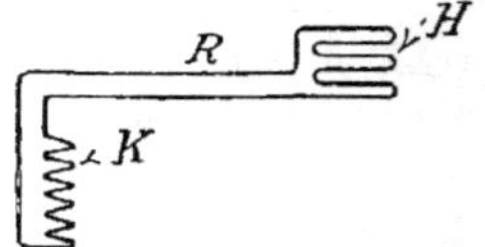

FIG 62 —Diagram of a high temperature hot water heating system

[1] "Safety Devices," from SCHMIDT, "Sicherheitsvorrichtungen," *Gesundh -Ing*, Np 3, 1914

The boiler is of the form shown in Fig. 63 in which two or more groups of pipes may be placed. A direct control of the heat output is impossible. It can be effected indirectly by permitting the water to flow through the coil by regulating a three-way cock and a bypass, thus diverting the water which is not needed in the coil. The expansion tank is made up of piping arrangement.

For computation relating to the design of hot-water heating systems see page 204 *et seq.*

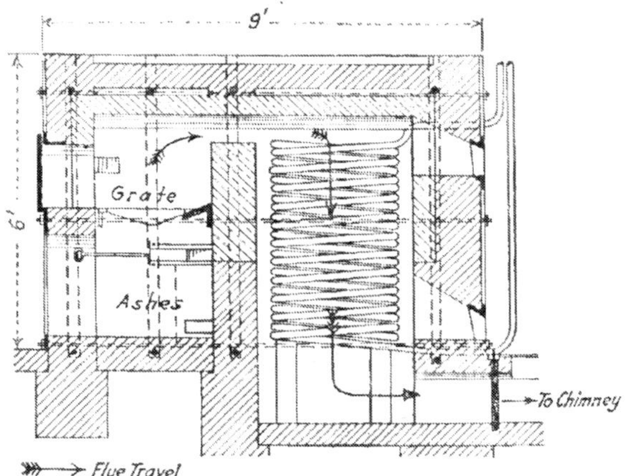

Fig. 63.—Boiler of a system similar to that shown in Fig. 62.

D. STEAM HEATING

High-pressure Steam Heating

1. Advantages, Disadvantages, Field of Application.

a. Advantages.—Low cost of piping installation, flexibility in the layout of pipe lines, and long-distance transmission of heat are the main advantages.

b. Disadvantages.—Among the disadvantages are the following: surface temperature of 250° F. and over, hence considerable burning of dust; excessive radiation effect; impossibility of regulating the radiators; noise from the circulating steam; water hammer in the pipe lines; constant attention to the steam traps; leakage of steam in piping connections; heat loss due to high temperature and therefore expensive operation; costly insulation; the necessity for official permits for the boiler installation.

c. Field of Application.—Due to these important objections high-pressure steam heating is not recommended for dwellings. It is sometimes used in factories, halls, etc., but even in these cases, unless other circumstances control it is better to use low pressure. For the use of high-pressure steam in warm-air heating, see page 93, and for steam hot-water systems, see page 88. The main application in heating practice is in the long-distance transmission of heat.[1] A special form of high-pressure steam heating is the circuit system used in America. A modification of the latter is the (Krantz) system shown in Fig. 64. Live steam from the boiler is fed to the turbine which in this case is of the extraction type.[2] From a suitable stage of the turbine (or from the receiver in the case of compound engines), bleed (extraction) steam is taken at about 45-lb. pressure and piped to the supply main. From this main several supply risers lead to different heating coils of ordinary pipe construction placed near the floor. Depending upon the outside tem-

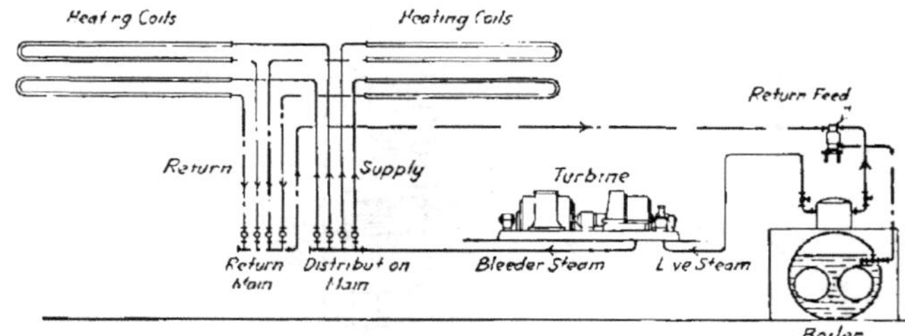

Fig. 64.—Krantz system.

perature one or more of the heating coils is put into service. The coils may be of different diameters so as to vary the heating surface.

The condensation flows through the return lines to the return main and to an automatic return trap from which it is returned directly into the boiler.

The Krantz system avoids the use of condensation return pump, replaces the control of the heating surface by group regulation, and makes extensive use of the heat in the condensation. In its application to a heating system a number of details must be considered so that each installation must be decided on its merits.

2. High-pressure Heating System Details.

a. Boiler.—Boilers used for high-pressure steam heating are the same as those used generally. For a discussion the reader is referred to the texts on the subject. In choosing the type of boiler, it should be noted that heating installations have a variable load and that they may require

[1] See District Heating, p. 96.

[2] Turbines from which steam is taken from intermediate stages are called "extraction" or "bleeder" turbines. The bleed steam is taken from that stage which has the required pressure suited to the purpose in hand.

large amounts of steam, especially for heating up. Superheat is of a restricted value for heating purposes and increases the cost of maintenance of the piping system. When superheat is used to insure dry steam, it should be limited to from 100 to 200° F. In general there is little need for steam pressures in excess of 120-lb. gage. It is usually inadvisable to use higher pressures than this, but if such is expedient, the pressure may be lowered by means of reducing valves.

b. Piping Connections and Special Apparatus.

(1) Steam-water Separator.—If superheat is not used, the steam from the boiler should be dried. One of the numerous forms of steam separators used for this purpose is shown in Fig. 65. The water is separated from the steam by impinging on the vertical baffles and also by abrupt changes in the direction of flow. In computing the steam piping resistance, the resistance of the separators should be considered.

Sendtner,[1] Deinlein,[2] and Hencky[3] have shown that steam contains only 1 per cent of moisture in the form of fine spray, while the remainder of the moisture is in the form of globules to be separated by means of orifice plates.

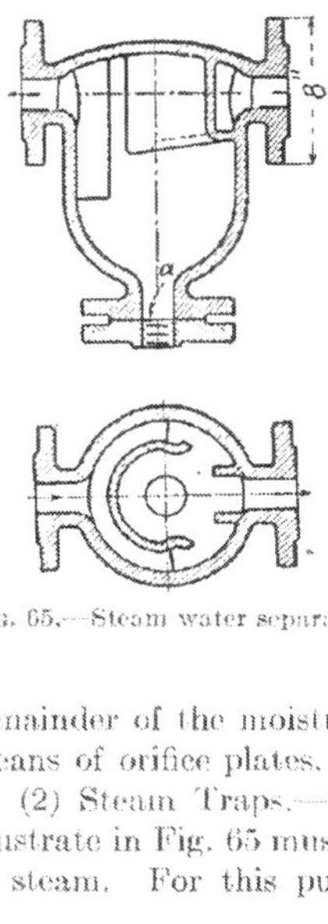

Fig. 65.—Steam water separator.

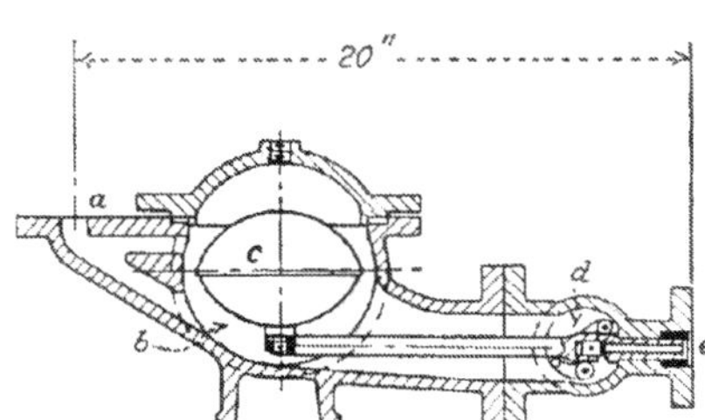

Fig. 66.—Steam trap (float type). (*Drayer, Rosenkranz & Droop, Hannover.*)

(2) Steam Traps.—The water which collects at *a* in the separator illustrate in Fig. 65 must be drained in such a way that there is no escape of steam. For this purpose traps are used. Since the loss of steam

[1] "Determination of Moisture in Steam," from Sendtner, "Die Bestimmung der Dampffeuchtigkeit mit dem Drosselkalorimeter und seine Anwendung zur Prüfung von Wasserabscheidern," Dissertation, Munich, 1910; *Mitteilung Forschungsarbeiten Ver. deut. Ing.*, Heft 98 and 99, Berlin, 1911.

[2] "Research on the Influence of Boiler Load on the Moisture in Steam," from Deinlein, "Versuche über Abhängigkeit der Dampffeuchtigkeit von der Kesselbelastung," *Zeit. Bay. Rev. Vereins München*, p. 135, 1913 and p. 203, 1914.

[3] "Moisture Content of Steam," from Hencky, "Die Grösse des Feuchtigkeitsgehaltes von Wasserdampf," *Zeit. Bay. Rev. Vereins München*, pp. 165 and 175, 1920.

through traps has an important bearing on the economy of high-pressure heating systems, they will be considered in some detail.

Traps are divided into float traps, bucket traps, expansion traps, and traps without movable parts.

Figure 66 shows a trap of the first type. The condensation enters at *a* into the chamber *b* in which there is a float *c* which is connected by means of a lever to the valve *d*. As the condensation fills the float chamber *b*, it lifts the float which opens the valve *d*. The steam pressure behind the water then forces it out through the outlet *e*. As the water level in the float chamber drops, the float drops with it and by means of the lever it closes the valve *d*.

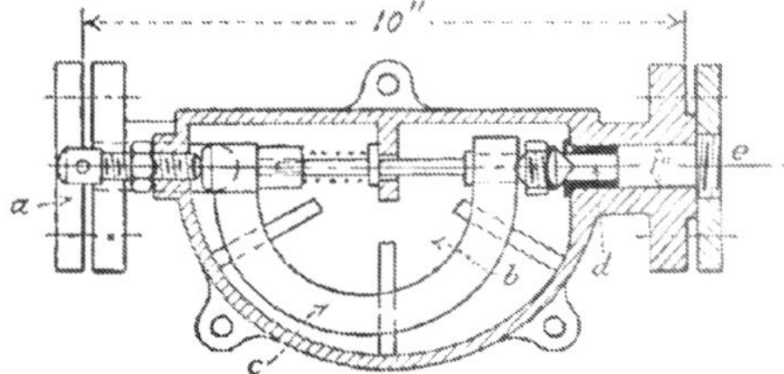

Fig. 67.—Steam trap (expansion type). (*Jager, Rothe & Nachtigall, Leipzig.*)

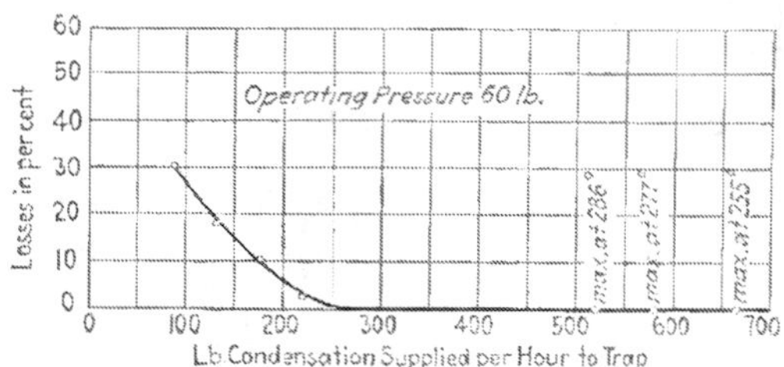

Fig. 68.—Graph of steam losses in a trap.

An expansion trap is depicted in Fig. 67. The mixture of steam and water enters at *a* and is quite hot when it enters the chamber *b*. This contains an expansion member *c* filled with a volatile liquid of a low boiling point. The needle valve *d* is closed by the hot condensate. As cooling occurs, the spring contracts, valve *d* opens, and the steam pressure forces the condensate out at *e*.

The trap designs shown often have large steam losses, especially at low loads. Figure 68 was plotted from tests made at the Research Laboratory, Charlottenburg,[1] for a type not shown here. It will be noted that the steam losses begin with an hourly condensation of 260 lb. per hour

[1] *Bull.* 2 of the Research Laboratory, Charlottenburg, published by R. Oldenbourg, Munich and Berlin, June, 1919.

and that the losses increase rapidly with decreasing load. The maximum discharge capacity with the condensation temperature at 255° F. (starting-up period) was 660 lb. per hour. With rising temperature (286° F.) it decreased to 520 lb. per hour.

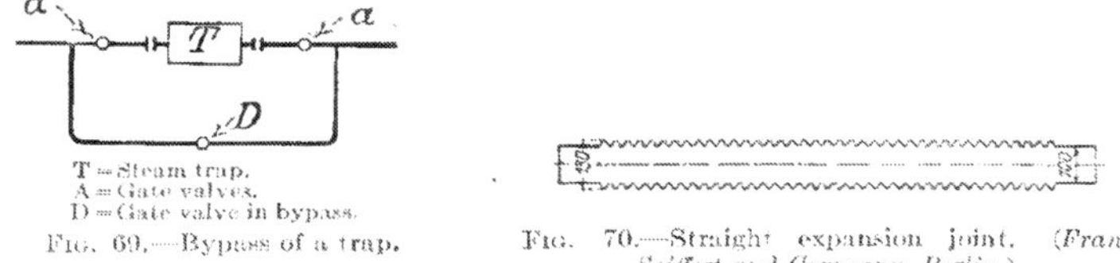

T = Steam trap.
A = Gate valves.
D = Gate valve in bypass.

FIG. 69.—Bypass of a trap.

FIG. 70.—Straight expansion joint. (*Franz Seiffert and Company, Berlin.*)

Due to operating conditions traps are subjected to considerable variation of load. With certain methods of installation, venting may become difficult so that special venting valves are frequently used. In view of the daily attention needed by traps, they should be installed to make them easily accessible. A bypass arrangement shown in Fig. 69 should be installed so that repairs are possible without interruption to the service.

(3) Piping and Piping Connections. Flanges are the usual means of connecting the larger sizes of pipe, particularly when high pressures are used. In this connection the standard American practice is to be followed. Welding may also be used, but in its application care must be exercised. When welded construction is used, flanges may be desirable about 50 ft. apart to permit repairs. Corrosion seldom takes place in steam lines, but it is usual in lines carrying condensate. For this reason the latter are frequently installed with brass piping.

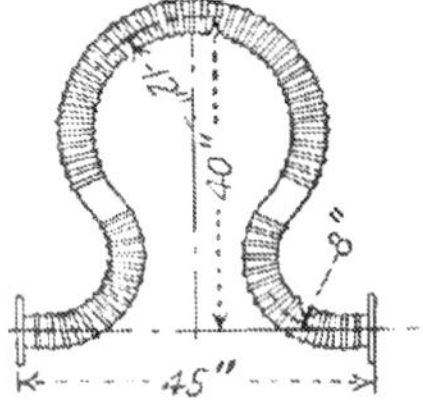

FIG. 71.—Curved expansion joint. (*Franz, Seiffert & Company, Berlin.*)

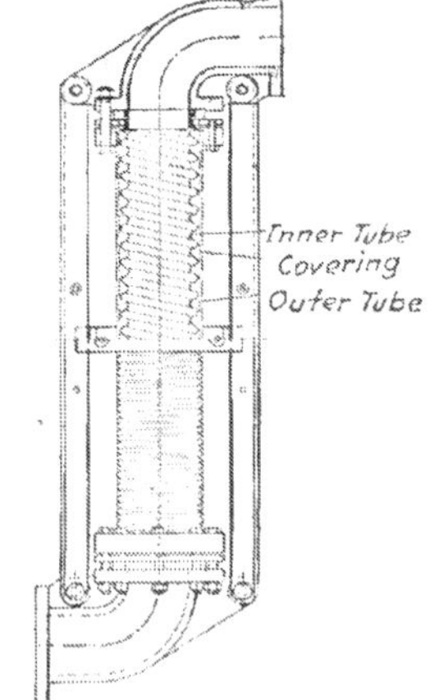

FIG. 72.—Metal hose connection. (*Metalschlauchfabrik-Pforzheim.*)

(4) Pipe Anchors and Expansion.—It is important that piping should be anchored and supported with due regard to operating conditions.

It may be noted that expansion may take place rapidly when a large valve is opened quickly and that the expansion may be nearly 2½ in. per 100 ft. of pipe. It is therefore desirable on long lines to use a small bypass valve through which small amounts of steam may be turned into the line for warming up prior to opening of the main valve.

Figure 70 illustrates a straight and Fig. 71 a curved expansion joint. A metal hose connection is illustrated in Fig. 72.

(5) Pipe Lines.—Pipe lines should always be pitched in the direction of flow. In Fig. 73 a separator should be located at *a* and a trap at *b*.

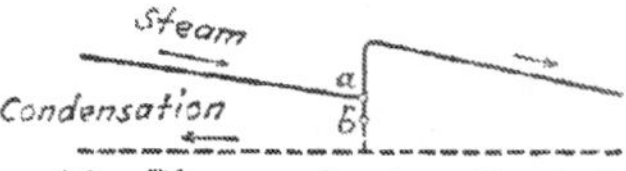

Fig. 73.—Diagram of a steam line drain.

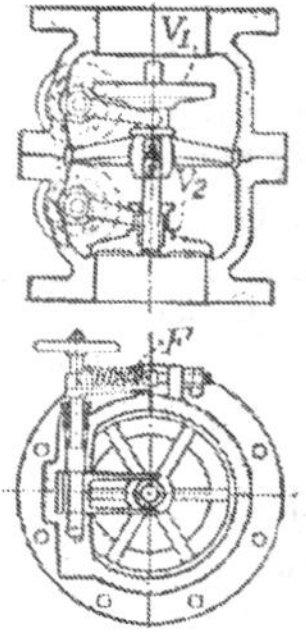

Fig. 74.—Quick-acting steam valve. (*Schaffer & Budenberg-Magdeburg.*)

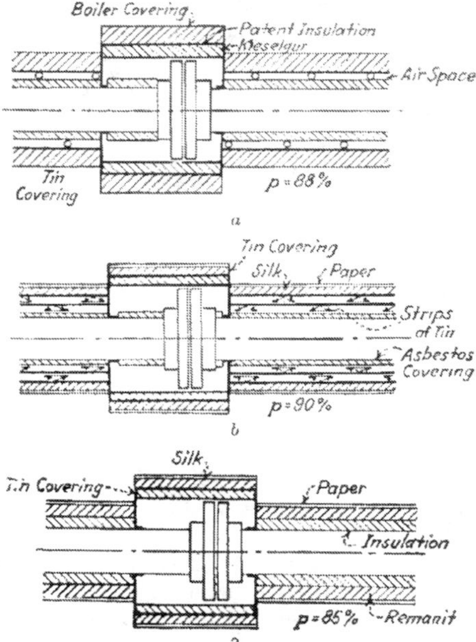

Figs. 75*a*, *b*, *c*.—Various methods of insulating high pressure steam pipe and flanges.

For speedy cutting off of the steam supply, quick-acting valves are used. Figure 74 shows one design. It will be noted that it has two valve discs V_1 and V_2, which shut either side of the pipe line in case of a break in the system. Springs F are located on the outside where they are open for inspection and adjustment of the tension. Electrically operated valves with remote control from any desired point (boiler room, distribution center, entrance of conduit, etc.) are also used.

The computation of high-pressure steam systems is given on page 230.

(6) Insulation.—It is important that excellent insulation be provided for high-pressure steam lines, since the surface temperature of the pipe is high, and as a consequence the heat losses are large. Flanges also should

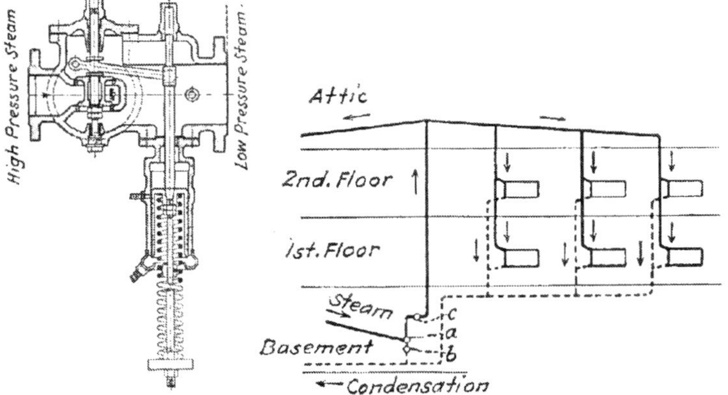

Fig. 76.—Reducing valve. (Gebr. Sulzer, Winterthur-Ludwigshafen.)

Fig. 77a.—Overhead distribution, high pressure steam heating system.

be provided with removable coverings. The materials mentioned on page 43 are unsuitable for coverings in direct contact with the pipe, since they would be injured by the heat. In this case special coverings are used in contact with the pipe. The insulating materials shown in Figs. 75a, 75b, and 75c have been tested[1] and the efficiency has been determined.

(7) Reducing Valves.—The high-pressure heating systems are operated with absolute pressures of about 30 lb. It is therefore necessary to reduce the line pressure from, say, 120 lb. to that indicated. For this purpose pressure-reducing valves are used designed to operate

[1] "Tests of Insulating Materials," from Eberle, "Versuche mit Isoliermitteln," *Zeit. Bay. Rev. Vereins*, pp. 105, 117, 129, 139, 151, 194; 1909; see also "Heating and Ventilating Equipment of Factories," from Hüttig, "Heizungs und Lüftungsanlagen in Fabriken," published by Otto Spamer, Leipzig, 1915.

with either a weight or a spring (Fig. 76). Care should be taken to insure retaining a small degree of superheat after the reduction of pressure to deliver dry steam.

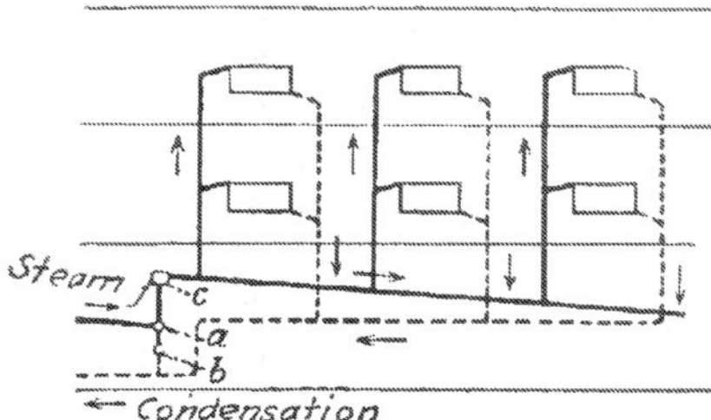

Fig. 77*b*.—Basement distribution, high pressure steam heating system.

(8) Layout of Heating Mains.—A good layout of piping is the overhead distribution system shown in Fig. 77*a*. The steam enters through the separator *a*, having a trap *b* for the drip as shown, and thence to the reducing valve *c*. The steam supplies the various risers by means of the upper distributing main. In this scheme the flow of condensation is in the same direction as the steam flow.

The layout illustrated in Fig. 77*b* is not as desirable, since the supply risers do not admit of the same degree of drainage and since the drip is opposed to the flow in the risers. Nevertheless the latter is frequently chosen owing to the lower cost of installation. The design of these piping systems is considered on page 230.

c. Radiators and Their Control.—Ordinary pipe is readily made up into good heating surface and permits easy venting. The various forms of radiators shown on page 45 *et seq.* may, however, also be used. Radiators which do not vent well must be supplied with vent valves or ordinary pet cocks and may require daily attention. At the present time automatic air valves have come into general use and give satisfactory results. They also act as air-inlet valves. Figure 78 illustrates the larger form of air valve used on mains, long runs of pipe, indirect stacks, etc. When in operation, the steam reaches the expansion member which contains a volatile liquid and which expands by virtue of the pressure generated and closes the valve. On cooling, the expansion member contracts and allows the air to reenter. Detailed computations of heating surface are given on page 230.

Fig. 78.—Air vent.

In Table IX the values of the heat transmission coefficient K are given for high-pressure steam radiators

Control of the steam supply in the usual form of radiator cannot be used in high-pressure work since even very small openings permit sufficient steam flow to fill the radiator The control therefore is effected by completely opening or closing the usual form of valve Two such valves should be installed on each radiator, one on the supply and the other on the return, since the return is also filled with steam (see Fig 79) The valve a on the return end of the radiator may be replaced by an automatic trap

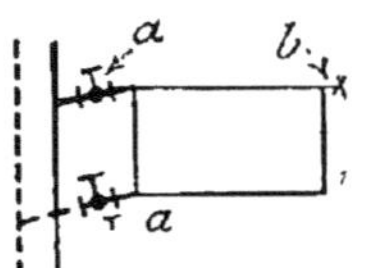

Fig 79—Connections of a high pressure steam radiator

d Returning Condensation.—The condensation flows into a sump near the boiler and is returned by means of boiler feed pumps If this solution is not feasible because of local conditions, the condensation is allowed to drain into an extra pit from which it is forced, frequently by an automatic pump, into a small elevated tank and from it into the boiler sump At times automatic boiler return feeders are used in place of feed pumps

Where for some reasons it is undesirable to return the condensation to the boiler, the water should be cooled off to below 100° F before entering the sewer The use of a sufficient length of cast-iron pipe to accomplish this is recommended before allowing the water to enter the vitreous drain pipe

Low-pressure Steam Heating

Low-pressure steam heating systems operate with the following pressures

	0 5–1 5 lb	for horizontal runs up to 500 ft
	2 lb	for horizontal runs up to 1,000 ft
Over	2 lb	for horizontal runs up to 1,500 ft

1. Advantages, Disadvantages, Field of Application.

a Advantages Compared with High-pressure Steam Heating—The surface temperatures are about 212° F so that the burning of dust is reduced considerably There is possibility of controlling the radiators When designed and installed properly, the operation is noiseless Complicated traps are eliminated with resulting simplification of attendance Tight connections are possible because of low pressure The heat losses are lower, with consequent economy in operation No permits are required for boiler installation The foregoing are reasons of the superiority of low-pressure over high-pressure steam heating

b Advantages Compared with Hot-water Heating—When the individual radiators or the entire system is shut off, the heating surface cools rapidly Consequently low-pressure steam is adaptable to rapidly changing heat requirements There is little danger from freezing, a fact of more importance in the United States than in Europe Slight danger does exist in low-pressure steam systems when starting up in very cold rooms, but no material damage can result The cost of installation is low. Comparative operating expenses that are reliable have not been established

c Disadvantages Compared with Hot-water Heating—Surface temperatures are higher; therefore the system is hygienically less desirable There is greater intensity of radiation The system is unreliable for centralized regulation Heat losses are larger and increase proportionally with lower outside temperatures There is possibility of disturbances in the return lines

d Field of Application—It was previously stated that low-pressure steam heating is less desirable than hot-water heating for residences and similar buildings For hygienic reasons it is preferable to use hot-water heating in homes On the other hand low-pressure steam is to be considered whenever rapid heating or speedy cooling off is desirable As a result it is suitable for theaters, assembly and amusement halls, business offices, churches, lecture auditoriums, for schools under certain conditions, and for the operation of steam air-heating systems

2. Forms of Installations.

a Reduction from High-pressure Steam—If low-pressure steam installations are supplied from a high-pressure steam system, there must be a reduction of pressure. One of the many different reducing valves is shown in Fig. 80 The high-pressure steam enters at *A* and flows through a strainer between the flanges at *B*. It then passes through a balanced double-seat valve *C* whose position is governed by the float *D* in a mercury chamber If the steam pressure rises on the low-pressure side, the float *D* and the mercury in the container *E* are depressed, the valve throttles the steam supply, and the pressure on the low side drops. Simultaneously with the falling of *D* the displaced mercury passes through *F* and the hollow rod *G* into the container *H*. If the low-pressure side drops below the pressure for which it is set, the mercury flows from the high position in *H* to *E*, raises the float, and opens the valve *C*, and then the operation is repeated The balls *J* and *K* are small safety valves to prevent mercury from being thrown about The extent to which the low pressure may be set (within certain limits) is by raising or lowering the container *H* With careful attention this reducing valve operates very satisfactorily and will throttle from 100 to ¾ lb in one step.

b Exhaust Steam from Power Plants—Exhaust steam from engines may be applied without difficulty Steam from the low-pressure cylinder is piped to a suitable receiver at a back pressure of from ¾ to 3 lb From the receiver the steam operates an ordinary low-pressure steam system. The subject is treated in greater detail under the subject of Exhaust Heating (p. 105).

c. Low-pressure Steam Boilers.—Low-pressure steam boilers in Germany are those which are set so as not to exceed a working pressure of 7 lb. per square inch. A standpipe (see p. 76) is used to limit the pressure permitted. In America pressures of 15 lb. per square inch are allowed for low-pressure work.

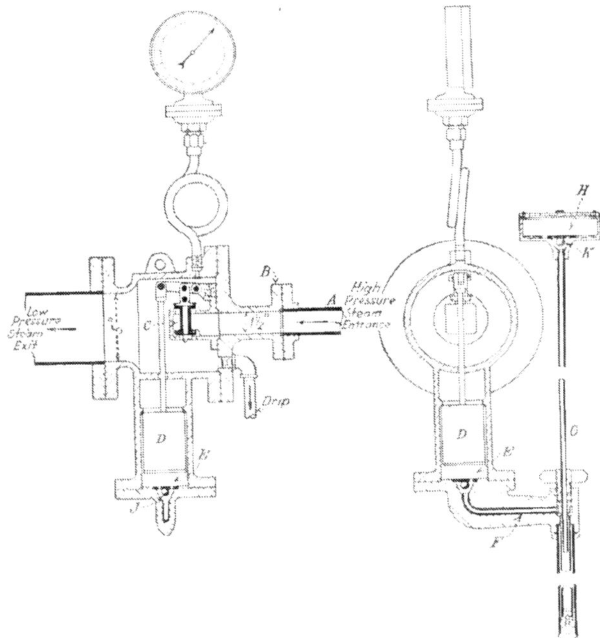

FIG. 80.—Reducing valve. (*F. Kaeferle, Hannover.*)

Steel boilers are used both for water heating and for steam heating. In the smaller installations vertical boilers are common while for the larger systems horizontal types are used.

Steel boilers of small and medium sizes are being supplanted by cast-iron sectional boilers. In recent practice the boilers for steam and water are the same except that the boiler trimmings are changed to suit. Consequently the designs shown in Figs. 21 to 29 may also be used for steam systems. It should be noted that because of the agitation within the boiler large quantities of water are at times carried over with the steam. Formerly steam headers (as shown at *A*, Fig. 81) were installed on the boilers, and the steam was taken from the outlet on top. The newer designs are provided in the boiler sections with means of effective

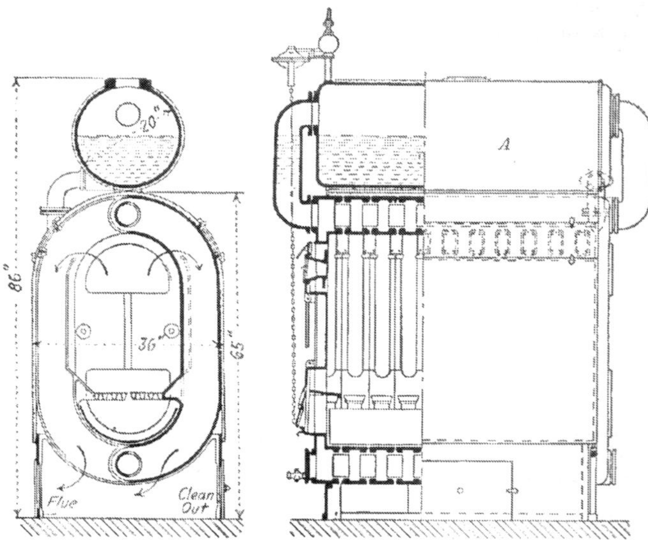

Fig. 81.—Strebel boiler with steam dome.

separation so that the installation of special steam headers becomes unnecessary. The boiler is therefore lower, an advantage in installation. Tests have shown such types to be very effective in that the amount of water carried over was very slight under average load, while at light loads a small amount of superheat was found.[1]

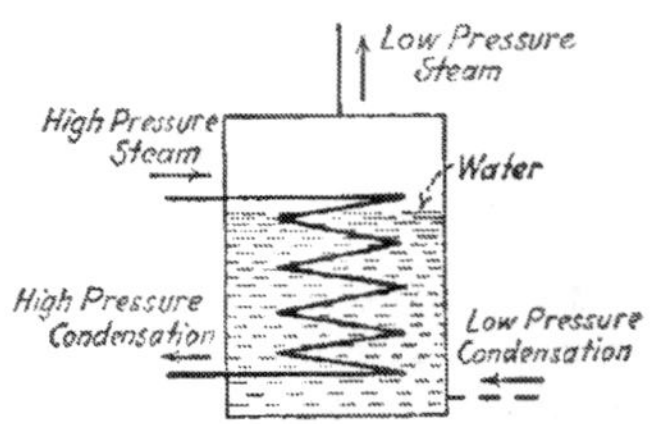

Fig. 82.—Steam heated steam boiler.

The height of the boiler room depends upon the size of the boiler and also upon whether dry or wet return lines are used. These considerations are given attention on page 77 in connection with the layout of the piping.

d. Steam-heated Steam Boilers.—In Fig. 82 is illustrated a type of steam boiler used particularly in chemical establishments for cooking, washing, etc. It has a fixed water level. If the steam pressure on the

[1] Research Laboratory Bulletin 17, April 1914.

low-pressure side rises, water is forced through the low-pressure condensation line into an elevated tank. As a consequence the water level in the boiler falls and exposes some of the high-pressure heating surface, rendering it thereby less effective. As soon as the pressure on the low side drops, the water returns from the elevated tank and the cycle repeats.

3. Boiler Equipment.

The equipment of steam boilers is different from that of water boilers and will therefore be considered in detail.

Pressure Gage.—Each boiler is equipped with an indicating pressure gage showing the range for normal operation.

Gage Glass.—Boilers must be provided with try-cocks to show desired water line and a gage glass indicating the water level.

Damper Regulator.—An important requirement for good operation is sensitive and accurate maintenance of the steam pressure. Figure 83

Fig. 83.—Damper regulator. (*American Radiator Company, New York.*)

shows a type of diaphragm regulator. An increase in the steam pressure causes the thrust rod to rise, actuating a lever which by means of rods or chains operates the air damper at the boiler. When the pressure drops, a weight tilts the lever, opening up the air supply to the grate. Adjustment for a desired steam pressure is obtained by setting the movable weight and by adjustment of the chain. The same regulator can also be used to control a cold-air check damper in the smoke-pipe. The check damper is opened by the regulator, after the air inlet is closed.

Standpipe.—In Germany official regulations call for the installation of a standpipe which must be open to the atmosphere and must be without a valve. The standpipe should be not more than 15 ft. high and the lower end must be connected to the water section of the boiler. In

America the pop safety valve is the standard safety device used. This is set at a pressure not to exceed 15 lb. per square inch (Fig. 84).

For details regarding filling, emptying, flue dampers, testing, attendance, firing, chimneys, etc., the reader is referred to pages 32 to 40.

a. Piping Connections.—Regarding the use of pipe, piping connections, supports, expansion, and insulation, reference should be made to pages 42 to 43.

b. Piping Layout.—Similar to hot-water heating systems, the overhead and basement distribution methods are used. In the former case the steam is piped to the attic in one or more large risers and distributed there to the various downfeed risers to the radiators. This arrangement depicted in Fig. 85 has the following advantages: good steam circulation

Fig. 84.—Pop safety valve. (*Crane & Company, New York.*)

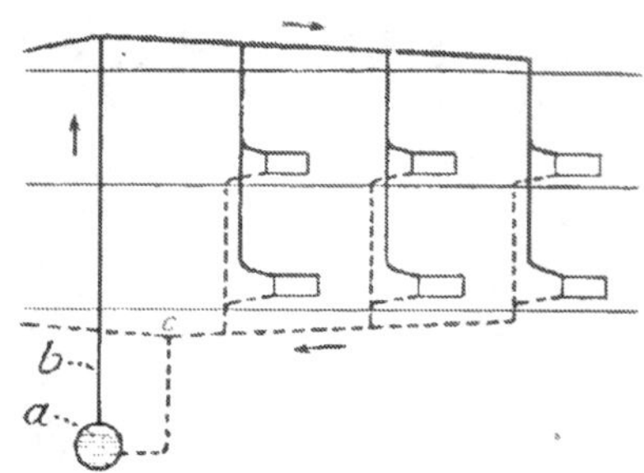

Fig. 85.—Overhead distribution, low pressure steam heating system.

in effectively dripped risers; flow of condensation in the direction of the steam flow; positive operation and practically noiseless action when carefully designed. The evident disadvantages are higher cost of installation and operating expense because of the heat losses from the attic piping. This layout of piping is used in extensive installations. The point *c* where venting takes place must be higher than the highest water line plus the equivalent head determined by the highest operating pressure and whatever factor of safety might be added.

The usual rule is to use basement distribution mains. They may be laid out with wet or with dry returns. Both types are illustrated in Figs. 86 and 87. The advantages are lower installation costs and lower operating expense. In general the heat losses from the basement mains are of some benefit to heating the house. A disadvantage is that the steam in the risers flows in opposition to the returning condensation. The steam main must be dripped frequently and the risers must be designed with a low-pressure drop (see p. 247). In the dry return system shown in Fig. 86, it will be noted that the riser S_2 is dripped and trapped through

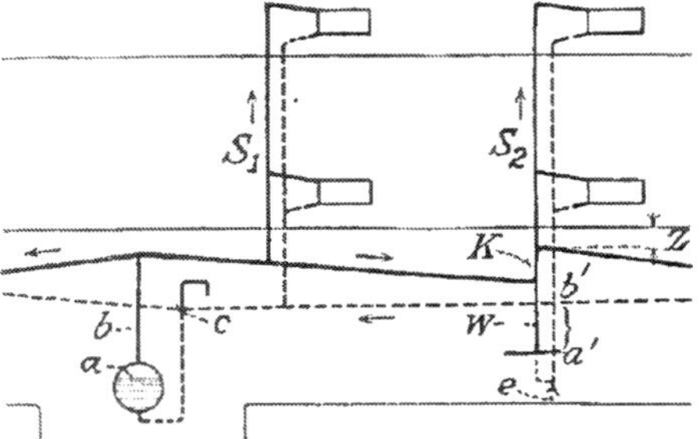

FIG. 86.—Basement distribution, low pressure steam heating system with dry return.

a water seal W. This is a simple U-tube in which the condensation flows off into the dry return at the point b'. If the steam pressure at this point is equivalent to a column of water of height $a'b'$, the steam will cause the water level to stand at a'. For every amount of water falling to a', a corresponding amount enters the return line at b', since the

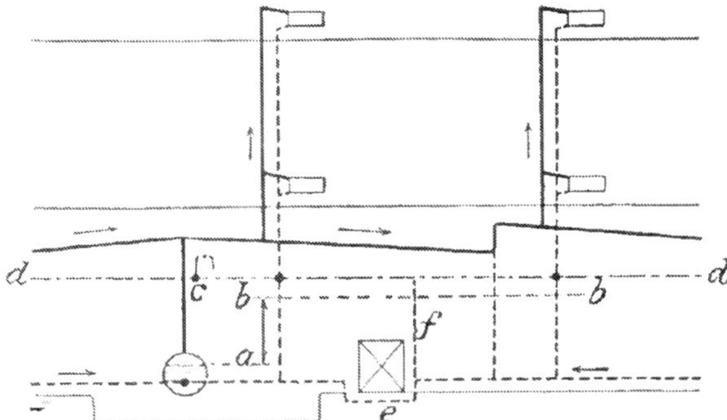

FIG. 87.—Basement distribution, low pressure steam heating system with wet return.

pressure differential $a'b'$ must be maintained. The seal is always made somewhat deeper than the head corresponding to the operating pressure, so that the disturbing blowing of the seal is avoided. Loop seals require no attention except that they must not be permitted to freeze. A cock e permits drainage. The point c (central venting) is higher than the water line corresponding to the sum of the hydrostatic heads *ab* and a safety factor

Figure 87 is a diagram of a basement distribution system with wet returns The latter are always below the boiler water level,[1] in fact up to the level *bb*. It was considered that wet returns are better than dry returns from the viewpoint of corrosion. According to the latest experiences there is no difference in the corrosive action, since the same effect takes place in both. In all cases, as it is often explained, it is not a matter of actual rusting that occurs, but it is a process of dissolving the iron. The condensation formed contains in the nascent state the ability to take up more oxygen than that which corresponds to the chemical formula H_2O On cooling, the excess oxygen is released and forms oxide of iron Since the latter is soluble in water, the piping is gradually dissolved Electrolysis also plays an important part

When wet returns must dip under doorways, they require a special vent line as shown at *f* in Fig 87 In the case of outer doors the piping placed under the floor is likely to freeze Dry returns in general should be installed with sufficient pitch to prevent trapping of water. This usually occurs in the T-connections (Fig 88) where leaks are found first. It is possible that the drip from above assists mechanically

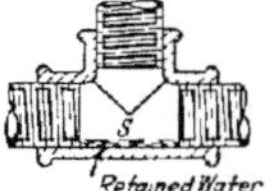

Fig 88—Trap water in "T" of return

With unfavorable water and drainage conditions the condensation lines may show signs of failure within a few years, particularly when low-pressure steam is reduced from high-pressure steam The use of copper for iron is a remedy

The height of the boiler room depends upon the piping layout For Fig 86 it is as follows

	Inches
Height of water line *a*	59
Operating pressure, 2 lb	55
Safety factor to *C* (including head due to friction)	12
Pitch of return = 0 06 in per foot of run, or 0 06 × 150 ft	9
Pitch of supply to ell at *K* = 0 06 in per foot of run, or 0 06 × 75	5
Boiler room height in clear	140 = 11 ft 8 in

The lowest ceiling height is obtained when overhead distribution is used A greater height is needed with basement mains and wet returns, and the highest ceiling is required with basement mains and dry returns

c Piping Computations—The determination of pipe sizes for steam and return lines is taken up in detail on page 245 The methods given insure a rapid and accurate calculation

[1] In this connection *a* is the highest water level, *ab* the head corresponding to the operating pressure, and *c* the central venting by means of a special vent line to which all condensation lines are connected

d Radiators and Enclosures.—The hot-water radiators described on page 45 *et seq.* may also be used for low-pressure steam systems. Also the references as to location of radiators, effect of enclosures, influence of the form and surface, air velocity,[1] etc. apply to low-pressure steam systems in estimating the transmission coefficient K. The general formulas for computing heating surface are discussed on page 182. The required surface for low-pressure steam heating is found from the equation

$$F = \frac{W}{K(t_d - t)}$$

where

F = the surface of the radiator in square feet.
W = heat requirement in B.t.u. per hour.
K = heat transmission coefficient in B.t.u. per square foot per degree Fahrenheit per hour.
t_d = steam temperature in degrees Fahrenheit.
t = required room temperature in degrees Fahrenheit.

Values of K are given on page 300 *et seq.* for various types of radiators.

A type of low-pressure steam heating was introduced in Germany under the name of "Milddampfheizung." As shown in Fig. 89 it consists of sectional radiation normally full of air and unvented. Steam is supplied through the lower nipple to one side to induce a circulation of the air-steam mixture. The mixture has a lower temperature than that of the steam at the pressure used. It is claimed that radiators operated in this way have the same advantages as warm water radiators and are more economical because of the lower installation costs. A rather complete investigation[2] at the Research Laboratory, Charlottenburg, revealed the following:

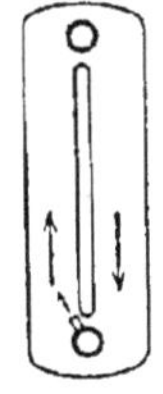

FIG. 89.—Diagram of "Milddampfheizung."

(1) The inner and surface temperatures and the heat transmission of air-steam radiators are dependent to a considerable extent on small pressure changes.

(2) The damper regulators for low-pressure steam heating in general use introduce changes in pressure such that the control of the surface temperature and the heat emission of the radiator are practically impossible.

(3) Shutting off radiators in the system produce changes of pressure in the remaining radiators and affect their surface temperatures so as to make control of room temperature difficult.

[1] The influence of the steam velocity is negligible within the limits used in practice.

[2] "Investigation of Air and Steam Mixtures in Low-pressure Steam Heating," from WERNER, "Untersuchungen uber Luftumwalzungsverfahren bei Niederdruckdampfheizungen," Bulletin 19 of Research Laboratory.

(4) Variations in boiler pressure (central control) do not cause proportional heat emission from radiators, consequently room temperatures are not under control

(5) The heat transmission coefficient does not remain constant but drops considerably with increasing mixture of air

From the foregoing it will be observed that the air-steam mixture in radiators is not equivalent to hot-water heating from either the hygienic or technical viewpoint

e. Useful Heat Output of Radiators —In recent years a new method of judging radiators has gained weight. Before that time the B t u input of a radiator was measured[1] and set equal to the total output. It was then believed that this latter figure gives a true indication of the heater's value In other words, it has been assumed that the radiator's efficiency is 100 per cent Now it becomes apparent that 100 per cent efficiency is as impossible on a heater as elsewhere. Science began to study the distribution of warmth in a heated room and there is no doubt that this has led to important discoveries

Supposing a radiator (*a*) transmits all its B t u emitted up to the ceiling and another radiator (*b*) directs all its output into the lower part of the room in which we live It is obvious that radiator (*b*) is more valuable for human purposes than (*a*)

The idea that heat on the ceiling is transmitted to the floor of the room above and therefore becomes effective for the warming of the house as a whole does not hold true as explained by the author in detail at the Convention of the A S H & V E, held in Buffalo in 1926 There is no longer any doubt that, concerning human comfort, a radiator's efficiency is not 100 per cent but is the quotient of the heat made useful for human purposes in the room in which the radiator is installed, to the total output (or input) of the radiator

The author, at the Conventions in Buffalo and St Louis (papers published in the Journals of the A S H & V E of November, 1925, page 501 March, 1926, page 163, and November, 1926, page 731) has pointed out one way in which the "useful output of a radiator" can be measured, hereby adopting a relative method of testing radiators in addition to the absolute method heretofore only recognized Even if details of this method should be changed later on, the principle seems to gain power and the actual figures will hardly be materially changed, an advantage of relative methods in general

The thoughts are clear and simple The sole purpose of a radiator in human housings is to give comfort and not a certain amount of condensation or a showing of a certain amount of square footage, or non

[1] By condensation cooling effect of water or watt consumption using an electrically operated heater

weight. Comfort calls for the delivery of warmth in the place where we live. This is the lower part of the rooms and not the ceiling. Warm feet and cool head, the requirement of health, is the goal at which we have to aim. The best radiators are not those which have the greatest actual heating surface or the maximum condensation (highest coefficient) but those which give maximum comfort with minimum condensation. In other words, maximum comfort must be obtainable with minimum installation and operating expense.

f. Regulation.—In low-pressure steam heating systems it is of considerable importance that steam is prevented from entering the condensation lines and that the latter carries nothing but air and water. If steam blows through, it gives rise to the following effects:

(1) Radiators are prevented from venting quickly and completely, and air is trapped in the radiator causing it to remain cold.

(2) Steam in the condensation lines causes radiators to heat from the returns even though the inlet valves are closed.

(3) Steam in the condensation lines forces the water ahead and causes it to impact at abrupt changes in direction. Since water is practically incompressible, its impinging upon the iron or other collected water introduces noises and water hammer.

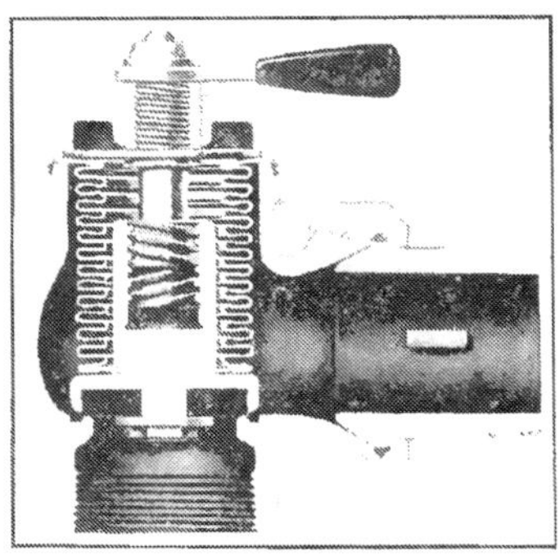

Fig. 90.—Steam radiator valve. (*American Radiator Company, New York.*)

The entrance of steam into the returns is prevented by adjustable valves and radiator traps.

(1) Adjustable Valves.—The valves used for steam systems are similar to those described on page 57 for water systems. An example of a steam valve is illustrated in Fig. 90. It has a quick-opening movement with a stem-locking device to make it tight against leakage into the radiator when closed. Individual radiators can be set after installation by loosening a lock nut on the dial and moving the indicator stop to the desired position. This limits the turn of the handle which prevents the valve disc from being raised further from the seat. Accurate and fine adjustments are made possible by a cone-shaped nut projecting into the inlet.

Cocks may also be used, but owing to their tendency to stick, they are not recommended.

Compared with hot-water valves, the necessity of minimum resistance to flow in steam valves is not as essential, since sufficient steam pressure is always available.

In regard to the design of equalizing valves the same conditions apply for steam systems as for water systems (see p. 57). They require that the relation for throttling and hand manipulation of the lever be as close to the ideal relation shown in Fig. 91 as possible and moreover that the

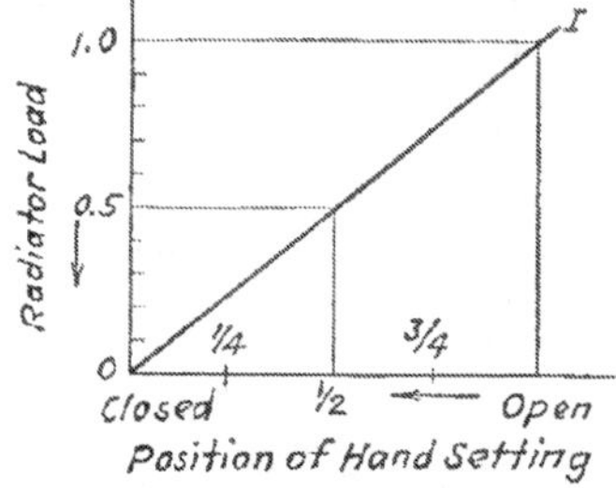

Fig. 91.—Graph of ideally regulated steam value.

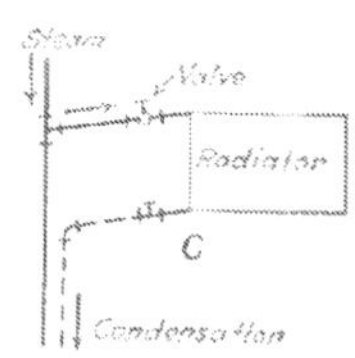

Fig. 92.—Double connections on a low pressure steam radiator.

hand control be independent of the equalizing features[1] (see also Figs. 55 to 56).

The adjustment is made by the steam fitter in the following manner: The boiler is held at the operating pressure, and the adjustment on each valve is throttled to the position where no steam enters the return lines. This condition may be determined more easily if a tee is placed at *c* (Fig. 92) in the return line. Frequently the entire radiator is not used, and the adjustment is made so that a small section remains cold even under the maximum pressure. The unheated part prevents blowing through with varying operating pressures.[2]

Fig. 93.—Vacuum trap. (*Warren Webster, Camden, New Jersey.*)

(2) Radiator Traps.—Radiator traps serve the same purpose as the traps for high-pressure systems. They are available in a large number of designs, of which one of the more common forms is shown in Fig. 93. They are installed at the return end of the radiator.

[1] "Investigation of Controlling Devices for Steam and Hot-water Heating Systems," from Ambrosius," Untersuchungen an Regelvorrichtungen für Dampf- und Wasserheizkörper," *Mitteilung der Anstalt* 25. Bulletin 25 of Research Laboratory.

[2] Relative to blowing through, "Pressure Relations in Low-pressure Heating Systems," from Frenckel, "Druckverhältnisse in Niederdruckdampfheizungen," *Mitteilung der Anstalt* 32, Bulletin 32, of Research Laboratory, Charlottenburg, published by R. Oldenbourg, Munich and Berlin, 1921.

The air passes freely through the trap, but as soon as the steam reaches the expansion member (bellows, etc.), the trap closes. When the water of condensation strikes the actuating member, the lower temperature causes the valve to open, permitting easy passage of the water.

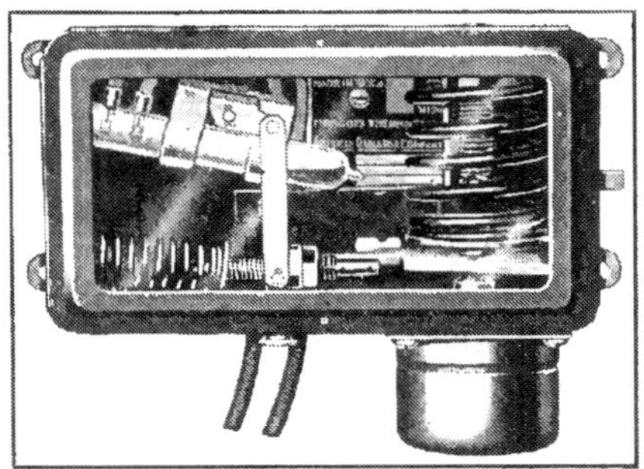

Fig. 94.—Mercoid temperature regulator. (*American Radiator Company, N. Y.*)

g. Automatic Regulation.—As previously mentioned (p. 58) for low-pressure steam, both direct and indirect control are used. The electric temperature regulator shown in Fig. 94 belongs in the latter class. An

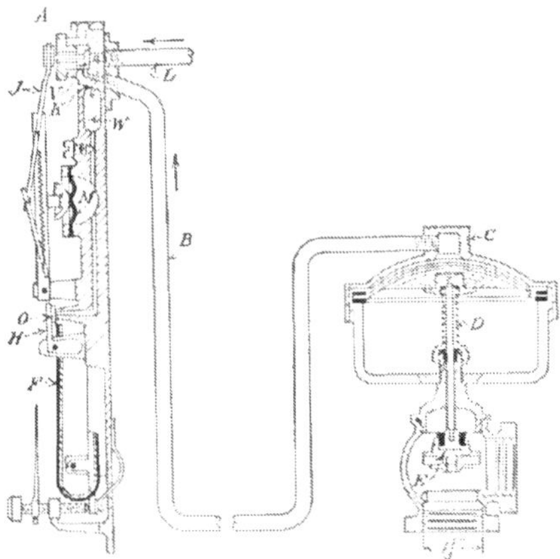

Fig. 95.—Air pressure room temperature regulator. (*Gesellschaft fur selbs-tatige Temperaturregelung, Berlin.*)

expanding member, metal bellows in this case, is actuated for immediate or remote control, by temperature or pressure changes, making or breaking the electric contact in the mercury switch.

The air-pressure regulator shown in Fig 95 operates on a different principle Compressed air enters the thermostat at L In the position shown the valve V is closed The air passes through the channel K to a cotton filter W and thence through a small passage terminating in a small tapered opening at O This opening is closed in the position shown, by means of the spring F and the lever H Compressed air forces the diaphragm M open and closes the valve V by means of the lever J. This releases the pressure (previously established) in the pipe B leading to the diaphragm valve where a spring D opens the valve E The air behind the diaphragm which escapes through the line B discharges around the loose stem of the valve V

When the room temperature rises, the spring F, which is sensitive to temperature, expands and, at the room temperature for which it is set, presses against the lever H in a manner to expose the opening O The compressed air discharges and permits the diaphragm to move inward through the operation of a spring, the lever J opens the valve V which closes the opening about the loose stem of the valve V Compressed air then enters directly from L to B, and C forces the large diaphragm downward against the action of the spring D and closes the radiator valve E

The use of automatic regulation and thermostatic control has greatly increased with the introduction of oil- and gas-fired boilers and with the improved types of coal-burning boilers having long firing periods Since they represent an additional investment, they should receive attention from time to time to enable them to function as intended

h. Return of Condensation —In most cases the condensation flows back to the boiler by gravity Where the gravity return is not possible, receivers (condensation tanks) are used into which the water flows by gravity and from which it is pumped directly into the boiler or into an elevated tank From this tank the water is fed to the boiler by gravity

i. Centralized Control —Even with the most careful design of the system, central regulation in proportion to the outside weather requirements by means of changes in the operating pressure is possible only within certain limits [1] In this respect steam systems are not the equal of hot-water heating

E. VACUUM HEATING SYSTEMS

Vacuum heating systems are those which operate at pressures below atmospheric This class includes the usual steam system in which the radiators serve as precondensers If the surface condenser of a steam plant (engine or turbine) is cooled by means of air and the air is then used as a heating medium, such system is also included in the present classification (see Exhaust Heat Utilization) When the surface condenser of a steam plant acts as a heater for a hot-water system, such arrangement

[1] See FRENCKEL footnote, p 83.

may also be called a vacuum heating system (see Exhaust Heating Systems)

The original American vacuum systems, frequently installed in tall buildings, are similar to the low-pressure steam systems in the return line of which a vacuum is maintained The vacuum up to as much as 15 in is created by means of a special pump whose suction is connected with the return line In addition each radiator and each vent are provided with a special trap operating by some form of expansion principle

Systems known as vapor, atmospheric, etc are modified vacuum systems using special fittings such as traps, fractional valves at the radiators, vent traps, centralized venting, return traps at the boiler, etc These serve to maintain an extremely low pressure, from a few ounces above atmospheric to a partial vacuum in the return lines After the air in a closed system, is expelled by means of the steam and the steam pressure is then allowed to drop, a vacuum will be maintained in the supply and return lines for long periods depending upon infiltration of air Nevertheless a vacuum pump with all accessories for automatic operation is also necessary. These various systems have been developed by individual manufacturers and have been applied to all kinds of buildings from residences to the tallest of office buildings

Up to the present, vacuum heating in Germany has been used only in connection with steam power plants In this case its utilization is very economical if the steam output of the machines corresponds to the average steam input to the heating system On cold days the vacuum is reduced to enable heating with higher steam temperatures Since extreme weather prevails for but a few days, the drop in economy of the steam plant is not very important Under certain conditions on cold days additional heat is supplied by a live-steam connection. To insure satisfactory results careful consideration must be given to the proper application in each particular case It should be noted that when vacuum systems are used in connection with reciprocating engines, the influence of back pressure on steam economy is less than in the case of turbine equipment

In Fig 96 is shown a diagrammatic layout of a vacuum system with extraction steam supply [1] The live steam after doing work in the high-pressure cylinder is exhausted to the low-pressure cylinder At *d* there is a pressure regulator At this point steam is extracted from the high-pressure exhaust which after passing through an oil separator is delivered to a manifold, from which place it is distributed for various uses The manifold is connected so that it may receive live steam direct from the boilers A valve *a* on a line from the manifold may be opened to supply steam to the vacuum system, should it be necessary

[1] Illustration from "Vacuum Steam Heating System," from SCHULZE, "Die Vakuumdampfheizung," *Mitteilung Warmestelle Ver deut Eisenhuttenleute*

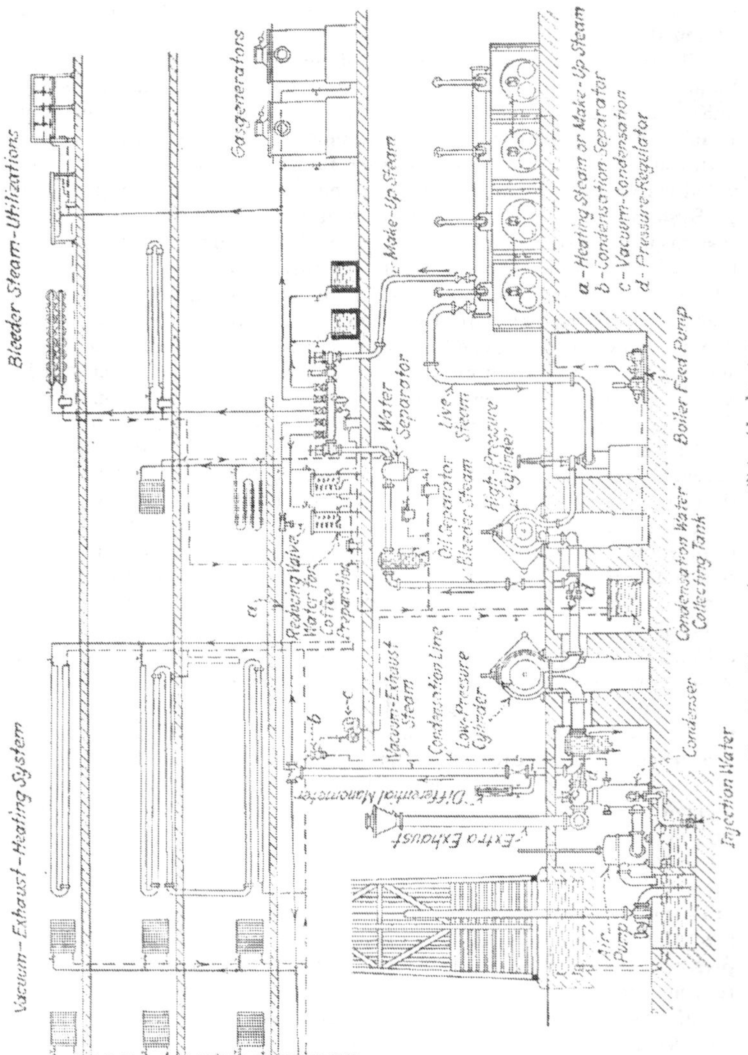

FIG. 96.—Diagram of a system utilizing bleeder steam.

The steam passes to the low-pressure cylinder, where additional work is performed. It is then carried to the injection condenser passing through an oil separator and a normal pressure regulator from which the steam passes into the heating system. In the drawing the following features are to be noted: the special exhaust line, the cooling tower, the air pump, and the differential manometer, the latter indispensable as an indicator of the vacuum.

The pressure regulators serve a number of purposes. On cold days they reduce the amount of vacuum carried in the system, and so permit operation at higher steam temperatures. In mild weather the regulators are set so that the steam flows into the heating system in the required amounts. The heating system then acts as a condenser. A further function of the regulator is to insure the steam supply to the heating mains despite the variable on the engines. It also serves to shunt excess steam to the condenser.

F. COMBINATION HEATING SYSTEMS

The combination systems will include steam hot-water heating systems and steam water heating.

Steam Hot-water Heating

1. General Arrangement.

The hot-water heater *K* (Fig. 97) consists of a number of tubes immersed in water. Steam is supplied through the pipe *a* and the con-

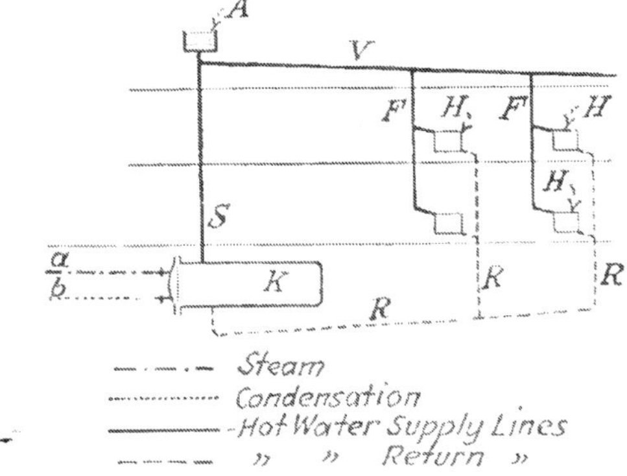

Fig. 97.—Diagram of a steam hot water heating system.

densation returns through the pipe *b*. The water contained in *K* is heated by the steam and rises through the supply main *S* to the expansion tank *A* and then to the distributing main *V* into the supply risers *F* which distribute it to the hot-water radiators *H*. The cooled return flows through the main *R* back to the heater *K*.

It is evident that the steam hot-water system described is nothing more than an ordinary hot-water heating system in which the boiler is not heated directly by means of a fire but by means of steam This heat supply may be high- or low-pressure steam, hot water, or oil

2. Field of Application.

The combination system described may be used in such cases where steam is available on the property, and it is desired to use hot-water heating for certain of the buildings This may be the case for hospitals where steam is needed for cooking, washing, sterilizing, etc , but the sick rooms are to be heated with hot water (see District Heating, p 100, and also Exhaust Heating, p 105) Another case may be that of a tall building, hotel, or other institution in which hot-water heaters are located at different levels with independent hot-water heating systems, but all of which receive their heat supply from the steam boilers Another application is that of dwellings erected on factory sites enabling such to operate with exhaust steam during week days and using a coal-fired boiler to provide heating when the plant is otherwise shut down

3. Installation.

The hot-water distribution system of the steam hot-water heating system is computed in the same way as the usual hot-water heating system The details for piping and radiators are considered on pages 42 to 59 On the other hand the steam lines must be considered as a part of a steam system whether high- or low-pressure steam is used In this respect what is mentioned on pages 64 to 79 applies in this case the heater, however, requires special consideration.

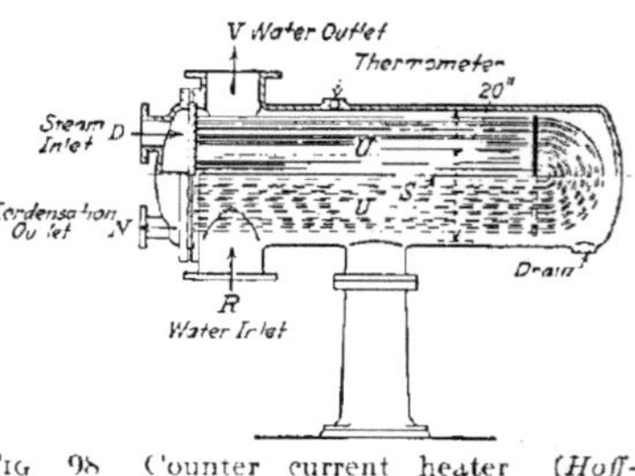

FIG 98 Counter current heater (*Hoffmannwerk, Leuben-Dresden*)

4 Hot-water Heaters.

a Countercurrent Apparatus —A countercurrent heater is shown in Fig. 98 Steam enters at D and flows through the U-tubes, and the condensation leaves at N The water from the heating system enters at R, flows in the opposite direction to the steam flow due to the baffle S and leaves the apparatus through the supply outlet V The U-tube construction of the heater insures simple and positive provision for expansion Corrosion is avoided by using tubes made of copper The entire coil may be removed from the shell Regulation is accomplished by throttling the steam supply or automatically by means of a thermostat

The advantage of the heater shown in Fig 98 is that it heats the water quickly A disadvantage, however, lies in the rapid cooling due to the small water content To overcome this the following is often resorted to

After the heating system is supplied with the necessary hot water and the rooms are brought up to temperature, the water from the heater is largely bypassed to suitable storage tanks These are nothing more than storage reservoirs with large water content to insure continued heating after the steam supply is discontinued [1]

b. Steam Hot-water Boilers—These are used because their water-storage capacity assures a storage of heat after the steam supply is discontinued These heaters do not warm up as rapidly as the one previously described

c Safety Appliances for Steam Hot-water Boilers—The safety measures discussed on page 35 should also be used for steam hot-water boilers In this the reference to heating surface of the boiler should be changed and the direct heating surface of the heater substituted therefor

5 Computation of the Installation.

The design of that part of the system which pertains to the hot-water heating system is found on page 194. Similarly on page 230 *et seq* the computations for the steam end are considered, noting in addition that the tubes deserve attention The determination of the amount of radiation of the steam hot-water apparatus is treated under the heading of Heating Surfaces on page 178 *et seq*

Steam Water Heating

In this form of heating an effort was made to combine the advantages of water and steam heating by supplying apparatus which will heat quickly and still remain hot for a long time after being shut down This type of system has been used by the Zahlerheizungsgesellschaft of Vienna [2] In the latter special boiler equipment for careful regulation is installed and steam water-transformer sections are used for each radiator, making the latter water radiators Control from the boiler room enables a proportional change in heat output from the radiators, a fact which tends to economy in operation Large installations in Austria have operated successfully, but its adoption to the building conditions found in America will require extensive study

G. WARM-AIR HEATING

General

Warm-air heating refers to such systems where the air acts as the carrier of the heat required The air may be heated by means of flue gases, steam, or hot water The methods named may accomplish the result in three ways (1) fresh air supply, exhausting after use, (2)

[1] See also reference to safety measures for hot-water boilers, in ' Verband der Centralheizungs-Industrie," 3d ed , Berlin, 1921

[2] 'Steam Water-heating Systems" from BRABBÉE "Dampf-wasserheizungen," *Gesundh -Ing*, vol 16, no 32, p 305, 1924

recirculation of air within the space to be heated, and (3) combination of fresh and recirculated air.

In the recirculation system the air is used over and over again. As a consequence it becomes charged with dust as it passes over the heating surfaces. Recirculation is used for reasons of economy in operating costs, but if so used, means should be provided for conditioning the recirculated air.

If the air is circulated by difference in density of the warm and cool columns of air, it is known as gravity circulation. When fans are used to produce circulation of the air, it is called "forced" circulation. Gravity circulation always depends on wind and temperature conditions. Positive operation under all conditions can be obtained only by means of forced circulation.

Warm-air Furnace Heating

1. Advantages, Disadvantages, and Applications.

a Advantages.—Low installation costs, low operating expense, simple regulation, absence of radiators, no likelihood of freezing, rooms ventilated while being heated, and few and simple repairs are among the advantages of this system.

b Disadvantages.—Frequently its operation is influenced by outside temperature and wind conditions. It is difficult properly to apportion the air supply to each room. The relation between the heat and the air requirements are disturbed when several rooms are to be heated. There are physiological objections due to overheated furnace surfaces. Frequently too high temperatures are registered. Drafts are present.

c Applications.—While steam and hot-water heating has made inroads in the heating field, warm-air furnace heating is still used in installations such as in churches[1] where low cost is a decisive factor, in small homes particularly in rural districts, in stores, in community halls, etc.[2] The question of suitability of warm-air heating systems requires careful consideration in each case.

2 Design.

Warm-air furnace heating as a general rule operates with gravity circulation (without blowers). The separate elements of the installation will be discussed under the following sections.

a Warm-air Furnaces.—Fundamentally any furnace may be used for warm-air heating. For the large heat transmission required of

[1] The larger churches in general use low-pressure steam or hot-blast heating. See UBER, "Kirchenheizungen," published by Ernest & Sohn, Berlin, 1915, also "New Heating System in the Cathedral at Metz," from SCHMITZ, "Die neue Heizanlage des Domes in Metz," *Zentralbl. d. Bauverwaltung*, p. 169, 1917.

[2] "Applications of Calorifer Air Heating," from KORI, "Das Anwendungsgebiet der Kalorifer-Luftheizung," *Gesundh.-Ing.*, p. 616, 1922.

furnaces it is better that they be designed especially for the purpose Among the more important requirements are compact form, moderate surface temperatures, uniform distribution of heat, good circulation over all heating surfaces, proper provision for expansion, small number of joints, easy accessibility for repairs or cleaning, feasibility of ash and soot removal without entering air chamber, and automatic regulation

Extreme care should be exercised to prevent unburned gas (carbon monoxide) from entering the air chamber, and for the same reason a small opening in the smoke damper must be provided, which is not large enough to make effective control impossible Regulation of the fire by controlling the air supply to the grate is desirable

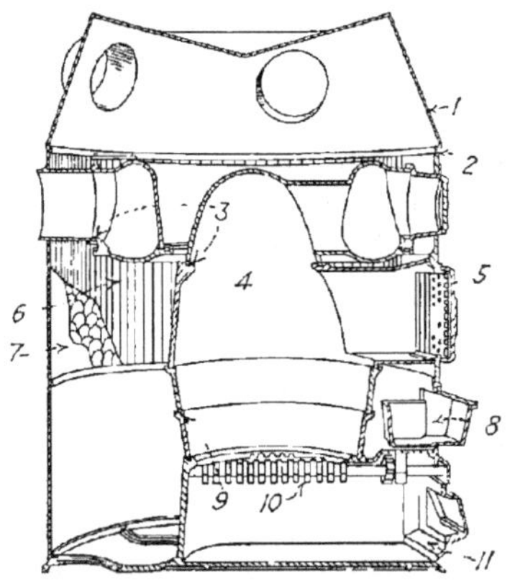

Fig 99 —Hot air furnace (*Fox Furnace Company, Elyria, Ohio*)

The air heating chamber should be located as low as possible to insure ample circulation Where large air chambers are built, they should be constructed along lines similar to the usual plenum chambers

Of the numerous designs of large furnaces the one in Fig 99 is typical Its different parts are

(1) Casing caps to provide full warm-air capacity for pipes leading to rooms

(2) Air- and dust-tight casing joints

(3) Double-seal cup joints throughout furnaces to prevent smoke and gases from entering the home

(4) Large radiating surface

(5) Large feed door for easy firing and use of large chunks of coal or wood

(6) Insulated upper casing to keep cellar cool

(7) Oversize casing to prevent overheating

(8) Large vapor pan to moisten the air

(9) Two-piece fire pot to provide for expansion and contraction

(10) Triangular grate bars to provide free air space

(11) Full-width ash-pit door for easy removal of ashes, air- and dust-tight

b Duct Layout—In warm-air furnace heating the ducts should be laid out exactly as in any ventilation system For treatment in greater detail the reader is referred to the section on Duct Installation (p 149) Since the ducts carry comparatively hot air, they should be insulated,

This applies particularly to extensive duct layouts where the heat losses may be considerable For this reason the heat-loss computations given on page 265 should be made in advance Due to the heat loss from the ducts gravity warm-air heating is unsatisfactory for extended buildings In small and compact houses the loss from the ducts may be useful when emitted in the right place, in which case the ducts are left uninsulated

Recirculation ducts should be planned carefully to include desirable physiological requirements and to avoid difficulties in construction They should be designed to require the minimum amount of fresh air For large spaces recirculation permits more rapid warming and also saves fuel

c *Inlet and Outlet Registers* —Statements made on page 136 relative to ventilating systems apply to warm-air registers Preferably the entering temperature of the air should not exceed 150° F

STEAM AND HOT-WATER AIR HEATING SYSTEMS

1. Advantages, Disadvantages, and Applications.

a *Advantages* —Easy control, absence of radiators, effective ventilation, establishment of pressure in rooms to prevent drafts, hygienic advantages if fresh-air or conditioned circulation is used are among the advantages of this system.

b *Disadvantages* —Operation with fresh-air circulation is expensive When a large number of rooms have widely differing requirement in heat and air supply, difficult operating conditions ensue

c *Applications* —Warm-air systems operated in conjunction with steam or hot water are used to advantage in heating and ventilating large single rooms, theaters, auditoriums, schools large factory plants, etc In these cases gravity circulation is replaced by blower systems In general for physiological reasons fresh air has been employed exclusively, although recent practice for reasons of economy uses a substantial percentage of recirculating air

2. Heaters and Regulation.

In the design of heating systems it is important at times to obtain a large heat emission in a small space This is made possible by using high air velocities and specially designed heaters The belief that the surface temperature of steam radiators is appreciably reduced at high air velocities has not been substantiated [1]

With steam radiators the contamination of the air by distillation of the dust particles diminishes with an increase in air velocities since the time of contact between the air and heating surface is but a fraction of a second The collection of dust on the heating surface when not in operation and the introduction of the vitiated air into the room on starting should be avoided For this reason alone high-pressure steam systems

[1] Research Laboratory, Charlottenburg, Bulletin, 3 1910

are not recommended. Other objections to high pressure are found in the difficulty of maintenance, expensive insulation, heat losses, attention to steam traps, venting, and noise in operation. If a steam system is installed, low pressure is preferable.[1] As far as the radiators are concerned, heating surfaces that will not harbor dust are to be desired. Vertical surfaces are therefore to be preferred.

For physiological reasons hot-water heating offers an excellent solution in warm-air heating installations. The use of hot-water heating throughout a building and steam in the ventilation system does not lend itself to simplicity of installation of boiler equipment and piping. Hot water has its objections, however, because of the possibility of freezing. In some cases it is possible to preheat the air by drawing the supply from ducts paralleling the chimney or passing through the boiler room, and so

FIG. 100.—Cross section through a Vento section. (*American Radiator Company, New York.*)

warming it sufficiently to prevent freezing in the radiators.[2] Danger due to freezing is less if the flow of water in the radiators is accelerated by means of pumps. In all cases neither the ventilation nor air heating must begin until normal circulation is established in the radiators. Special forms of radiators are designed for these systems some of which will be described in greater detail.

a. Vento Heaters.—The form illustrated in Fig. 100 has been in general use for hot-blast heating systems. It has a large amount of heating surface for the space it occupies and offers little resistance to the air flow.[3] To facilitate cleaning it may be erected in sections of two or more with space between for easy access.

[1] In drying systems the choice is sometimes reversed due to the greater drying capacity at high temperatures.

[2] A drop in temperature of the water below 45° F. may be indicated or prevented by automatic devices.

[3] "Discussion of Air Heaters in Particular Reference to the Rhombikus Type," from MARGOLIS, "Die Bewertung von Lufterhitzern unter besonderer Berücksichtigung des Rhombikus-Lufterhitzers," *Gesundh.-Ing.*, 1916; "Design of Heating and Cooling Coils," from GRÖBER, "Die Berechnung von Heiz-und Kühlrohren," *Gesundh.-Ing.*, 1920; "Design of Air heaters," from SCHMITZ, "Über Berechnung von Lufterhitzern usw.," *Gesundh.-Ing.*, 1920.

b. Tubular Air Heaters.—Heaters built up of tubes were formerly used to a large extent. In this case the steam surrounds the tubes while the air passes through them. The heaters should be erected with the tubes in a vertical position to insure against deposit of dust. This arrangement secures a high heat transmission (especially as whirling is created in the air motion), but its resistance to air flow is high. The tubular heater is less frequently used at present.[1]

c. Fin-surface Heaters.—Recently heaters of the type shown in Fig. 101 have been used where the surfaces are in only partial contact with the heating medium, *i.e.*, with only the steam tubes. On the tubes thin metallic flanges or fins are placed. These take the heat from the tubes by conduction and transmit it through extended surfaces to the air passing over them. This arrangement is based on the fact that the conductivity of heat from the steam to the metal is high and likewise through the metal but the transmission from metal to air is low and therefore requires a large surface. The output of fin-type heaters is high for the small amount of material and space required. This is equally true of heaters where rapidly circulating hot water is used in place of steam.

FIG. 101.—Fin surface heater with motor and fan on test.

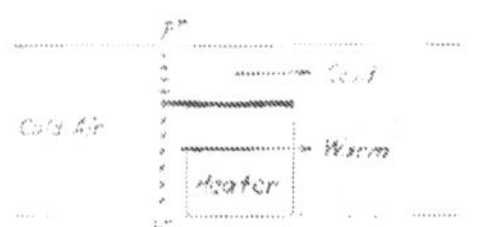

FIG. 102.—Temperature control by double mixing damper.

When fin heaters are erected with the fin surfaces in a horizontal position, they harbor dust. While the speed of the air through the heater over horizontal surfaces may cause them to be self-cleaning, this is not strictly the case when the heater is not in operation or under light load. Therefore the fins should always stay in a vertical position.

d. Heat Output and Temperature Control.—The control of the heat output is sometimes accomplished by means of the double-mixing damper *K* shown in Fig. 102. It is arranged so that when it opens the

[1] "Heat Transfer at High Air Velocities," from "Untersuchungen über Wärmeabgabe usw., unter Anwendungen grosser Luftgeschwindigkeiten," *Mitteilung der Anstalt*, Heft 3. Research Laboratory, Berlin, Charlottenburg, Bulletin 3.

cold air duct, it closes the warm-air duct simultaneously By gradually turning of the damper K (by hand or by remote control), any desired temperature is obtained

3. General Details

General details of warm-air systems are the same as for the usual ventilation system Reference should be made to page 136 *et seq* A difference lies in the fact that the air is supplied to the rooms at higher temperatures The heat losses from the ducts should be considered and insulation provided where necessary When rooms are widely separated, it may be more economical to subdivide the distribution system and use individual heating plants in the most unfavorable locations

4. Heating of Factories.

Factories may be heated by individual radiators or by warm air through a properly designed duct system Recent practice however, has developed along special lines In this method of heating, unit heaters H (Fig 103) supplied with steam are located as shown The air is kept in motion, part of it being recirculated and warmed in accordance with the heating requirements

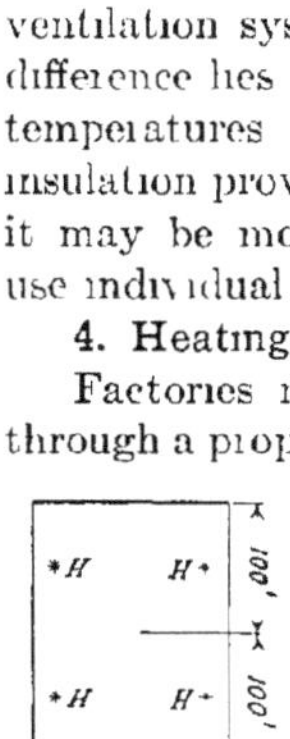

Fig 103 —Diagram of a factory heating system

In this case the following points should be noted

(1) As a general rule high-pressure steam has been used, which for reasons previously stated should be replaced by low-pressure steam

(2) Recirculation, of itself, cannot prevent drafts at the large doors and windows usually found in certain buildings A steam coil or other radiation should be installed at such places to counteract the tendency to produce drafts

(3) Heaters having horizontal surface are not desirable because of the dust they collect

In cases where ventilating equipment may be driven by means of steam engines or turbines and where the exhaust heat may be recovered in a heater, very economical operation is assured

The design of ventilation systems, particularly with reference to the heating elements, is considered on page 204.

H. DISTRICT HEATING

General

The use of air as a heating medium in district heating plants is commercially impracticable due to the enormous cost of this mode of transmission The present discussion will be limited to the use of water and steam systems District heating systems in general are large systems where several buildings or groups of buildings are supplied with steam or hot water from a central location

Advantages.

The handling of fuel and ash is eliminated at the consumer's premises. Smoke and soot are reduced to a minimum, due to the greater refinement in large power plant design. Installation at consumer's premises is simplified. There is safety from fires, with consequent reduction of insurance rates.

Applications.

This method may be applied to hospitals and similar institutions, buildings and groups of buildings, residence developments on a large scale, exhaust heating systems, and heating systems for entire blocks of buildings or groups of blocks.

District Steam Plants

The oldest and at the same time the largest heating system of this type is that operated by the New York Steam Company.[1] The downtown system of this plant has delivered so-called "Street Steam" since 1879 and at present has more than 45,000 hp. in two boiler plants. More than 1,400 customers are supplied, some of them 3,000 ft. away. The operating pressure is 125 lb. per square inch, and the steam is used for operating machinery, elevators, pumps, heating, hot-water supply, washing, and cooking. A sliding scale of prices is charged so that large consumption commands a lower unit rate. The downtown system shown in Fig. 104 is in demand because of the convenience and economy in saving valuable cellar space. About 300 such systems are in operation in the United States varying in size from small privately owned undertakings to very large public utility companies supplying heat outputs up to 1,000,000 lb. of steam per hour, equivalent to over 30,000 boiler hp.

Between the years 1885 and 1901 the first large district heating plant was constructed in Europe. It was located in Dresden, Germany, and was designed by Rietschel and Henneberg, of Dresden. In 1909 it had 25,500 sq. ft. of boiler heating surface, and 32 main and exhibition buildings were connected.[2] In 1911 the firm of Doerfel in Dresden purchased exhaust steam from power plants and supplied several public and other buildings with hot-water heating. This installation is noteworthy because of the contracts made both for the purchase and resale of heat. During the last 15 years many large steam heating plants were constructed.

[1] COMBE, "Central and District Heating," Dominion Fuel Board, Canada, 1924.

[2] TRAUTMANN, FESTNUMMER, *Gesundh.-Ing.*, June, 1909.

In regard to the boiler installation, piping, auxiliaries, and other details reference should be made to page 64 *et seq* and page 230 *et seq.* Some individual items may be noted.

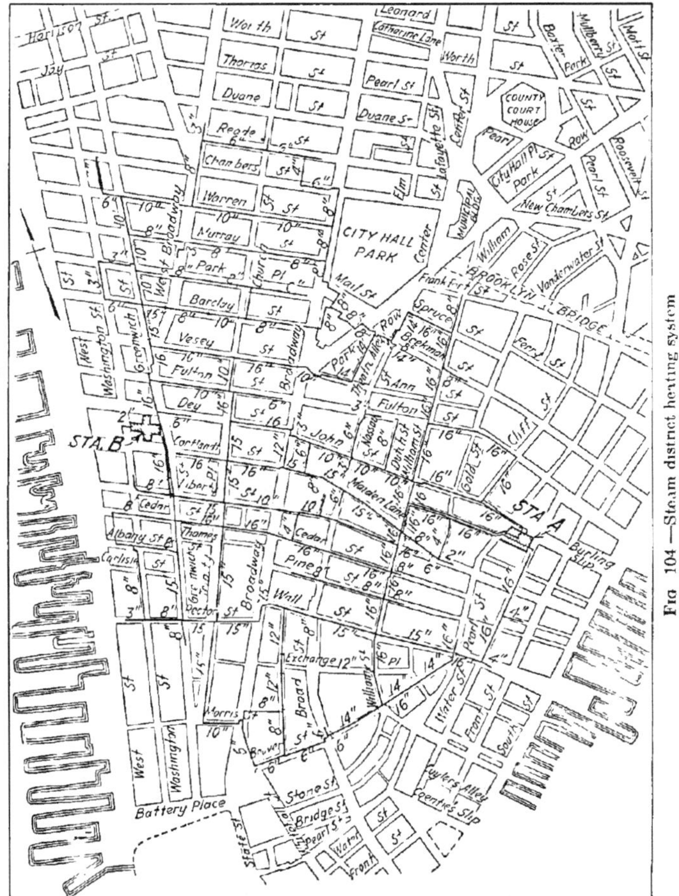

Fig 104 —Steam district heating system

1. Supplementary Steam Mains.

In the older installations the steam main was laid out in the form of a closed circuit, but this practice is now abandoned. In most cases two steam mains are installed which may be used independently of each other or may be interconnected at several points by suitable valve arrangements The one line is designed to supply the maximum demand, including allowance for possible extensions, while the second main is designed for a smaller load to be determined as follows: (1) the quantity of steam used for domestic purposes as for cooking, washing, sterilizing, etc , (2) the minimum winter steam used when that required for ventilation and such other load which may not be urgent is shut off The greater of the two quantities determined by the above fixes the pipe size to be used for a supplementary main.

2 Pipe Tunnels.

Steam mains should be erected in suitable conduits, a typical design of which is shown in Fig 105 The conduit should be at least 6½ ft. high for passage and should be provided with the necessary manholes. The spacing of manholes and the corresponding interchangeable pipe length should be such that flanged sections can be disconnected and removed In the design of the pipe tunnels, provision must be made for the proper expansion of the piping

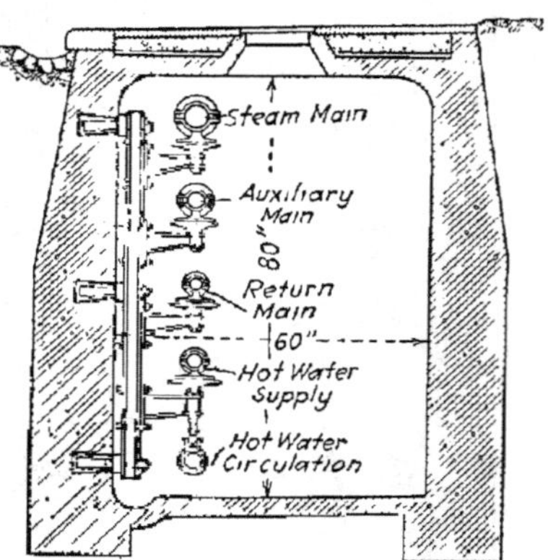

FIG 105.—Conduit with sufficient headroom for passageway (*R O Meyer, Hamburg*)

Special attention is desirable in the design of branch connections, since they should never obstruct the passageway The latter serves primarily to permit installation and insure inspection of the piping Absolute safety is not insured even with quick-closing stop valves In many cases the tunnels also convey electric cables and serve as a passageway between buildings, and also for the laying of electric cables, which not seldom add to the destruction of the iron pipes by electrolytic reactions

3. Steam Heating of Buildings.

If the buildings connected to the district heating system are to be supplied with high-pressure steam, there is nothing of special note to be considered beyond that discussed on page 64 If the buildings are to be operated at low pressure, a reducing valve as shown on page 70 is used at the service entrance From then on the system is treated as a low-pressure steam heating system described in detail on page 72

4. Hot-water Heating of Buildings.

When buildings are to be equipped with a hot-water heating system and fed from a high-pressure steam service main, it is done by means of steam water heaters described on page 89. The circulation may be maintained by means of gravity or in connection with circulating pumps. These systems are described on pages 21 and 61.

As will be seen when the distribution is accomplished by means of of high-pressure steam, the individual buildings may choose heating systems using high- or low-pressure steam or hot water.

Hot-water District Heating

A certain tendency in European practice is in the direction of forced-circulation hot-water systems (Fig. 106). Of the numerous advantages the following deserve mention: simplicity of boiler installation because

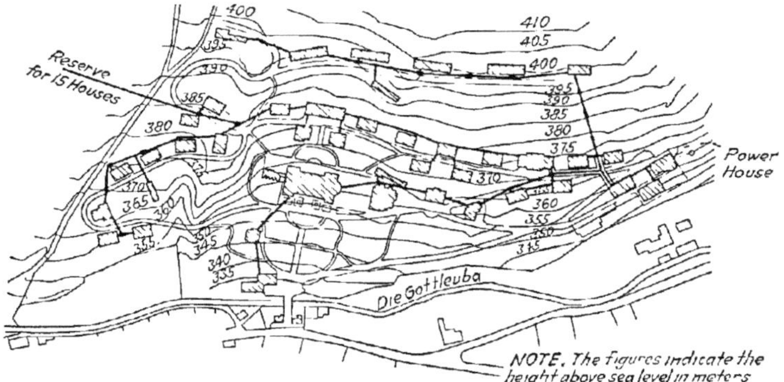

FIG. 106.—Hot water district heating plant, Gottleuba near Dresden. (*Jeblinski & Tichelmann, Dresden.*)

of simple accessories, smaller boiler loads and therefore longer life, decidedly lower maintenance of the piping due to the gradual temperature variations; ability to use smaller conduits instead of the larger pipe tunnels necessary with steam, simpler and fewer expansion joints, simple connection to hot-water heating system of a building, elimination of separators, steam traps, reducing valves, and the attention required for their upkeep, simple and certain operating control, smaller heat losses at maximum load and rapidly diminishing losses with higher outside temperatures, control of system from the boiler room, and possibility of using exhaust steam for heating even remote buildings. As to the first cost and operating expense, this will be decided by the economics of each individual case. In general the application of hot water district heating is the same as for district systems in general. It might be noted that hospitals and similar institutional buildings are usually designed for hot-water heating The type of layout is of particular importance In this the designer should locate buildings that require steam (kitchens, laundry, sterilizing rooms and possibly auditoriums, etc) close to the boiler plant so that long steam mains are eliminated Isolation hospitals are sometimes supplied with a separate plant (*e g*, gas-fired boilers); smaller cooking apparatus is electrically operated, and distant buildings are provided with some usual form of heating plant. In the preliminary layout provision is usually made in proportioning the piping system to accommodate future extensions.

1. Boiler Installation.

Reference should be made to discussion on page 25 Where possible, all exhaust steam should be put to use The amount of high-pressure steam for operating machinery, tools, cooking, laundry, etc should be determined For this purpose the usual forms of high-pressure boilers are selected with sufficient reserve capacity in the form of a spare unit This spare unit may also act as a reserve for the hot-water heating system by installing a suitable steam hot-water heater The quantity of steam available from the power plant at bleeder or exhaust pressure is to be determined. That portion of the steam which is constantly available should be used for those operations which need steam summer and winter, such as baths, domestic hot-water supply, water for cleaning, etc The excess steam available is diverted into the hot-water heating system To obtain a clear picture on these matters, curves must be drawn showing the power, lighting and various heat requirements, not only by month to month but by daily and hourly variations to locate peak loads The curves will furnish data from which the amount of thermal storage may be determined and from this the contents of the storage tanks and the available heat To accomplish this result, direct-fired boilers may be set up and, these in turn may be used in one of two ways In the first instance boilers of large storage capacity can be used so as to enable sufficient overnight storage of heat with banked fires and pumps in operation The other method is to install countercurrent heaters (p 89) laid out to heat up the system quickly and by the use of storage tanks introduced in the circulation system to absorb the excess heat above normal requirements

In all cases the basis for design should not be the heat requirement for extreme weather conditions (if 0° F) but rather the fair average of winter weather (say, 32° F) The heat requirement in severe weather can be maintained in most installations by forcing the boilers to their maximum capacity. In this way a considerable saving is made in reducing the installation expense and improving the general operating conditions

As a rule each individual case requires careful analysis to insure the best installation from the technical as well as the economic viewpoint

2 Pipe Lines and Pipe Conduits.

In general pipe lines should be designed with great care and the connections should be substantial If these details have been given proper consideration, it may be possible to dispense with tunnels which permit the necessary headroom for passage and limit the size to conduits Figure 107 shows an installation where this is the case Should repairs be necessary, the earth is excavated and the trench cover is removed to gain access to the damaged section. Many schemes are used to limit the excavating to short sections Conduits should be drained to prevent the moisture from injuring the insulation

The choice of piping layout for the buildings to be heated (whether one- or two-pipe) and the pumping head must be decided for each case These data are obtained from methods discussed on page 191. Effort should be directed to reduce the fixed and operating expenses which depend upon first cost, depreciation, amortization, labor, power consumption of pumps, supplies, and heat loss In general the pressure differential of each building is estimated at about 3 ft of water-column pressure and the entire pumping head not over 30 ft. of water (gravity head neglected). For other details (pumps, starting, regulation, etc) the reader is referred to the section on Forced Circulation (p 61)

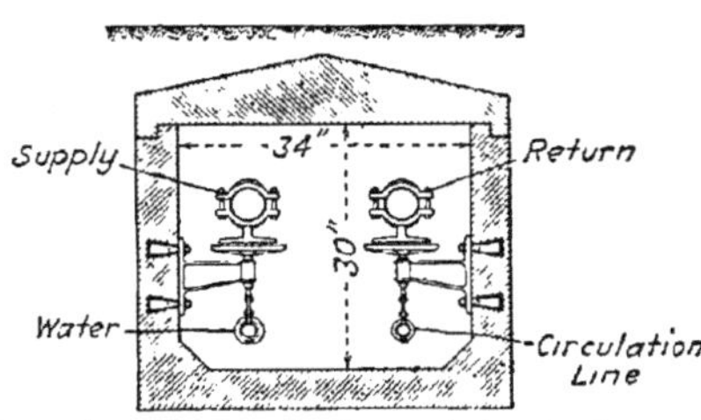

FIG 107 —Conduit for district hot water heating (*R O Meyer, Hamburg*)

3. Pump Connections and Expansion Tanks

In forced-circulation systems the pumps are usually installed in the return line Assume that K is the boiler (Fig 108) and that the flow and return mains of a district system are represented diagrammatically in the form of a circle. Let x represent the flow main shown by the solid line to the point M and the return main shown dotted returning to the boiler through the pump P The branch u_1 supplies the building G_1, and the return R_1 connects with the corresponding main The pressure differential between the two connecting points of these lines is the operating pressure required for the building G_1 A similar condition exists for the building G_2 It will be noted that the pump P is installed in the return main A static pressure h exists in the entire system and is determined by the maximum level of the water in the system The static pressure h is represented by the shaded ring in the diagram

Assume that the expansion line L, which leads to the expansion tank E, is connected to the flow main at the point A This implies that the pump exerts pressure up to the point A and that a suction exists in the mains from the point A back to the pump P. The form of the pressure line is determined by the friction of the pipe and fittings in the flow main. Every point on the flow main (in a clockwise direction) between P and A is subjected to an increase in pressure when the pump is in operation. For example, if a building is connected at the point x, the lowest radiator is subjected to the static pressure h and the excess pump pressure p', i.e., the total pressure H to which the radiator in question is subjected is $H = h + p'$.

The head due to the expansion tank is to be chosen and the mains are to be designed so as not to exceed the total head H of 100 ft. of water column pressure when cast-iron radiators are used. The same applies to cast-iron boilers since in this case also the boilers are subjected to the static pressure and practically the entire pumping pressure. American practice for standard house-heating hot-water boilers is 30 lb. per square inch total pressure.

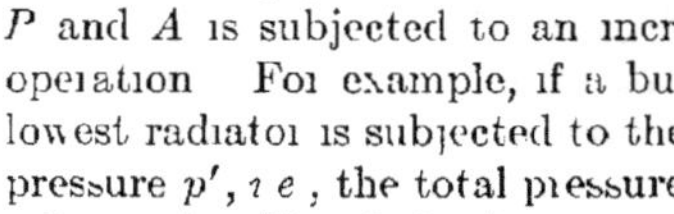

Fig. 108.—Pressure distribution in a forced circulation system.

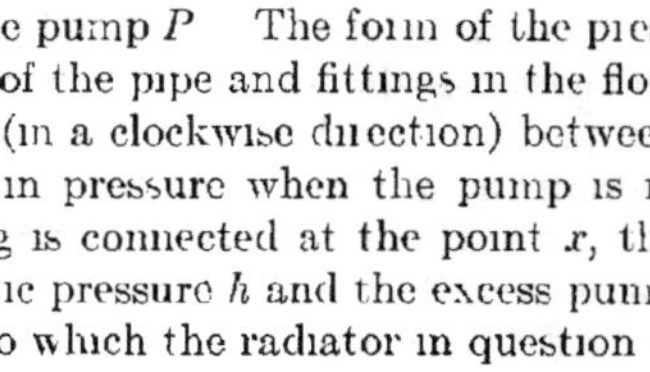

From the diagram it is apparent that the pump pressure beyond the point A (in a clockwise direction) drops below the static pressure. If therefore a building is connected at the point y, the pressure is reduced by an amount p''. This condition is likewise true at the pump. The static pressure h acting on the suction side of the pump is reduced by the amount p_2 when it is in operation. The position of the expansion-tank (point A) as well as the proportion of the main lines should be chosen so that p_2 is smaller than h thereby eliminating possible interruptions of the flow to the pump.

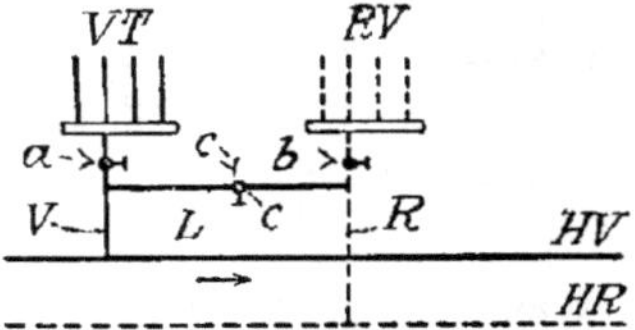

Fig. 109.—Connection of service mains to a building.

4. Distribution and Connections in Buildings.

The connection of the service mains to the buildings is simple. From the service main HV in Fig. 109 the branch supply line V connects the supply distribution point VT, whereas the return main R connects the return point RV to the return line HR of the service mains. By suitable adjustment of the valves a, b, and c, the crossover connection L makes it possible to control the flow between VT and RV A pressure gage between VT and RV indicates the operating pressure at any time If temperature differences between flow and return mains in the building are to be other than those existing in the service mains, the special apparatus discussed on page 105 is to be used

5. General Control.

Temperature control may be secured from the boiler room by changing the water temperature in the flow main or by mixing in the return main The effectiveness of this control depends upon the time taken by the water to reach the farthest radiators. This time depends upon the length of the circuit and upon the velocity of the water in the distribution mains With a velocity of 4 ft per second and a length of main of 3,000 ft a change in the boiler room will become effective at the farthest radiator about 12 min after the change is made

6. Control Panels.

One of the main advantages of district heating by means of hot water is the simplicity of the operating control This is obtained only if the necessary apparatus for attention and observation is placed conveniently and in a suitable location The control instruments include boiler-performance indicators, draft gages, starters and measuring instruments for the pumps, ventilating and control equipment, etc In general only the most essential instruments should be provided, and they should be maintained constantly in operating condition

7. Exhaust Heat. Utilization

Exhaust steam from engines and turbines as well as the exhaust heat from internal combustion engines, gas ovens, etc. may be used to good advantage in hot-water district heating systems Reference should be made to the section on Exhaust Heat Heating Installations (p. 105) for more complete treatment

8. High-temperature Hot-water District Heating Systems.

The usual hot-water systems seldom extend beyond 3,000-ft. radius The limitations are due to pipe sizes, heat losses, and power consumption of pumps It also depends to a large extent upon the kind of power used for the pump

If a heating system is designed to operate with a temperature drop from 195 to 165° F., the amount of water circulated may be reduced one-half on increasing the temperature drop to 70° F by raising the flow temperature to 235° F with a return temperature of 165° F Using the same pipe sizes in both cases, the power consumption of the pumps for the latter combination is reduced to about one-eighth If conversely, the power requirement remains the same for both cases, the pipe sizes may be diminished considerably An installation employing high-temperature drop was made by Rudolf Otto Meyer of Berlin in the extension of the Charlottenburg City Hall

The change from the high-temperature service mains to the building heating system may be made in two ways. In one case counter-current heaters (p. 89) are installed which are supplied with high-temperature water instead of steam. The "Reck" system of water heating which was employed in the Charlottenburg installation mentioned is simpler. In this system the flow main *V* (Fig. 110) is supplied by the service main *HV* with a valve *a* installed in the connection. The return *R* connects with the service return *HR* and also by means of the pipe *b* with the flow main *V*. The valve *a* is adjusted so that water from *HV* at, say, 235° F. is sufficient to maintain a temperature of 200° F. in the distribution main *V*. A corresponding amount of return water at 165° F. flows from *R* to the service main *HR*. The circuit in the building proper is from the main *R* through *b* into the flow main *V*, from which the various local circuits may operate as gravity or forced circulation systems.

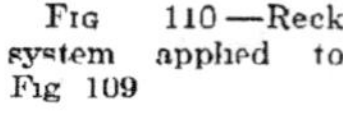

FIG. 110.—Reck system applied to Fig. 109.

The possibility of two methods of control are noted in that the temperature in the service mains may be varied or the regulation at *a* within the building itself is feasible.

Hot-water district heating systems may also be extended by installing separate boiler plants for blocks of houses. The boilers are fired with gas instead of coal or coke where producer gas is available in the desired location and in sufficient amounts. Oil-fired boilers are also used and good results obtained.

I. EXHAUST HEAT UTILIZATION[1]

GENERAL

Exhaust and bleeder steam[1] (steam extracted from intermediate stages of engines or turbines) may be used in one of the following ways: exhaust heat power applications and exhaust steam heating systems.

Exhaust steam heating systems as treated here include not only warming operations, but also ventilation, hot-water supply, heat for cooking, steaming, drying, etc.

[1] "Utilization of Exhaust Heat," from BRABBÉE, "Abwarmeverwertung," *Zeit. Werkstattstechnik*, 1912, "Exhaust Heat Utilization in Power Plant Operation," from SCHNEIDER, "Die Abwarmeverwertung im Kraftmaschinenbetrieb," 3 Aufl., Springer, Berlin, 1920 (complete bibliography included), "Finding the Most Economical Power for Factories," from URBAHN-REUTLINGER, "Ermittlung der billigsten Betriebskraft fur Fabriken," 2 Aufl., Springer, Berlin, 1920, "Selection, Design, and Operation of Power Plants," from BARTH, "Wahl, Projektierung, und Betrieb von Kraftanlagen," 2 Aufl., Springer, Berlin, 1920, "Exhaust and Bleeder Steam Utilization," "Maschinenfabrik Augsburg-Nurnberg, Abdampf-Zwischendampf-Verwertung," privately published, 1920, "Economics of Public Utility Heating Systems," from BEHRENS, "Gemeindliche Warme-wirtschaft," published by Weltwirtschaft und Technik, Berlin, 1920.

In the case of power installations where the exhaust steam is not required immediately, particularly for machines operating intermittently (rolling mills, steam presses, etc), it is discharged to storage tanks from which it is drawn for supply to low-pressure turbines Formerly these storage tanks required large masses of iron in their construction, while the later designs utilized water (Rateau accumulators) [1] Other types (Balcke storage tanks) were built in the form of gas holders in which condensation was prevented as much as possible by suitably insulating the holder as well as by using liquid seals [2] For recent designs of accumulators (Ruth apparatus), see *Report* of Kongress fur Heizungs-und Luftungstechnik, Munich, 1921 [3] The discussion of exhaust heat power plant installations is beyond the scope of the present text The use of exhaust for heating systems will be taken up in greater detail

Exhaust Heating Systems

From the heat balance diagrams in Figs 111 to 113 it will be noted that for the prime movers used at present about 60 per cent of the heat in

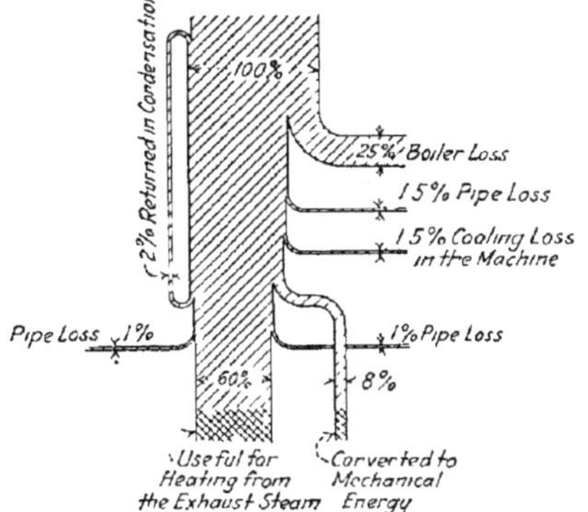

Fig 111 – Heat balance of a one cylinder engine

the fuel passes into the exhaust, cooling water and flue gases If this is allowed to escape into the atmosphere, the heat contained therein is wasted

In making power installations the following questions must be answered

[1] "Exhaust Steam Power Plants," from RUSTER, "Abdampfkraftanlagen," *Zeit Bay Rev Vereins*, 1909, "Turbines," from TREITEL, "AEG Turbinen," *AEG-Zeit Jahrgang* XIII

[2] A "New Exhaust Steam Accumulator," from KOPPLIN, "Ein neuer Abdampfspeicher," Gluckauf, 1911, "Maschinenfabrik Augsburg-Nurnberg, System Estner-Ladewig "

[3] Published by *R Oldenbourg*, Munich and Berlin, 1921

1 Is it possible to use exhaust or bleeder (extraction) steam, if of sufficient quantity, for heating, drying, hot-water supply, baths, or any other purpose?

2 In municipal or public utility plants, is it feasible to so arrange the plant that steam may be extracted at suitable pressures for process use in nearby plants?

3 Is it possible for baths, slaughter houses, chemical factories, breweries, industrial establishments in general, group dwellings, etc to be brought within the scope of a power plant so as to utilize bleeder steam?

4 Is it possible for bleeder steam to be applied for such special purposes as forcing growth in truck gardens, incinerating plants, frost protection of bath houses, or many other applications?

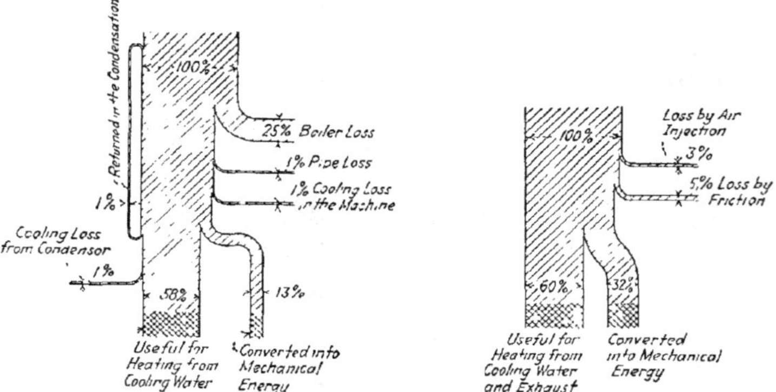

FIG 112—Heat balance of a 1,000 kw turbo dynamo

FIG 113—Heat balance of 100 hp Diesel engine

It must be remembered that applications of this sort should be made if possible to installations requiring heat throughout the entire year Heating systems will then take secondary place

After the answers to the above questions become known, the site of the power plant must be determined The plant should then be designed so that the equipment admits of extracting steam at the required pressures and so that its application to uses is economically justified If exhaust heat is to be used for process work, it is best to provide for this in the original installation In general the cost of installation may be increased thereby, but this may be offset by the savings otherwise effected In such plants where no provision was made for utilization of waste heat the economic aspect of the problem must be investigated for each particular case.

For approximate calculations it may be assumed that for non-condensing engines about 16,000 B t u per horse power is available for other uses and in the case of condensing engines 10,000 B t u is available

Exhaust heat applications often influence the choice in the type of prime mover. Whereas, in general, steam plants give the lowest cost of power when considerable use is made of the exhaust, the situation favors internal combustion engines when little use is made of waste heat It is important to note that the steam consumption of a single unit is not the deciding factor in an analysis but the operating expense of the plant as a whole In general the comparison between purchased power and the cost of generating it at an isolated plant may under extreme conditions be favorable to the isolated plant in sizes above 500 kw A desirable arrangement is one wherein the exhaust requirements of a plant are amply provided for by the isolated plant and where the excess over the power requirements of the isolated plant may be sold at a profit to neighbors.

As will be seen in the diagrams in Fig 111 to 113, exhaust heat utilization is possible in both steam and internal combustion engine plants

1. Exhaust Heat from Steam Power Plants.

It may be noted that both the steam engine and the turbine are adapted for exhaust steam heating From a technical viewpoint in heating practice the reciprocating steam engine is preferable, while for bleeder steam use the steam turbine is favored In deciding it must be remembered that steam taken from reciprocating engines carries over some oil which is difficult to remove entirely, while on the other hand the steam from turbines is practically free from oil Oil in the steam considerably reduces the heat transfer of the radiation, it is dangerous in the boilers, and it makes the steam useless for certain processes (dyeing, etc)

The operating principle of oil separators is impact, centrifugal force, or both Figure 114 illustrates a type which operates with both impact and centrifugal force According to experiments of Eberle,[1] oil separators may be considered effective if they reduce the oil in 1,000 lb. of steam to ⅕ oz

The use of exhaust or bleeder steam may be employed by: (1) direct use of the steam, (2) heating water by means of steam, and (3) heating air by means of steam or hot water

(1) *Direct Use of the Steam* (1) Exhaust Steam Utilization — Figure 115 illustrates the connection for an exhaust steam heating system The steam discharged from the engine passes through an oil separator and enters a compound regulator which controls the pressure on the heating side If the pressure is insufficient live steam is automatically

[1] "Tests of Oil Separators," from EBERLE, "Versuche mit Dampfentölern," *Zeit Ver deut Ing*, 1910, *Zeit Bay Rev Vereins*, 1911

admitted. In case of excessive pressure the regulator exhausts into the atmosphere.

If the exhaust heat of the machines always corresponds to the steam consumption of the connected system, about 50 per cent reduction in operating costs may be effected over the operating expense of condensing units with live steam heating. The combination of power and heating discussed here will become uneconomical only when about one-half of the exhaust steam is wasted or more than one-half of the live steam is to be added.[1]

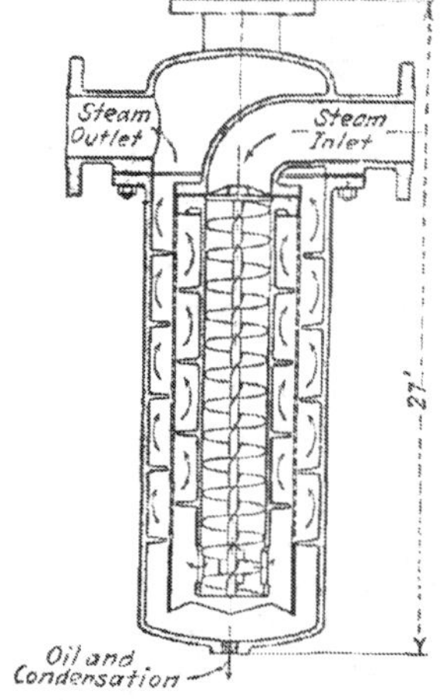

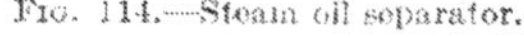

Fig. 114.—Steam oil separator.

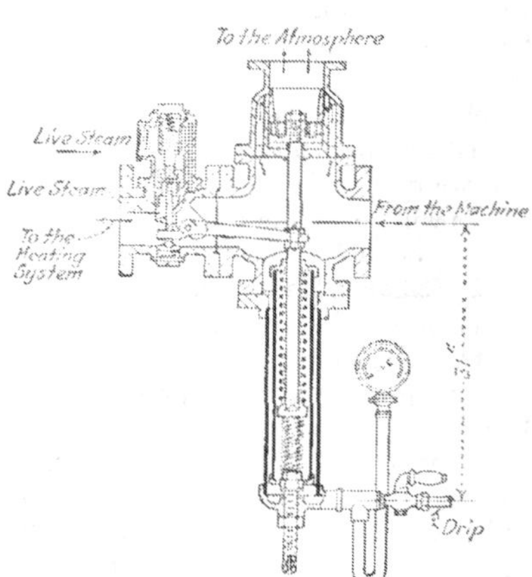

Fig. 115.—Compound regulating device.

Complete utilization of exhaust steam is then the case when the desired quantities of exhaust are available at the required pressure. In other cases steam may be extracted from the prime mover at the required pressure to supplement the use of the exhaust or to be used independently. The advantages of operating under the conditions set forth are: possibility of employing wide ranges of pressures independent of the load; automatic regulation for any desired supply pressure; automatic regulation of the machine for a fixed rotative speed; operation of the low-pressure prime mover as a usual free exhaust unit or as a condensing unit when its process

[1] For further details as to the specialties used in making installations, see Schneider, "Exhaust Steam Utilization in Power Houses." "Die Abwärmeverwertung im Kraftmaschinenbetrieb," published by Springer. 1920; see also, "Steam Engines and Heating Systems," from Deinlein, "Dampfmaschinen und Heizungsanlagen," *Zeit. Bay. Rev. Vereins*, 1908; also by same author a course of lectures at the Hauptstelle für Wärmewirtschaft, Berlin, 1920.

steam is not required; smaller reconstruction costs than those required for converting condensing units into back pressure units.[1]

A further discussion of details leads beyond the scope of the text. Bleeder steam is to be considered as high-pressure steam, so far as design and installation are concerned, and reference should therefore be made to that part of the text. Bleeder steam is used extensively for cooking, steaming, drying, hot-water heating in general, dye works, bleacheries, textile mills, paper mills, malting establishments, breweries, chemical industries, kaolin works, slaughter houses, baths, etc.

An example of a bleeder steam turbine installation is shown in Fig. 116. The turbine is of about 3,000-hp. capacity and runs at 3,000 r.p.m.

Fig. 116.—Bleeder turbine of about 3,000 hp. and an hourly steam consumption of 26,000 lbs. (*Maschinenfabrik-Augsburg-Nurenberg.*)

with a total steam consumption at full load of about 26,400 lb. of steam per hour.

(2) *Water Heating by Means of Steam.*—Steam engines and turbines are equipped with surface condensers of a type suited for the particular operating conditions. When maintaining the usual vacuum, the condensate temperatures range from 90 to 125° F. If it is necessary to use water at higher temperatures than these during the winter, the change is brought about by admitting air into the exhaust line, thus decreasing the vacuum. As a consequence, with the same load on the prime mover the steam consumption rises as the vacuum decreases. The relative steam consumption for single-cylinder engines is greater than for compound engines when the vacuum is decreased, being still

[1] See Hottinger, "Heating and Ventilating;" Woyl, "Manual for Hygiene," published by J. A. Barth, Leipzig, 1913.

greater in the case of turbines since their economy is dependent very largely on maintenance of a high vacuum. Higher water temperatures may also be obtained in a simple manner by means of coil heaters, the heating of which is done by means of steam bled from the turbine at some intermediate stage. The vacuum in this case need not be disturbed. Live steam may also be considered for this purpose. In cases of this kind the exhaust from feed pumps or other auxiliaries may be used to advantage. When the steam is free from oil as in turbine drives, the clean exhaust steam can be fed directly into the water to be heated.

The transfer of heat from the steam to the water is accomplished in steam hot-water heaters as described in detail on page 89. The water so heated is generally supplied to the various points of use by means of pumps so that the installation provides a common pump heating system from the water heaters. These were considered in detail on page 61.

FIG. 117.—Cast iron sectional heater utilizing Diesel motor exhaust gas. (*Gebr. Sulzer, Winterthur, Switzerland.*)

(3) *Heating Air by Means of Steam or Water.*—Bleeder steam, exhaust steam, or circulating water from surface condensers may be used for heating air. The types of equipment suitable for this purpose were discussed on page 90 *et seq.* It might be noted that similar equipment may also be used for precooling (precondensing), of which examples will be shown later.

2. Exhaust Heat from Internal Combustion Engines.

The jacket water from internal combustion engines, being clean and at a temperature of from 125 to 150° F., may be used for hot-water heating systems or hot-water supply. The exhaust gases from combustion motors range from 550° F. at half load to about 650° F. at full load. This heat may be also rendered useful by means of cast-iron sectional heaters[1] (Fig. 117). To prevent condensation of the moisture in the exhaust gases, cooling is not permitted below 260° F.

For producer gas engines the available heat in the exhaust may be taken as 2,000 to 2,400 B.t.u. per horsepower; for Diesel engines the values range from 2,000–1,400 B.t.u. per horsepower depending upon the load. Recalling figures previously given (p. 108) of from 10,000 to 16,000 B.t.u. per horsepower for steam engines, the advantage of steam engines

[1] "Exhaust Heat Utilization from Diesel Engines," from HOTTINGER, "Die Abwärmeausnutzung bei Dieselmotoren," *Zeit. Ver. deut. Ing.*, 1911.

over combustion engines for utilization of exhaust heat is apparent. In the case of combustion engines the total heat in the exhaust is dependent upon the load. With steam engines the use of bleeder or live steam in addition makes them independent in this respect.

3. Examples of Exhaust Heat Heating Systems.[1]

a. Exhaust Steam Heating.—Figures 118 and 119 show the fourth section of the heating system installed in the plant of the Gasglühlicht A. G. in Berlin. On main I there are almost as many radiators as on mains II, III, and IV. With the lowest outside temperatures, the build-

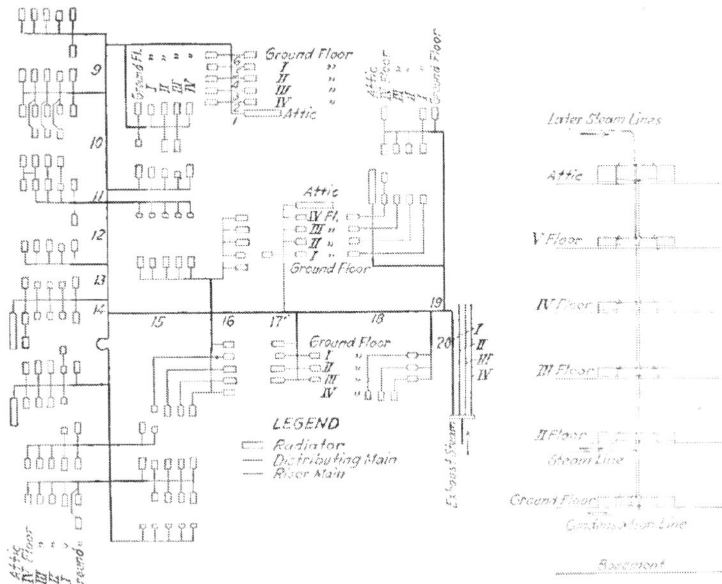

FIG. 118.—Diagrammatical layout of a heating system using exhaust steam. (*R. O. Meyer, Berlin.*)

FIG. 119. One of the risers shown for the system illustrated in Fig. 118.

ing requires about 8,000,000 B.t.u. per hour. Originally an exhaust steam system was provided which was to operate with a back pressure of 3 lb. per square inch on the engines. During the first winter, however, the building was not completely heated with 22 lb. per square inch, so that exhaust heating was abandoned and live steam heating had to be used.

[1] See also "New Power Plants," from JOSSE, "Neuere Kraftanlagen," published by Oldenbourg, Munich, 1924.

The installation was redesigned by the author[1] who corrected errors in piping sizes and in installation. After installing suitable traps, effectively insulating the mains, changing valves, and adding a new steam main to the fourth floor, the system was put in operating condition. The heat requirement of 8,000,000 B.t.u. per hour was provided at a back pressure on the engines of 1½ lb. per square inch.

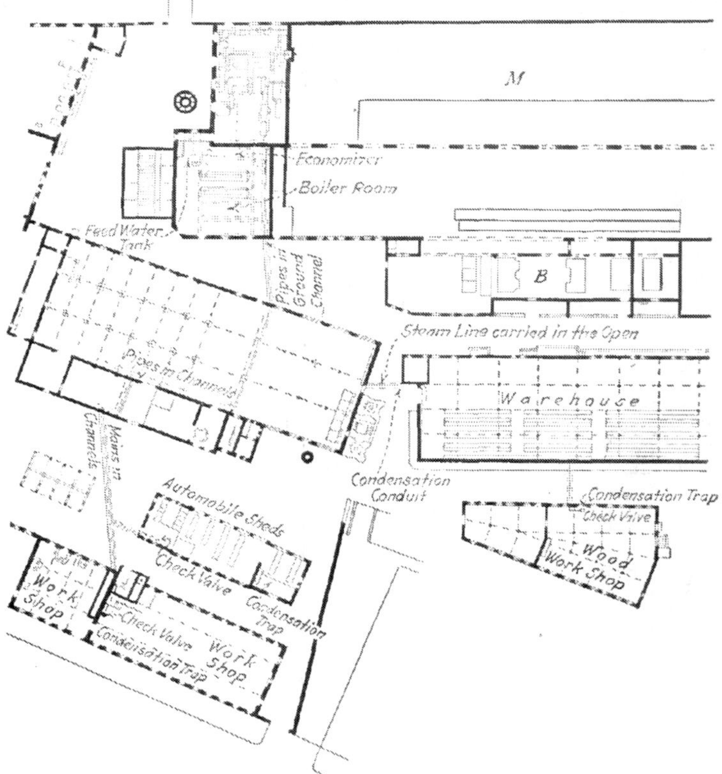

FIG. 120.—Factory heating with exhaust steam. (*Gebr. Sulzer, Winterthur, Switzerland.*)

By raising the back pressure, exhaust steam may be used effectively for even distant buildings. An example is shown in Fig. 120. This

[1] "Heating System in the Plant of the Auer Company," from BRABBÉE, "Die Heizungsanlage im Fabrikgebäude der Auergesellschaft. Ein Beitrag zum Thema, Wirtschaftliche Vorteile bei der Verbindung von Kraft-Heizbetrieben," *Gesundh.-Ing.*, 1909, p. 221.

shows a large group of factory buildings with numerous outlying buildings.

b. Exhaust Steam Hot Water Heating.—A residence situated 300 ft. from a paper mill was originally equipped with a warm-air heating system. This was later replaced by a system in which the air is heated by means of hot water.[1] For this purpose a surface condenser was installed

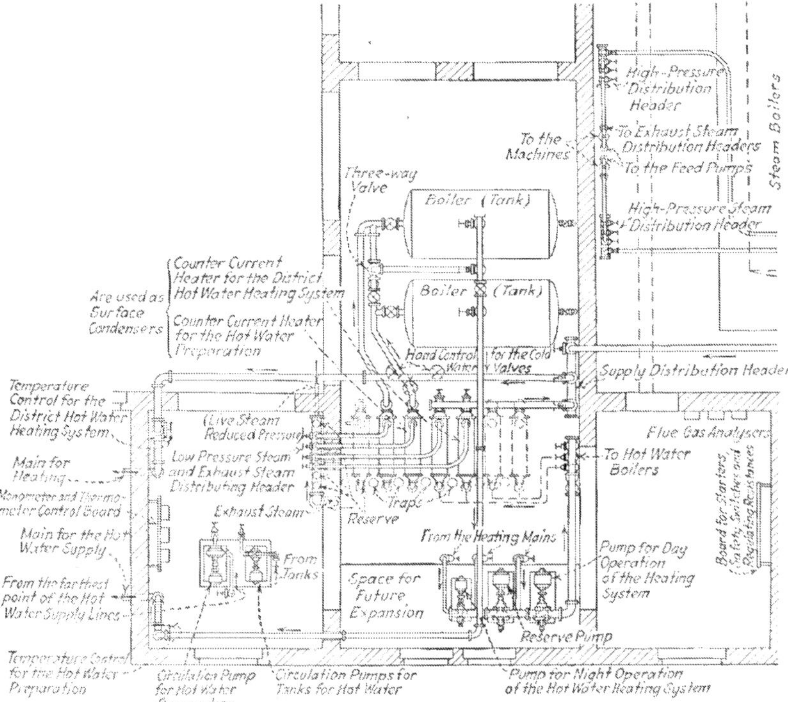

FIG. 121.—Exhaust water heating system for a large hospital. (*Beckem & Post, Hagen, i. W.*)

paralleling a jet condenser, by inserting an automatic change over valve. The hot water is supplied to the heating system by means of a pump. If the exhaust steam is insufficient or the engine is shut down, live steam is supplied to a surface condenser by means of a reducing valve. In mild weather the flow temperature of about 130° F. is maintained with

[1] "Hot-water Heating by Means of Exhaust Steam," from MEYERS, "Warmwasserheizung unter Ausnutzung der Abdampfwärme einer 100 pferdigen Kondensationsmaschine," *Zeit. Ver. deut. Ing.*, p. 244 *et seq.*, 1910.

the usual vacuum on the engine, and this flow temperature is sufficient for the heating requirements In severe weather a coil heater is used in addition so that with the use of live steam the temperature may be further raised to that required

Figure 121 illustrates a central hot-water heating system and a central domestic hot-water supply system using exhaust heat from engines The size of the steam and water boilers for this installation was found from diagrams showing the load for the day and night and for summer and winter operation Consideration was given to the steam requirements of the engines for cooking, laundry, heating, ventilation, and general steam supply An additional steam boiler with proper arrangement of water heaters serves as a reserve for the water boiler The engines and feed pumps are connected to the high-pressure manifold by double lines of pipe and also by means of a reducing valve to the exhaust steam manifold From the water boiler a line leads to the distributing header into which the circulating water from the surface condensers is connected so that the central heating system can be supplied from either the boiler or from the water heaters (coil heaters using exhaust steam) The hot water returning from the buildings enters the circulating pumps The first pump is used for day operation, the second for night operation, and the third for a reserve unit A return distributing header makes it possible to send the water either to the coil heaters or to the boilers, thus closing the circuit

The exhaust steam from the engines passes through the apparatus previously described (via separator, etc) and then enters the low-pressure distributing header (also supplied with a live-steam connection) To this are connected the surface condensers which act as countercurrent heaters, delivering the water heated therein to the heating system or to the hot-water storage tanks from which the central water supply is taken Two electric pumps, one of which is connected to the water storage tanks and the other to the farthest point in the domestic hot-water supply lines, maintain the required temperatures in the circulating system A well-designed control room with control panel equipped with current and pressure indicators, starting switches, resistances, gages, remote thermometers, flue-gas recorders, etc , complete the installation

c Heating Air by Means of Steam—In Fig 122 is shown an installation in which air is heated by means of steam The exhaust from the low-pressure cylinder connects with the steam inlet to the heater, the latter acting as a precondenser. The air passes through filters and then through a drum type washer, enters the suction side of the blower, and is forced through the air heater The hot air in this installation is used in a weave shop In severe winter weather the air heater is connected to the receiver of the compound engine and is operated with bled steam instead of the exhaust steam from the low-pressure cylinder.

d Exhaust Gas Heating.—As previously mentioned the cooling (jacket) water and the exhaust gases of Diesel and similar engines may be used for heating or industrial hot-water supply, whereby this water, flowing through cast-iron sections embedded in the exhaust gas ducts, may be further heated by the heat of the exhaust gases.

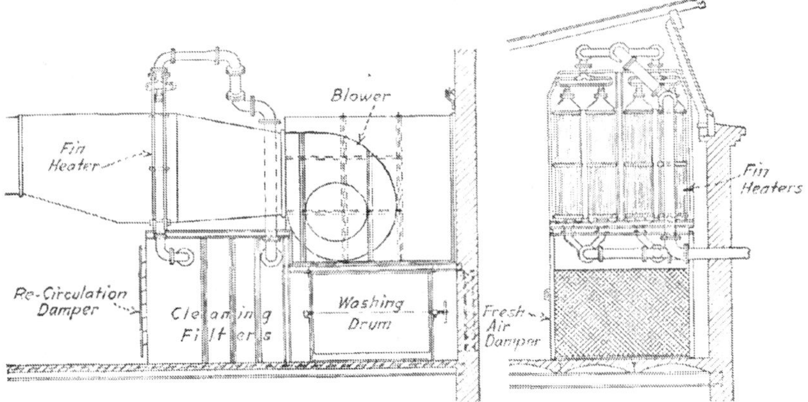

Fig. 122.—Steam air heater using exhaust steam. (*Junkers & Company, Dessau.*)

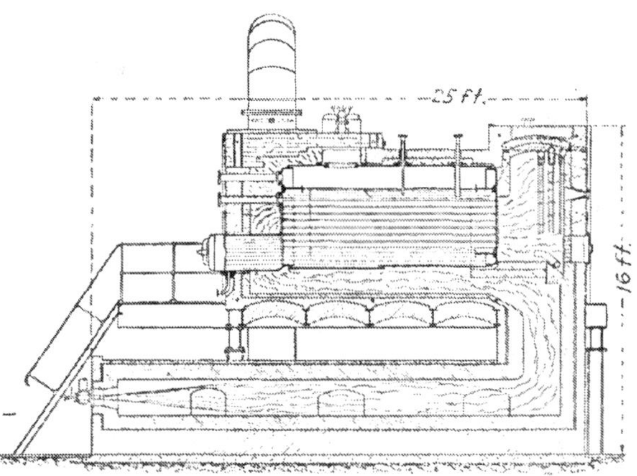

Fig. 123.—Steam boiler utilizing exhaust heat. (*Gebr. Sulzer, Winterthur, Switzerland.*)

4. Other Forms of Exhaust Heat Utilization.

Examples of numerous cases may be cited where exhaust or otherwise waste heat may be utilized, but the following will suffice for illustration.

Figure 123 shows a high-pressure steam boiler heated by means of the exhaust gases from a smelting furnace. The exhaust gases flow through the tubes and finally enter the chimney. The particular boiler shown has about 230 sq. ft. of heating surface and operates at a working pressure of about 120 lb per square inch.

5. Examples Showing the Economy of Exhaust Heating.

a In the installation shown in Figs 118 and 119, when changed from high-pressure steam to exhaust steam, the saving was approximately 660 tons of coal a year. This saving was exceeded, due to the fact that additional steam engines have been changed over to use their exhaust in the same way.

b The exhaust steam hot-water heating system mentioned under *b* (p 114) compared with live steam shows an annual saving of about 165 tons of coal a year despite the smallness of the installation

c In the hospital at Gross-Lichterfelde (Berlin) the utilization of exhaust heat from the engines was inaugurated when alterations were made, with the result that savings of 4,400 tons of coal annually were obtained [1]

d In the steam plant (300 hp) of the Munchener Neusten Nachrichten, 1,320 lb of steam per hour were required for heating purposes In this case the consumption was:

23 8 lb of steam per kilowatt-hour with heating
22 7 lb of steam per kilowatt-hour without heating

By use of the bleeder steam consuming only 1,320 lb per hour, about 1,650 tons of coal were saved annually

e The Stuttgart Public Bath Company[2] erected its own electric power plant, sold the power in the year 1910 for 1 43 cents per kilowatt-hour and used the engine exhaust for heating the hot-water supply. In the plant about 16 per cent of the fuel is useful in the form of power and 65 per cent in the form of heat. In spite of the relatively small output of 500 to 600 hp the annual savings of 2 750 tons of coal were effected

f In the construction of the third section of the new hospitals in Munich an electric power plant of 2,200 hp was installed, the exhaust of which was sufficient to heat the buildings. The excess power generated at this plant was fed to the city supply. The annual saving in coal was 8,850 tons

[1] "Exhaust Steam Utilization in the Hospital at Gross-Lichterfelde," "Abdampfausnutzung zur Warmwasserbereitung und Heizung im Kreiskrankenhaus in Gross-Lichterfelde," *Gesundh -Ing* , 1908

[2] "Combined Bath and Electric Generating Station," from "Verbindung von Bad und Elektrizitatswerk," *Zeit Bay Rev Vereins*, p 20, 1909, "New Steam Plant of the Stuttgart Bath Company," from EBERLE, "Die neue Dampfanlage der Stuttgarter Badegesellschaft," *Zeit Bay. Rev* Vereins, p 96, 1910.

When favorable results like the foregoing can be obtained with such comparatively small installations, it is likely that similar economies may be effected if the larger public utility plants were operated in conjunction with district heating systems.

Radical improvements in the operating economy of modern boilers or power plants is hardly to be expected It appears therefore that the future economic development must consider the generation of power concomitant with the distribution of heat

Many of the applications here noted were discussed 15 years ago [1] At that time but few European engineers were seriously occupied with these questions At present the problem is sufficiently serious to demand the attention of the foremost engineers

It will be appreciated that the heating engineer of today must be equipped with a thorough understanding of the fundamental principles involved coupled with an extended experience in the application to commercial problems

[1] Utilization of Exhaust Heat from Brabbée "Abwarmeverwertung," Zeitschrift Werkstattstechnik, 1912

PART II
VENTILATION

SECTION I

NECESSITY OF VENTILATION

A. INTRODUCTION

The need of ventilation arises from the fact that the air in occupied rooms undergoes contamination. This is caused by:

Heat and moisture emission from the human body, illumination, processes, etc

Carbon dioxide emission from people, illumination, etc.

Generation of impurities, odors, etc

Dust

The degree of the possible contamination of the air supply may be appreciated when it is considered that people assimilate 25 lb of nourishment in a gaseous form and 6 lb in the form of solids and liquids in 24 hours Though there is general public feeling against consumption of contaminated food and uncleanliness in its service, a much graver danger in breathing uncleaned air is tolerated

It is essential to educate people so that these important relations are understood better and so that they will learn how harmful it may be to neglect air nourishment Hygiene and sanitation for air conditioning are as important as for water conditioning and its removal of waste matter from houses and cities

The influence of good ventilation in factories is reflected in the improved performance of the workmen As the public authorities demand suitable atmospheric conditions, certain buildings become better equipped. In public halls, theaters, hotels, etc unhealthful conditions frequently prevail

The ventilation of schools in Europe is not given sufficiently serious attention In America numerous ventilating codes have been enacted, and some thorough investigations of ventilation as applied to schools have been made [1] A recent resolution of American Medical Society[2] in this connection indicates that no definite standards have been agreed upon It may be observed that children appearing healthy up to their sixth year will upon entering certain schools become pale, weak, and disinclined to study. Efforts to combat tuberculosis of the lungs offers

[1] "Ventilation," *Report* of New York State Commission

[2] Reprinted in *Heating Ventilating Mag*, January, 1926

its greatest resistance in human beings between the ages of 6 and 10 years Much of this difficulty may be removed by adequate ventilation in the school room From observations by the author over a number of years it was found that in well-ventilated lecture rooms the capacity of the audience attains a maximum, and the effort required of the lecturer is reduced to a minimum [1]

B HEAT AND MOISTURE EMISSION FROM PERSONS AND ILLUMINATION

General

The blood temperature of human beings in a state of health is about 98 5° F This temperature remains constant within narrow limits if there is no disturbing influence in the health of the individual It is a surprising fact that this temperature remains constant independent of wide variations in atmospheric conditions This result is secured by processes in which the heat from the oxidation of food is dissipated by the body in a way to leave the blood temperature constant

The body dissipates heat by radiation, conduction, and evaporation The temperature and the humidity of the surrounding air must be such as to permit the bodily heat emission at a sufficient rate. Heat removal from the body may take place by accelerating the evaporation of body perspiration from the skin For this reason fans almost immediately produce an improvement in conditions without a fresh air supply. If the temperature of the room were equal to the blood temperature and the air completely saturated with moisture, a serious disturbance to health would occur which would finally result in death

From the foregoing it is clear that the moisture content of the air as well as the temperature of the room and the air motion is essential to the comfort of the occupants

There is an increase in the moisture content and the temperature due both to occupancy and to sources of illumination and these are important factors affecting air conditions [2] Evidence of these important facts was obtained from the experiments of Flugge and his coworkers and published in 1905

A further series of valuable experiments on the effects of air motion, humidity, and temperature has been completed by the Research Laboratory of the American Society of Heating and Ventilating Engineers, and published from time to time in the *Journal* of that society (1924–1925)

[1] "New Auditorium Ventilation System at the Research Laboratory, Charlottenburg," from Brabbée, "Über die neue Horsaalluftung in der Versuchsanstalt," *Gesundh -Ing*, p 441, 1915

[2] *Deutsche Vierteljahrschr offentl Gesundheitspflege*, 1890 Flugge, Paul, Ercklenz *Zeit Hyg Infektionskrankh*, 1905

Heat Emission from Occupants and Sources of Illumination

From the experiments of Pettenkofer and Rubner the hourly heat emission of occupants may be taken as follows:

For adults at hard physical labor........... 560 B.t.u. per hour
For adults at moderate physical labor....... 480 B.t.u. per hour
For adults at rest........................ 400 B.t.u. per hour
For children up to 10 years of age.......... 200 B.t.u. per hour

A part of this heat is dissipated by means of conduction and radiation and the rest by evaporation.[1] For fully occupied rooms the radiation component is diminished because of the counterradiation from nearby equally warm sources of radiation. Experimental evidence on this subject is meager, but according to Rietschel[2] the results may be taken as follows:

For rooms with average occupancy (residences, offices, etc.)................................ 300 B.t.u. per adult
For densely occupied rooms (theaters, halls, schools, etc.)........................... 200 B.t.u. per adult

For children of about 10 years of age one-half of the above values are to be taken. Rooms are densely occupied when there is one person for each 10 to 12 sq. ft. of floor area.

The heat emission from sources of illumination[3] is given in the following table:

Form of illumination	Hourly consumption for 1 Hefner candle	Hourly heat emission for 1 Hefner candle in B.t.u.
Arc lamp	1.1 watts	4.0
Tungsten filament	1.2 watts	4.0
Carbon filament (16 Hefner candles)	4.5 watts	16.0
Acetylene	0.0212 cu. ft.	22.0
Gas mantle (vertical)	0.0742 cu. ft.	26.0
Argand burner	0.353 cu. ft.	200.0
Bray burner	0.459 cu. ft.	268.0
Kerosene lamp	0.0073 lb.	144.0

[1] According to Rubner if the heat dissipation by means of radiation from the body is reduced or eliminated, this heat dissipation takes place through conduction (as in the case of a water bath) or through evaporation (as in a hot-air bath, by evaporation of the moisture formed by perspiration).

[2] Rietschel, *Gesundh.-Ing.*, 1913.

[3] Wedding, *Deutsche Vierteljahrschr. öffentl. Gesundheitspflege*, 1901.

In the heat emission given above is included all the heat in the light, water vapor, and products of combustion. Since the heat equivalent of the light is relatively small in gas and similar illuminants, and since the heat in the vapor, though quite variable, is also unimportant, the values given may be considered correct for all practical purposes. For open flames it is presupposed that the products of combustion mix with the air in the room.

Moisture Given Off by Individuals and Sources of Illumination

The evaporation of moisture from individuals varies with the relative humidity and the temperature of the surrounding air and also with the age, feeding, and bodily activity. According to Rietschel[1] the evaporation may be taken as shown:

In rooms with average occupancy[2]..... 600 grains per hour per person
In rooms with dense occupancy........1,200 grains per hour per person
For children up to 10 years of age these values are reduced to one-half.

The moisture emitted by illuminants may be assumed as follows:

Gas lighting for one cu. ft. of gas................ 1 oz. of moisture
Candles for 1 Hefner unit........................ 154 grains per hour
Kerosene for 1 Hefner unit...................... 90 grains per hour
Alcohol light for 1 Hefner unit.................... 60 grains per hour

It should be noted, of course, that there is no moisture emission from electric lights.

C. EMISSION OF CARBON DIOXIDE FROM INDIVIDUALS AND ANIMALS

Due to breathing[3] there is a chemical change in the air, with the following analysis by volume:

	Dry air when inhaled, per cent	Average air when exhaled, per cent
Oxygen	20.96	16.4
Nitrogen	78.00	79.0
Argon	1.01	1.0
Carbon dioxide	0.03	4.6

The nitrogen and argon content remains practically constant, the oxygen content is reduced about one-fifth, while the carbon dioxide con-

[1] Rietschel, *Gesundh.-Ing.*, 1913.
[2] See Heat Emission from Occupants, p. 123.
[3] This refers to breathing by means of the lungs and considers that through the skin as unimportant.

tent is increased more than a hundred times It can be shown that with continuous breathing without room ventilation death will eventually result due to an increase in the carbon dioxide content rather than a reduction in the oxygen supply.

Air with a carbon dioxide content of 1 per cent may be breathed for long periods without danger, but with volumes from 5 to 10 per cent for short periods only The permissible carbon dioxide content of the air in the room is limited by Pettenkofer[1] to 0 7 and by Rietschel to 0 7 to 1 5 parts per thousand From the foregoing it is seen that the limits set are not determined by the poisonous effect of the carbon dioxide but by the fact that, according to Pettenkofer, the unmeasurable poisons taken up under Poisonous Effluvia increase with increase in the dioxide content with the latter being measurable

According to Pettenkofer[2] and Scharling[3] it may be assumed that for ordinary room temperatures the emission of carbon dioxide is as follows

Adults (on the average)	0 7 cu ft per hour per person
Children up to 10 years	0 35 cu ft. per hour per person

Furthermore according to Fischer[4] the carbon dioxide formed by combustion (computed for 32° F) may be taken as follows.

0 6 cu ft per cubic foot of illuminating gas
25 cu ft per pound of kerosene
22 cu ft per pound of wax

The figures assume that the combustion gases mix with the room air

D. POISONOUS EFFLUVIA (AMMONIAE, ANTHROPOTOXIN)

Poisonous effluvia are a number of generally unknown products given off by individuals either through the skin or from its surface Among them are fatty acids such as capron and capryl acids These products have an unpleasant odor noticeable even in small quantities[5] In this category belong gases originating internally from food products In addition there is the contributory factor of the ammoniac due to putrefaction of the mucus from the throat and mouth and the decay of the teeth

In the year 1880 it was thought that the cause of the unpleasant characteristics of the exhaled air had been found in the so-called anthro-

[1] PETTENKOFER, *Ann Chem*, 1862, 1863
[2] *Zeit Biol*, Bd II
[3] LEHMANN, "Handbuch d physiol Chemie," 1854
[4] FISCHER, "Jahresbericht d chem Technologie," 1883
[5] WEYL, "Handbuch der Hygiene," Bd IV, p 259

potoxin (poisons in exhaled air) Hermans,[1] Jensen,[2] Rauer,[3] and Formanek,[4] among others, have demonstrated, however, that the experimental methods were open to criticism and that the results were unreliable Later Wolpert[5] again attempted to support this viewpoint of poisonous exhalation but his work was refuted by B Heymann.[6]

Neither have the theories of Weichardt[7] on the so-called kenotoxins been supported, particularly as a result of the work of Inaba [8]

The theory that poison is present in the breath is not accepted at the present time, although it cannot be ignored that the exhalations affect breathing, in time causing sluggish blood circulation, headaches, and other ill effects. While these contaminations of the air cannot be measured at present, according to Pettenkofer's proposal the ill effects are restricted by setting a limit on the allowable carbon dioxide proportion of the air supply.

E. DUST

Dust when inhaled irritates the mucous membranes and makes them susceptible to the attack of various disease germs. Furthermore, the presence of disease germs (pneumonia, tuberculosis, etc) has been shown in room dust Dust in rooms is not removable by ventilation and should therefore be prevented by vacuum cleaning of rugs and furniture rather than by beating or dusting All properly designed ventilation systems will provide the least number of dust particles in the air supply (air washing)

If the surface temperature of the radiators exceeds 150 to 175° F the dust coming in contact with the heating surfaces will decompose The resulting products irritate the mucous surfaces of the eyes, the mouth, and the throat and are the cause of numerous and unjustified complaints against the dryness of the air [9]

[1] HERMANS, *Arch Hyg*, 1882
[2] JENSEN, *Arch Hyg*, 1890
[3] RAUER, *Zeit Hyg Infektionskrankh*, 1893
[4] FORMANEK, *Arch Hyg*, 1900
[5] WOLPERT, *Arch Hyg*, 1903
[6] HEYMANN, *Zeit Hyg*, *Infektionskrankh*, 1905
[7] WEICHARDT, *Arch Hyg*, 1908
[8] INABA, *Zeit Hyg Infektionskrankh*, 1911
[9] NUSSBAUM, *Hyg Rundschau*, 1905, VON ESMARCH, *Hyg Rundschau*, 1905, HERBST, *Hyg Rundschau*, 1905.

SECTION II

REQUIRED AIR CHANGES

A. GENERAL

In the former editions of the original text the determination of the required air changes was based on limits of the carbon dioxide content of the atmosphere In the last edition the air change was based on the following requirements:

Temperature
Humidity
Carbon dioxide
Rules from experience

Recent studies show that these measurements (with the exception of those under Rules from Experience) are not entirely adequate for the reasons given below. Nevertheless the more important methods of a calculation are treated in detail on page 257 according to methods prevailing at present.

The American Society of Heating and Ventilating Engineers has adopted the Synthetic Air Chart as a measure of ventilation requirements The quality of the air is based upon a weighted combination of the factors of temperature, humidity, air motion, dustiness, odor, bacteriological content, and carbon dioxide content [1]

B. TEMPERATURE REQUIREMENTS

On page 3 the room temperatures to be maintained in different kinds of rooms are listed. These values should therefore be made the basis for computing the air changes for a specified room temperature that is not to be exceeded

Designing for a fixed temperature only is open to objection. In Europe a room temperature of 68° F. may be assumed sufficient for winter weather with normal air changes It was frequently found in the author's well-ventilated lecture room that the occupants believed the temperature much too low at 68° F in summer weather Indeed on very warm humid summer days, the room temperature had to be increased to 75° F to insure comfortable atmospheric conditions Similar conditions were found to be the case in the new ventilation system of a representa-

[1] See "A S H & V E Guide," Chaps XIV and XV, 1925–1926

tive office building in New York,[1] when on hot sultry days a temperature difference of 10° F. between outside and room temperature was sufficient for comfort, whereas at greater differences the room felt cold. The required room temperature depends largely upon the amount of air movement and the degree of relative humidity within the room. Thus for a greater air motion a higher temperature must be maintained. At present there is a lack of information on the influence of heat storage from the cold walls and the effect of counterradiation. Moreover, it is not known whether all the air in the room takes part in the air currents produced by mechanical ventilation nor to what extent the exhaust air temperature is affected by it. Precise data for computation purposes are therefore lacking.

C. HUMIDITY REQUIREMENTS

The factor of humidity is not sufficient in itself to form a basis for determining the required number of air changes. It must be recognized that the permissible limits of the relative humidity are very wide. According to Flugge a relative humidity of 25 per cent is sufficient with otherwise good atmospheric conditions, while relative humidities of 60 to 70 per cent may be sustained. Kisskalt[2] also determined that the surrounding walls of the rooms, depending upon their construction, have the property of absorbing large quantities of moisture and, as a consequence, influence the resultant computations (see p. 258).

In this connection reference should be made to the experiments of the American Society of Heating and Ventilating Engineers on temperature, humidity, and air motion.[3] The charts accompanying this work show the wide range of combinations which may be endured and the conditions considered comfortable.

D. CARBON DIOXIDE REQUIREMENTS

The carbon dioxide as an indicator for the required number of air changes gives considerably smaller values than the heat or moisture factors. On this account it is applicable in rare cases only (see p. 260).

E. HEAT CONTENT OF AIR

As previously mentioned, personal comfort depends primarily upon the relation of temperature, moisture content, and motion of the air. The total heat content per pound of air represents a value which varies

[1] *Heating Ventilating Mag.*, October, 1926.

[2] "Influence of Temperature and Wind Pressure on Air Changes," from KISSKALT, "Der Einfluss von Temperatur und Winddruck auf die Selbstluftung," *Gesundh.-Ing.*, p. 853, 1913.

[3] *Trans.* A. S. H. & V. E., 1924; see also "A. S. H. & V. E. Guide," 1925–1926.

with both temperature and humidity content. The following figures are of interest:[1]

Air temperature, degrees Fahrenheit	Relative humidity, per cent	Total heat content, B.t.u. per pound
65	68	17
68	55	17
72	43	17

It will be noted from the above that, for the temperatures and the corresponding relative humidities found acceptable in practice, the total heat per pound of air is constant. It was suggested that the results of the table be made the basis for design. This, however, is not entirely practicable since the disadvantages noted under Heat and Moisture Emissions and Carbon Dioxide Emissions apply also in this connection.

F. PRESSURE REQUIREMENT

Dietz[2] extended the work of O. Krells in establishing a pressure factor for ventilation by stating that the air supply required in operation was only determined "by the amount of air needed to build up the pressure for the air change" (but at the minimum 1,000 cu. ft. per person per hour).

G. AIR CHANGES DETERMINED BY EXPERIENCE

It is customary in practice to use past experience as a guide in estimating the required number of air changes. Roughly this depends upon providing a given number of cubic feet of air per minute per person, and when this is not known, upon assigning a given number of air changes per hour. If J is the cubical content of the room, L the required volume of air, and five air changes per hour are desired, then

$$L = 5J \text{ cu. ft. per hour}$$

Formerly it was the opinion that if more than five air changes per hour were used in ventilation, it would give rise to drafts. In the auditorium of the Research Laboratory, Charlottenburg, however, it was demonstrated that where careful planning of the inlet and outlet registers is introduced in the design, ten or even more air changes per hour may be used without discomfort to the occupants. In this case it is presupposed that the ventilation current is from the floor towards the ceiling (see p. 150).

[1] "Humidity of the Air," from MARR, "Die Feuchtigkeit der Luft," *Gesundh. Ing.*, 1915, pp. 73–90.

[2] DIETZ, Compendium of Heating and Ventilation. "Lehrbuch der Lüftungs-und Heizungs Technik," R. Oldenbourg. Munich and Berlin, 1920.

Drafts affect occupants in various degrees Numerous observations made in the auditorium mentioned showed that drafts begin to be annoying if the air at room temperature near a person enters with a velocity exceeding 40 ft per minute. At lower temperatures, particularly when air is introduced from above, the discomfort appears long before this velocity is reached. Even unnoticeable air currents may lead to physiological disturbances, according to Rubner,[1] when the air temperature is sufficiently low Based on numerous tests the following tables are given for estimating the air changes required for adequate ventilation.

1. Rooms of Known Occupancy.

Assembly rooms, theaters, auditoriums, schools, etc

	Cubic Feet per Hour per Person
In winter	700–1,000
In summer	1,400–1,800

but not more than ten air changes per hour

	Cubic Feet per Hour per Person
Prisons:	
Sleeping cells for prisoners	350
Single cells	500–700
General jail quarters	350

2. Rooms of Unknown Occupancy.

	Air Changes per Hour
Living rooms and rooms of a similar nature	1–2
Halls and stairways:	
Used frequently	1–2
Used infrequently	½–1
Hotel rooms	3–5
Locker rooms, dressing rooms, etc	2–3
Bathrooms	1–2
Toilets	3–5
Kitchens	up to 40

[1] "Insensible Air Currents," from RUBNER, "Über insensible Luftstromungen," *Arch Hyg*, 1907, see also FUSWILER, "The Neutral Zone in Ventilation," *Jour* A S H & V E, January, 1926

SECTION III

MEANS OF INSURING AIR CHANGES

Before proceeding with the discussion of the means for insuring air changes, it will be desirable to investigate the pressure relations in a closed room

A. PRESSURE DISTRIBUTION IN A CLOSED ROOM

Consider a closed and absolutely air-tight room R (Fig 124), which is to be ventilated by a volume of air L In this case it makes little difference whether the air L is forced into the room (pressure ventilation) or whether it is drawn out of it (exhaust ventilation) In practice, however, the surrounding walls are never perfectly tight, and as a consequence pressure and exhaust ventilation are quite different in their application. For example, suppose the room shown diagrammatically in Fig 125 is heated to a temperature higher than the surrounding air.

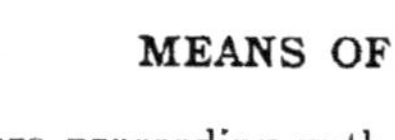

FIG. 124 —Diagram of room ventilation

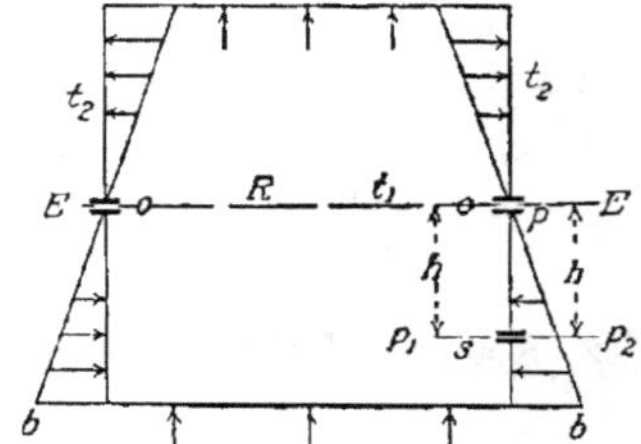

FIG 125 —Pressure distribution in a heated room

Let t_2 be the outside temperature and t_1 the inside where $t_1 > t_2$ If openings o are assumed at the midheight of the room, an equalization of pressure takes place in the plane of these openings The plane EE is the neutral zone, and its pressure is assumed $= p$ lb per square foot

The pressure existing at a plane a distance h in feet below EE is increased to p_1 where

$$p_1 = p + hs_1$$

and in which s_1 denotes the density in pounds per cubic feet at a temperature t_1 On the outside of the room the pressure has increased from p to p_2 lb. per square foot according to the equation

$$p_2 = p + hs_2$$

where s_2 is the density of the outside air in pound per cubic foot at a temperature t_2

Since $t_1 > t_2$ and as a consequence $s_1 < s_2$, there results the condition that

$$p_2 > p_1$$

Thus at planes below the neutral zone the outer pressure is greater than the pressure within the room. The increase, moreover, is such that it reaches a maximum at the floor. Above the neutral zone the effect is reversed and the pressure within the room is greater than the pressure on the outside at the same level. The pressure variation is shown in Fig 125. While the line *ab* appears as a straight line, it is in reality a logarithmic curve.

As long as the room *R* is air tight, the excess or deficiency of pressure cannot have any effect on the air movement. In practice, however, rooms do not have openings at the neutral zone, but rather due to the porosity of the walls and other crevices their functioning is similar to the opening at *o* in the plane *EE*.

Due to air leaks of various kinds in the usual construction of rooms without air ducts, the neutral zone is located about midway between floor and ceiling. Above the neutral zone the warm air leaks outwardly, while below it the cold air enters the room. The latter effect is that which gives rise to the disturbing drafts, particularly at the window cracks. This is one reason why radiators are placed beneath windows.

If, for example, the room *R* were an orchestra pit in a theater, every door opened would be the source of a draft. On the contrary, if the room were a kitchen, then gases above the neutral zone would diffuse throughout the house.

It is therefore desirable that in rooms where drafts are to be avoided (theaters, auditoriums, schools, etc.), a positive pressure should exist within the room to insure outward leakage, and as a consequence the neutral zone should be at the floor level or below it. In rooms where objectionable odors are created (kitchens, locker rooms, toilets, etc.), a minus pressure should prevail for the entire height of the room to induce adequate ventilation, and as a consequence the neutral zone should be at the ceiling or above it.

It is desirable to choose the excess (positive) pressure sufficiently low to prevent unnecessary heat flow from the room. Likewise the minus pressure should be selected with a low differential. In most cases it is sufficient to choose the neutral zone on either the floor or the ceiling. For special cases such as theaters, kitchens, with open passages, etc., a plus or minus pressure of 0.02 in. of water-column pressure is used in the design. It should be noted that even with so low a pressure the effect on large openings, such as fire curtains in theaters, is considerable. For example a surface of 1,000 sq. ft. corresponds to a total pressure of about 100 lb.

The shifting of the neutral zone downward for a plus pressure within the room is effected

(1) By installation of a low-set air-inlet duct of low resistance, *i e*, large size, and a high-set exhaust outlet of comparatively high resistance

(2) By means of pressure ventilation, *i e*, by introducing outside air by means of blowers, and restricting the exhaust outlet

A shifting of the neutral zone upward may be effected

(1) By installation of a low-resistance (large) exhaust duct at the upper level of the room and at the lower an air inlet of comparatively high resistance

(2) By means of exhaust ventilation, *i e*, drawing the room air by the use of exhaust fans and increasing the resistance to the entrance of outside air

Definite levels for the neutral zone are not possible if the surrounding walls permit considerable infiltration of air To insure the desired pressure distribution, it is essential that walls, ceilings, windows, etc be of an air-tight construction and that a definite amount of power be available to provide the necessary pressure to guarantee effective ventilation On the other hand if it is desired to use the walls themselves for ventilation, they should be built of materials pervious to air

The effects just discussed increase with increasing inside temperatures Thus it is possible in churches to have proper atmospheric conditions when unoccupied and intense drafts when occupied In this case particular attention should be paid to the creation of air currents due to occupancy, and as a consequence it is often erroneous to install a heating surface underneath pews

If the room temperature is lower than the outside pressure, variation again takes place but in a reverse direction from those shown in Fig 125

B. NON-MECHANICAL ROOM VENTILATION

Aside from the infiltration through windows, skylights and doors, the amount of air passing through a room depends upon the porosity of the building materials In this connection Lang[1] and Gosebruch[2] have made experiments and found that

$$L = \frac{Fc(p - p_o)^*}{e}$$

[1] "Non-mechanical Ventilation and Porosity of Building Materials," from LANG, "Uber Naturliche Ventilation und Porositat der Baumaterialen," Stuttgart, 1877

[2] "Porosity of Building Materials" from GOSEBRUCH, "Über die Durchlassigkeit der Baumaterialen," Dissertation, Berlin, 1897

* "The Influence of Temperature and Wind Pressure on Gravity Ventilation" from KISSKALT, "Der Einfluss von Temperatur und Winddruck auf die Selbstluftung," *Gesundh -Ing*, p 853, 1913, has shown, on the other hand, that infiltration depends only slightly on the temperature difference if at all, but increases or decreases respectively with the wind pressure In the latter case it is immaterial whether the pressure is positive or negative

where

L = quantity of air in cubic feet per hour
F = wall surface in square feet
c = porosity factor, *i e*, the quantity of air which passes through each square foot of wall surface 1 ft thick per hour with a pressure difference of 1 in of water column
$p - p_0$ = pressure difference on opposite sides of wall in inches of water column
e = wall thickness in feet

Lang found the factor c to average as follows

Stone	$c = 0\ 0339$
Brick	$c = 0\ 0550$
Fire-clay bricks, glazed	$c = 0\ 0$
Fire-clay bricks, unglazed	$c = 0\ 0397$
Mortar	$c = 0\ 00192$
Concrete	$c = 0\ 0706$
Portland cement	$c = 0\ 0375$
Plaster, cast	$c = 0\ 0112$
Oak	$c = 0\ 00192$
Pine	$c = 0\ 276$

Other values of c are given by von Thielmann[1] as follows

Pumice stone	$c = 272\ 8$
Hollow tile	$c = 1\ 586$
Pressed brick, hard burned	$c = 0\ 137$
Portland cement	$c = 0\ 156$

From the values of c it is apparent that the air infiltration through building materials is slight It is further decreased by the use of wall paper or plaster and is also decreased by the absorption of moisture of the outer building walls Since one air change per hour has been observed for usual living rooms, this change must be accounted for by the accidental leakages rather than by infiltration due to the porosity of the building walls Furthermore, since the air infiltration is largely dependent upon the wind velocity and in general is insufficient from a hygienic viewpoint, this phase of the subject will not be considered in greater detail Nevertheless nonmechanical gravity systems of ventilation are important in rooms where mechanical systems of ventilation are not provided

The purpose of any ventilation system is to insure the required air change under specified conditions As a consequence the amount of infiltration should be known to the designing engineer The porosity

[1] "Porosity of Building Materials," from VON THIELMANN, "Die Luftdurchlässigkeit von Baumaterialen," *Gesundh -Ing*, 1915, p 265

of building materials decreases (according to Lang) in the following order

Limestone
Crosscut pine
Stucco
Concrete
Pressed brick
Portland cement
Sandstone
Oak (cut with grain)
Plaster

The order of the reduction of the porosity of the walls due to surface coatings is as follows

Whitewash
Sizing
Oil paint (non-porous when fresh)
Waterglass (non-porous in time)

The natural air change of a room may be determined as follows (Fig 126) With the aid of candles, gas flames, or by chemical means a fairly high carbon dioxide content is produced in the room atmosphere All sources of carbon dioxide generation are then removed and the air of the room is thoroughly mixed At suitable time intervals z_1, z_2, and z_3, an analysis of the carbon dioxide content p_1, p_2 p_3 is made The air change L in cubic feet per hour is given by the equation

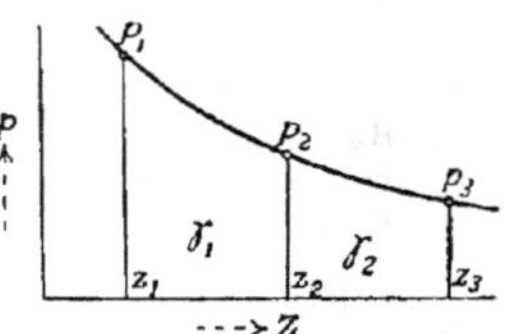

FIG 126—Diagram for calculating the ventilation of a room by measuring the CO^2 content

$$L = \frac{J}{z_2 - z_1} \log_e \frac{p_1 - a\frac{s_1}{s_0}}{p_2 - a\frac{s_2}{s_0}}$$

in which

J = volume of room in cubic feet
a = carbon dioxide content of outer air
p_1, p_2, etc = carbon dioxide content of room air
s_1 = average density of the room air in pounds per cubic foot during the interval z_2 to z_1 in hours
s_2 = average density of the room air in pounds per cubic foot during the interval z_3 to z_2 in hours
s_0 = average density of the outside air at the time of test.

A purely mathematical treatment, though based on assumptions not sufficiently verified, was attempted by Hencky.[1]

C. VENTILATION BY GRAVITY CIRCULATION

Gravity ventilation systems comprise such systems which depend upon the difference in density between warm and cool air to induce circulation. Installations may be designed to operate without special heating of the exhaust air.

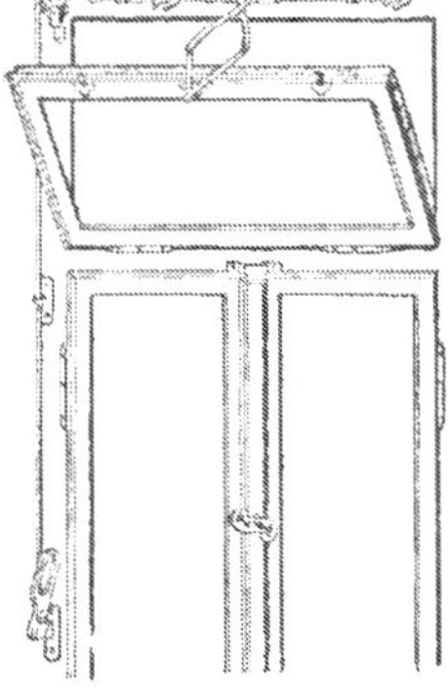

Fig. 127.—Transom. (*Furstenberg, Berlin.*)

Gravity Circulation without Exhaust Air Heating

1. Air Supply.

Window ventilation is the simplest method known due to the directness of the connection with outdoor air. It is not to be recommended, however, for general application owing to the lack of control of air velocity due to winds and the objections due to dust, rain, etc. Transoms as shown in Fig. 127 are widely used for window ventilation. Draft diverters as shown in Fig. 128 are also often applied.

The effect of drafts may be diminished and in some cases eliminated if the entering air is heated by passing it over a radiator. Objection due

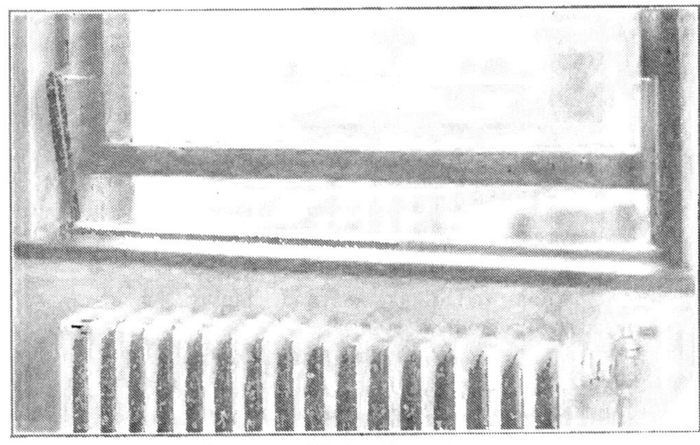

Fig. 128.—Window draft diverter.

[1] "Loss through Plane Walls," from Hencky, "Die Wärmerverluste durch ebene Wände," R. Oldenbourg, Munich and Berlin, 1921; also Arnold, "Hauch-Schreider- und Porenlüftung," *Gesundh.-Ing.*, p. 252, 1921.

to rain or snow may be overcome by suitable protection to the inlets It is difficult, however, to counteract the effect of wind velocity. Air-inlet dampers should be designed so that entering air is directed towards the ceiling

2. Discharge of Exhaust Air.

Similar to that for fresh-air supply, the simplest arrangement for discharging the used air is to employ a direct connection with outdoors, but this in general cannot be recommended Exhausting to a nearby room or hallway, if the latter is connected with outdoors, is somewhat better This arrangement may be desirable when the exhaust air is discharged into a central location and when the annoyance of transmission of sounds is not serious

The system of ventilation described is entirely dependent upon outside wind and temperature conditions When these are unfavorable, a reversal of the air current in the ducts is possible In hospital work the spread of contagious diseases is likely under such adverse circumstances As a consequence when funds are not available to install a proper ventilating system, hospitals would be safer if open window ventilation were used in connection with window draft diverters

Gravity Circulation Induced by Heating the Exhaust Air

The operation of gravity ventilation systems is improved by heating the exhaust air As shown in Fig 129 a simple radiator may be installed for the purpose It is also possible to use the heat of the chimney gases for this purpose when the exhaust ducts are placed adjacent to the chimney duct and separated from it by cast-iron plates With regard to ducts and auxiliary apparatus reference should be made to page 96, and for the detailed design to page 265

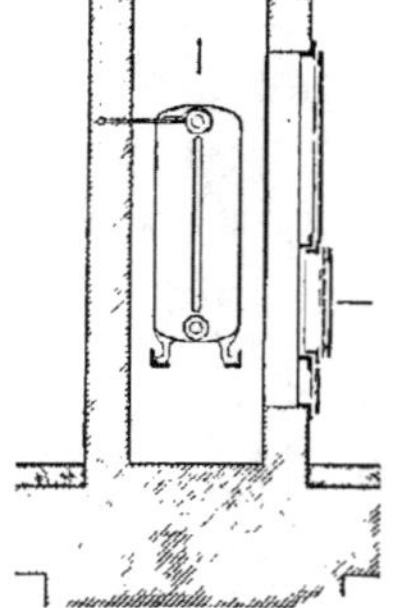

Fig 129 —Heating exhaust air by a radiator

Even the ventilation systems last described are subject to the influence of outside wind and temperature conditions For this reason each installation should be analyzed for its individual volume and pressure requirements Hence such simple ventilation systems must be restricted to comparatively unimportant rooms having small occupancy In some recent developments the question has been shown to be of considerable importance,[1] and further investigations are under way.

[1] *Report* of the New York State Commission on Ventilation, 1923, E P Dutton & Company, Winslow, "Fresh Air and Ventilation," E P. Dutton & Company, 1926

D. VENTILATION SYSTEMS WITH FORCED CIRCULATION

Ventilation systems employing forced circulation are those which use motive power additional to that inherent in gravity circulation or as a substitute for it.

The following means are employed for the purpose: (*a*) roof ventilators using wind pressure, (*b*) small mechanical fans, and (*c*) blower systems using larger fans employing pressure or exhaust.

Fresh Air and Recirculated Air in General

Forced-circulation systems in general should be used where a constant supply of fresh air is required in a room. Some systems are designed in connection with heating plants, and all the air is recirculated. For hygienic reasons total recirculation is not recommended because of the

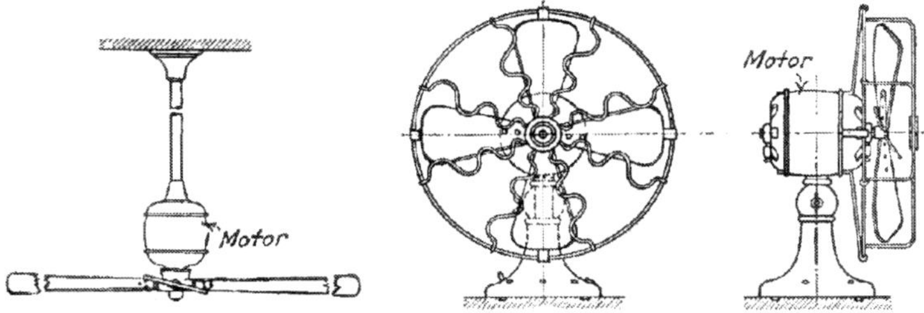

Fig. 130*a*.—Ceiling fan. (*Siemens, Schuckert, Berlin.*)

Fig. 130*b*.—Table fan. (*Siemens, Schuckert, Berlin.*)

contamination due to occupancy and other products of combustion. Such systems are chosen frequently for reasons of economy (see also p. 90). A compromise is often made by recirculating and conditioning a part of the room air and mixing it with air drawn from outdoors. Indeed this is now the general practice.

Fans of the type shown in Figs. 130*a* and 130*b* are used in ventilating railway cars, ships, etc. and are installed without the aid of ducts or heating elements. Since they merely keep the air in motion, they simply add to comfort by accelerating the evaporation from the body by air movement.

Forced Circulation Due to Wind Pressure

Wind pressure in ventilation is utilized by the following means: (*a*) suction cowls or deflectors and (*b*) pressure cowls. On ships they are commonly known as ventilators.

1. Suction Cowls.

If a plate P be superimposed on a vertical duct K as shown in Fig. 131, a wind current directed against the plate will be unable to enter the duct and will give rise to a suction due to a vacuum beneath the plate. Suction cowls are made up of a number of plates in a way to insure a suction independent of the direction of the wind. In Fig. 132 is shown a polar diagram for a particular design of cowl in which the radii show the several directions of the wind and the vector length the corresponding volume of air exhausted. A good design of a suction cowl does not necessarily require a high maximum exhaust effect but rather a uniform exhaust capacity over the entire range of wind directions. Several types of cowls were tested in the Research Laboratory, Charlottenburg.[1]

Types of cowls shown in Figs. 133 are used extensively in railroad work.

To find the effectiveness of suction cowls, it is essential to determine the required volume of air in addition to the suction pressure. For this

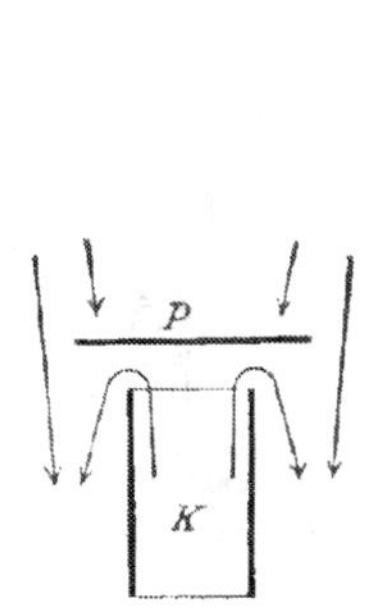

Fig. 131.—Stationary suction cowl.

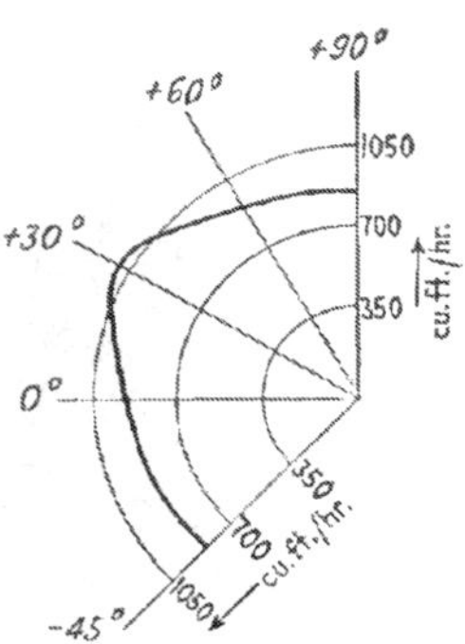

Fig. 132.—Polar diagram of a cowl showing capacity for different wind directions.

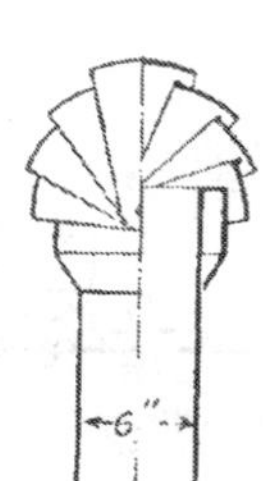

Fig. 133.—Stationary suction cowl. (*David Grove, Berlin.*)

reason in the tests at the Research Laboratory, Charlottenburg, the differential pressure was included and also the tightness against entrance of rain.

Since the volume of air moved by suction cowls depends upon the velocity and the direction of the wind, stationary plants in general should not depend upon them for ventilation purposes. Their most important use lies in preventing the wind from reversing the direction of circulation in the duct system. On the other hand, suction cowls are used extensively on trains and ships though their value is often over-estimated.

If the duct outlets of a low building are adjacent to a high wall, suction cowls are useless. In this case the wind in impinging upon the high wall produces a pressure and so reverses the direction of the circula-

[1] *Bull.* 2 of the Research Laboratory, Charlottenburg, R. Oldenbourg, Munich and Berlin, 1910.

tion. Under these conditions leading the outlet of the duct above the wall is the only solution, if fans are not to be used.

Revolving-head suction cowls in spite of all precautions frequently fail to operate and in that case may create a pressure within the duct instead of a suction. Since the ventilators themselves offer resistance to the air movement, their free area should be made large, and changes in direction of flow and in the cross-section should be as few as possible.

In the tests mentioned the following results were observed:

a. The effect of suction cowls (or pressure cowls) depends largely upon the wind velocity.

b. The volume of air handled is approximately proportional to the first power of the wind velocity (linear relation). The suction (or pressure) produced is proportional approximately to the second power of the wind velocity. The capacity is approximately proportional to the third power of the wind velocity and decreases markedly when the resistance to be overcome increases.

c. The relative performance of the various designs is independent of the wind velocity.

d. The relative value of the suction cowls changes with the magnitude of the air resistance connected, an important item in the choice of the design.

2. Pressure Cowls.

Pressure cowls (Fig. 134) must always be turned so that the opening faces the wind. In the author's experiment station a number of these were tested for the amount of air supplied, as well as the pressure at which it was supplied. The results were plotted on polar coordinates similar to Fig. 132. Pressure cowls are used on ships, and as in the case of suction cowls, their performance is frequently overrated. When the pressure or draft produced is known, the duct system may be computed in a manner similar to other ventilation systems (see p. 265 *et seq.*).

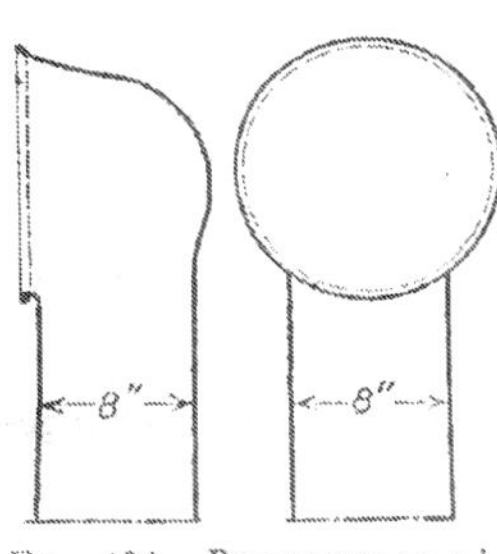

FIG. 134.—Pressure cowl. (*Norddeutscher Lloyd.*)

MECHANICALLY OPERATED BLOWERS (SMALL SIZES)

At times small blowers are operated by means of springs, water power, etc. Since the air resistance in most ventilation projects is of some importance, the use of the small blowers is restricted to kitchens, etc. In case the water power is used, the waste water may be made available for washing. Steam jet blowers are also installed but are objectionable on account of the noise produced.

Ventilation by Means of Blowers

1. Advantages, Disadvantages, and Uses.

a Advantages —Blower systems are independent of all outside temperature and wind conditions They can maintain predetermined pressure conditions They permit of effective air filtering, washing, humidifying, heating, cooling, etc They admit of a choice of duct areas within certain limits They may be connected to several rooms with a simple duct system

b Disadvantages —Blower systems are expensive to install and they incur operating expenses As previously mentioned, the necessity of providing adequate ventilation for certain buildings is so generally accepted that the refusal to operate them owing to the expense can hardly be justified It is certainly necessary that every installation should be designed so that the operating expense is a minimum This expense should be recognized and included in the budget If this is not done, a desire to economize may affect the functioning of the ventilation equipment even to the extent of disuse In this case the effect may be worse than if no ventilation equipment were installed originally. In the following discussion the reduction of the operating expense of the ventilation equipment will be considered paramount.

c Uses —Mechanically operated systems are used for places of assembly such as theaters, auditoriums, schools, offices, and industrial plants

2. Design of the Ventilation System.

To avoid discontinuity in the treatment the computations involved are considered on page 265 *et seq* The maximum outside temperatures on which to base ventilation can be considered as 80 to 90° F for all-year operation (100° F if a cooling system is incorporated), or if ventilation is used only during winter, 40 to 50° F

The lowest outside temperature can safely be taken as 10° F, as for lower outside temperatures a reduction in the number of air changes is proper

3. Accessibility for Cleaning Equipment.

An important requirement of a ventilating system is that it should admit of thorough cleaning To comply with this requirement it may introduce certain difficulties Considering that those parts which do not admit of easy access become dirty after a short time, that this condition prevails for years, and that all air supplied to occupants must pass over the uncleaned surfaces, the necessity of these measures is apparent Those charged with the responsibility for design or installation should see that adequate provision is made for the required cleanliness Cleaning does not need to be done daily, it is sufficient that it be done say once yearly in the duct system In many existing installations thorough cleaning is impossible, and therefore the installation is neglected as a whole.

a Air Supply (1) Fresh-air Sources.—The fresh-air supply of a ventilation system should be taken if possible from a location protected against wind, dust, smoke, etc with inlet registers in a vertical and not in a horizontal plane In densely located buildings the air supply need not be taken at the ground level but at some suitable elevation on the side of the building or above the roof of it In systems having low operating pressure differences it may be necessary to provide two air inlets in opposite directions to eliminate the influence of the wind In general, care should be exercised to insure simplicity and good arrangement of apparatus and so provide excellent operating conditions A suitable air filter is better than several air inlets with their complicated duct and damper systems

A grille should be provided at the fresh-air inlets to prevent entrance of leaves, animals, etc A damper should be installed immediately behind the grille to permit closing the inlet when the system is out of service and to prevent dust from entering This damper should not be used for regulation, since it interferes with proper distribution in the duct system

(2) Dust Removal Since dust causes a contamination of the air (see p 126), it is essential that the air supply to the rooms be kept as free from it as possible. For this purpose the apparatus to be described may be used

(*a*) Dust Chambers.—Dust chambers are enlargements in the duct system so that the air velocity becomes low due to the enlarged cross-sectional area of the duct This allows the heavier particles to settle The resistance of such a chamber is small, but it should be noted that, if abrupt changes of cross-section are made, the velocity head of the air is lost to a considerable extent The design of dust chamber should be such that further contamination of the air may not take place within the chamber itself To this end the walls, floor, and ceiling should be constructed of tile or any construction insuring a hard smooth surface. Plastered walls finished with several coats of oil paint and subsequently varnished are suitable Such surfaces may be washed with a damp cloth If the walls are continually wetted, a tile construction is to be recommended

The dust chamber must be protected against air or water from the ground The chambers should be lighted, passable, and easily accessible They ought to be arranged so as to be out of the path of general traffic and should not be used for storage or passageways Dust chambers can remove only the heavier particles and not the fine dust

(*b*) *Dust Collectors.*—Dust collectors are intended to remove the dust from the air passing over them Since they are used almost exclusively in gravity circulation systems, their resistance should be small. The filters are made of rough material stretched on frames and arranged as shown in Figs. 135*a* and 135*b*. Their cleaning effect is relatively small.

(*c*) Air Filters.—In all through filters the air flows by virtue of the pressure difference on both sides and cleans it in passage. The operation of a filter is to be judged in two senses: its cleaning effect and its pressure loss. It is to be noted that the effects are opposite in their characteristics and that a filter of high cleaning efficiency is accompanied by a large pressure drop. With respect to the latter, see page 262.

Cloth frame filters are often used. The filtering materials, available in various grades of roughness and transmission capacity, are spread over frames. In case of cotton filters the cotton is held between thin vertical metal grilles. These filters are cleaned by beating them or by vacuum cleaning. In the latter case provision is made for a passageway in front of the filter. Filters should be accessible for cleaning, and in the case of high filters it is important that the filter cloth should be easily removable. The resistance of the filter must be noted from time to time so that when it reaches a predetermined value it is taken out of service for cleaning.

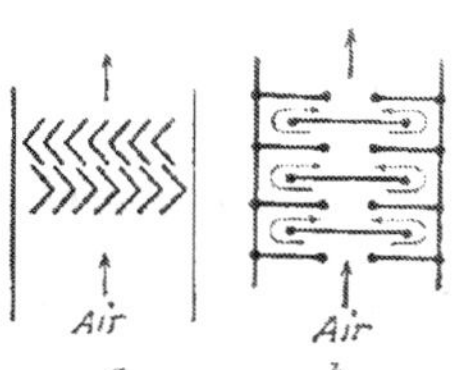

FIGS. 135*a*, *b*.—Dust collectors.

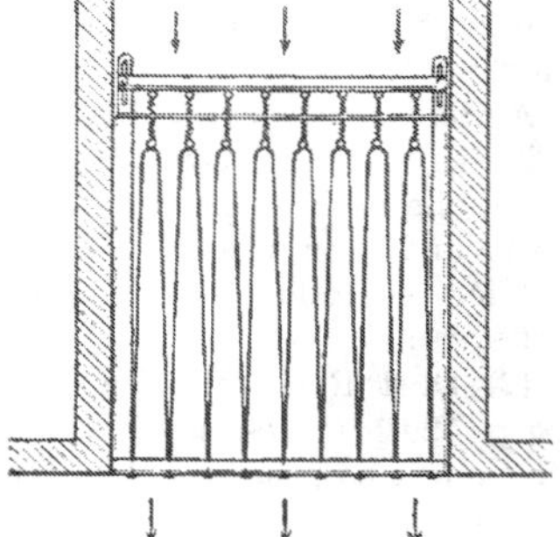

FIG. 136.—Pocket filter. (*X Haberl, Berlin.*)

The pocket filter shown in Fig. 136 is well liked by some because of its large filtering area and its small space requirements. The frames should be made interchangeable and easily removable so as to insure good cleaning.

Filters may be made of stone, coke, wood shavings, peat, straw, etc,. placed between wire netting. The use of these agents has the disadvantage that small particles of the filtering material may dislodge and enter the air stream. This type of filter sometimes uses water sprays, but the practice should be abandoned. The use of small stones worn smooth from abrasion in river beds is recommended, since they expose an easily cleaned surface. An arrangement of this form is used in the Research Laboratory, Charlottenburg and also in the American Radiator Building, New York City.

Combination units customary in America include heating, washing, humidifying, and drying. A typical diagrammatic layout is shown in Fig. 137 where the tempering (preheating) coils are at *A*, spray chamber at *B*, eliminator (water separator) at *C*, and the blower at *D*. A heating coil not shown is placed on the suction side of the blower to bring the air to the desired temperature. Figure 138 gives a full view of such an installation.

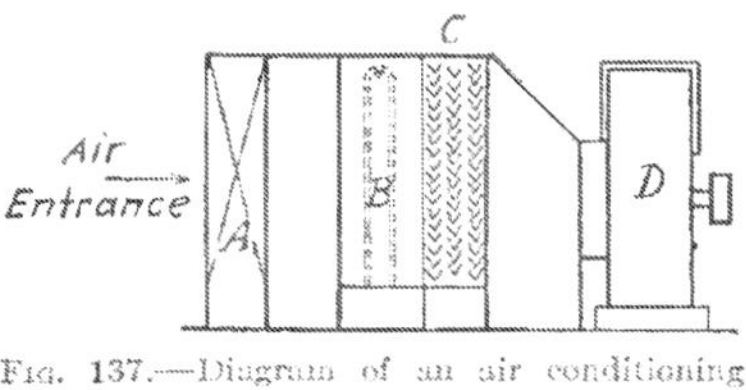

Fig. 137.—Diagram of an air conditioning apparatus.

(3) Ozonizing the Air.—Opinions, in the past, on the use of ozone were divided.[1] At the outset it should be noted that ozonization of the air can never take the place of ventilation. Furthermore, it has been found that adding ozone to the air must be done with great care and only in very small quantities.

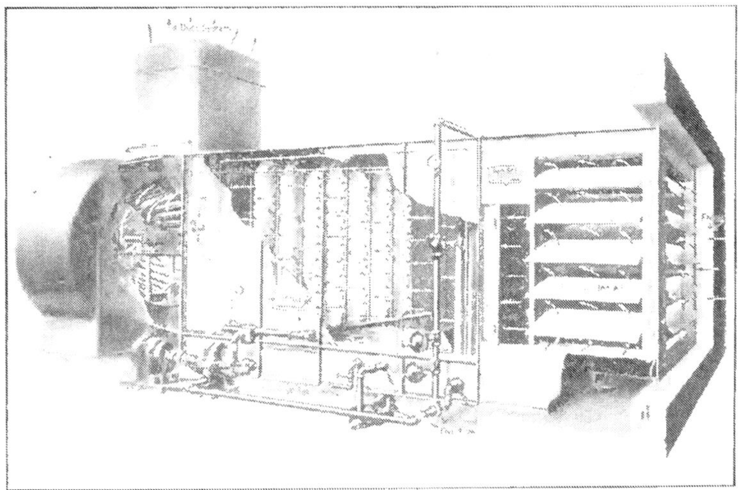

Fig. 138.—Air conditioning apparatus.

[1] "Improving Air in Rooms by Means of Ozone," from Lubbert, "Über die Gesundheitsshädlichkeit der Luft bewohnter Raüme und ihre Verbesserung durch Ozon," *Gesundh.-Ing.*, 1907; Experiments on Air Improvement," from Erlander and Schwartz, "Experimentelle Untersuchungen über Luftverbesserung," *Zeit. Hyg. Infektionskrankh.*, 1912; "Experiments on Deodorizing," from Kisskalt, "Versuche über Desoderierung," Ebenda. "Application of Ozone in Ventilation," from Konrich, "Zur Verwendung des Ozons in der Lüftung." Hallet, E. S., "Forced Ventilation in St. Louis Schools," *Domestic Engineering.*

(4) Reheating the Air.—Air may be heated by the various types of apparatus described on page 93. Particular reference should be made also to the difference noted between steam and hot-water heating surface as well as to the temperature regulation. Frequently, the air to be warmed passes through preheaters and reheaters which are separated, and the apparatus for humidifying, cooling, and drying is placed between them.

If the air is fogged, it is to be removed by heating the air above the saturation point. Systems in which fog removal is necessary therefore require the introduction of warm air, a fact which is frequently overlooked. At times efforts are made to remove the mist by drawing air from the room or introducing cold outside air through windows, neither of which can eliminate the foggy condition.

(5) Humidifying, Cooling, Drying.—While it is common to assume that a relative humidity of from 50 to 60 per cent is necessary for human comfort, experiments of Flugge and his coworkers have changed this viewpoint. It is known at present that rooms having a comparatively dry atmosphere are healthier than those in which a relatively high humidity prevails. Moreover, a relatively high humidity is conducive to the transmission of lung diseases. It may be assumed, when using clean air, that 30 per cent relative humidity is sufficient and that under the most unfavorable conditions, relative humidities of from 60 to 70 per cent may be endured.

Complaints of excessively dry air may not be due solely to the dryness but may be due to the irritation of the mucous membranes caused by the disintegration of dust particles after contact with the hot heating surfaces. In rooms which are continuously heated, constant ventilation is taking place due to the pressure conditions which prevail under such circumstances. This is the cause of the drying of furniture, flowers, etc. Where central heating is employed and the rooms are heated continuously from fall to spring, excessive drying is likely due to the frequent condition of high room temperatures. Excessive drying becomes more apparent when ventilation is provided in addition to the heating.

In certain trade processes where drying must be prevented as in textile plants, suitable provisions are made by installing air-conditioning equipment to insure proper temperature, relative humidity, and air motion. In art galleries, furniture show rooms, etc., ventilation systems are not in general use because of the drying effect. When a substantial system of humidification is installed, however, this objection does not hold.

At times humidifiers are made by placing pans of water at radiators or otherwise attached to them. The humidifying effect by such means is slight unless such containers are operated electrically or mechanically. A modern design is shown in Fig. 139, which is to be connected to an electric plug and has been found to give good results. Adding a certain pine extract to the boiling water greatly increases the comfort afforded by this device.

Central humidifiers are located at suitable places in the main air duct. In some cases steam-heated evaporating pans are used in the heating

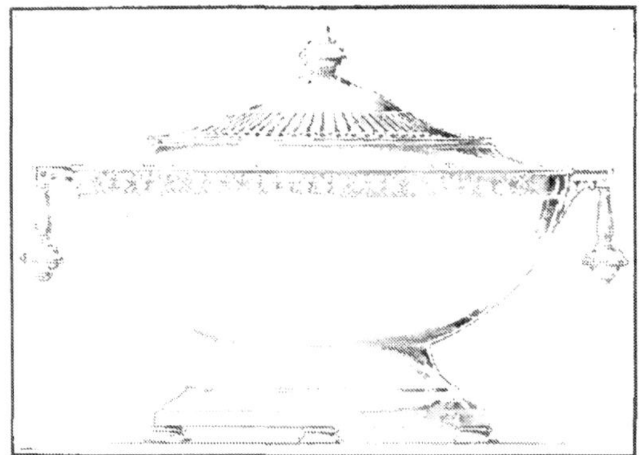

Fig. 139.—Humidifier. (*American Radiator Company, N. Y.*)

chamber (Fig. 140). The apparatus described on page 144 for air washing is also used for humidification.

(6) Fans, Blowers, Exhausters. (*a*) Construction.—A distinction is made between disc fans and centrifugal fans.[1] Disc fans propel the air in an axial direction in the main, but experience shows that air is discharged also in the plane of the wheel and that reverse currents appear

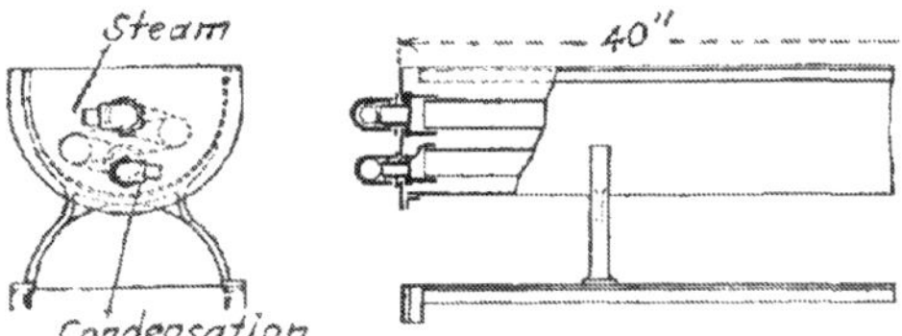

Fig. 140.—Steam heated evaporating pan. (*Rietschel-Henneberg, Berlin.*)

near the axis. The efficiency of these fans decreases rapidly with increasing resistance. On this account disc fans are used only for small outputs. In most cases fans of this type are direct connected with motors as shown in Fig. 141 and are installed in outer walls of windows. An interesting design is the fan shown in Fig. 142. It consists of a steam

[1] Berlowitz, *Gesundh.-Ing.*, p. 141 *et seq.*, 1921, made some important proposals in classifying various types of air-propelling apparatus.

turbine-driven rotor and stationary guide vanes. Tests of this design were reported from the Research Laboratory, Charlottenburg.[1] While

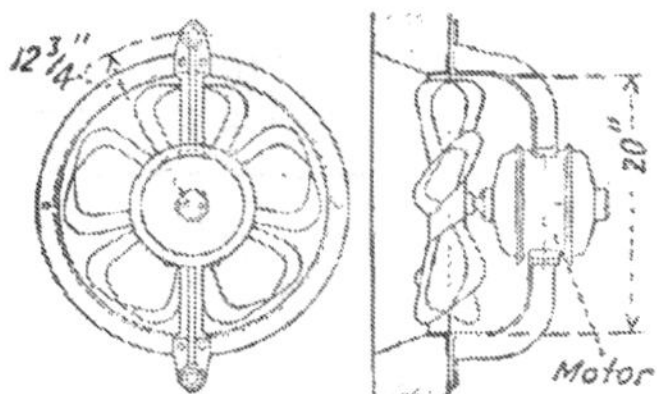

Fig. 141.—Meteor fan. (*T. Frohlich, Berlin.*)

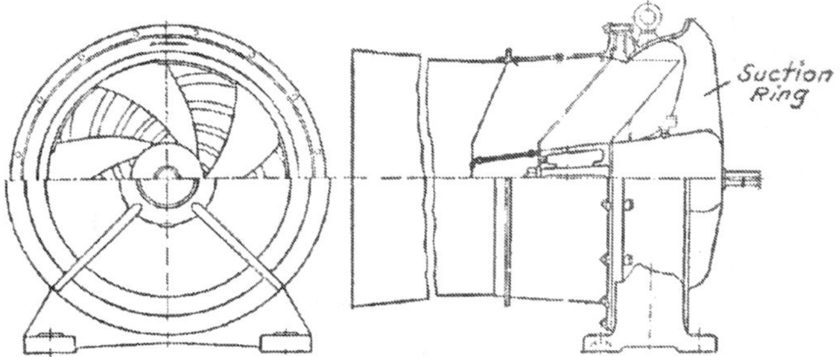

Fig. 142.—Schlopter fan. (*Siemens-Schuckert-Werke, Berlin.*)

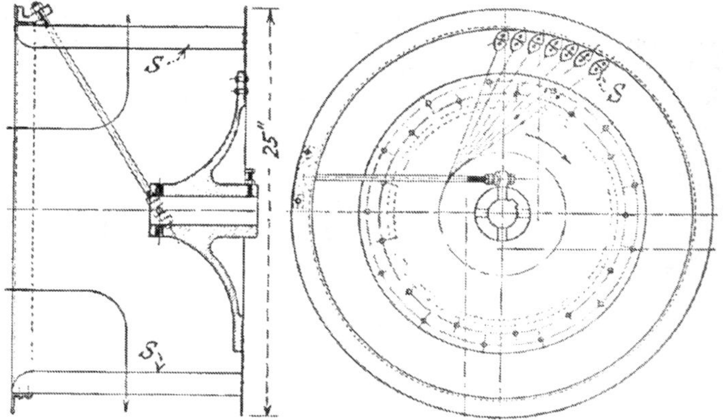

Fig. 143.—Centrifugal fan. (*T. Frohlich, Berlin.*)

this type has several advantages, the noise produced in operation makes it unsuitable for ventilation of buildings.

[1] *Bull.* 18 of the Research Laboratory, Charlottenburg, (Mitteilung der Anstalt 18). R. Oldenbourg, Munich and Berlin, 1914.

Centrifugal fans known as blowers discharge air at right angles to the axis in the plane of the wheel and as a result are sometimes not easy to apply to certain layouts They operate satisfactorily and produce comparatively high pressures As a general rule the pressure steps should not be increased too much at a time since the fans may introduce an objectionable humming noise The fan wheel of one of the usual designs is shown in Fig 143 and consists of a great number of short nearly radial blades S secured to the wheel as illustrated The power consumption of a blower and the variation of output under different operating conditions are of considerable importance in design For a given output the power requirement of a fan installation is dependent upon the efficiency. If a value $A = c\frac{Q}{\sqrt{h}}$ be plotted as abscissa and the efficiency η be plotted as ordinate in the graph in Fig 141, an important curve for design purposes will result In the equation

A = equivalent orifice
Q = volume of air in cubic feet per hour
h = total pressure in inches of water column.
c = a coefficient

In this connection the following points are to be noted

Tests to determine the efficiency of domestic-fans were not made until recently It is admitted, however, that there are numerous difficulties involved In Germany the Verein Deutscher Ingenieure published rules for testing fans and compressors, and in these rules the testing of fans was placed on a standardized basis While many special designs of fans were not investigated, the common forms have been tested and the results are available Blowers for large outputs should be subjected to individual test, particularly when they are to be used for important applications Fans are designed to have a high average efficiency over the range of load for which they are intended rather than a peak efficiency which may be misleading

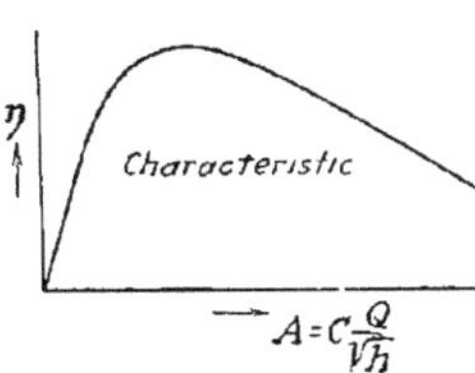

FIG 141 —Characteristic of a fan

Attention should be paid to proper design, correct selection of equipment inspection of manufacture, and installation with the result that the operating expense will be reduced to a minimum [1] Where systems are shut down due to exceeding the estimated operating expense, it might have been better from a ventilation viewpoint not to have installed the apparatus in the first place

[1] "Investigation of Ventilating Systems," from BRABBÉE-BERLOWITZ, "Untersuchungen an Luftungsanlagen," *Zeit Ver deut Ing*, 1910

(*b*) Operation and Regulation.—Fans are generally driven by electric motors either by direct or by belt drive. The direct drive has the advantages of small space requirements, cooling of the motor by means of the air supply to the fan particularly if of the double-suction type, elimination of belt adjustment, and simplicity of testing before acceptance. On the other hand, belt drive has the advantages of using less expensive motors, simplicity in replacing driving motors, and less transmission of noise which frequently originates in the motor rather than in the fan. It is important to choose motors with wide variation in speed control, since this is required in ventilation work. In general it is better to control the speed of the blower rather than to control the volume of air supplied by dampers.[1] For this purpose a variation of from 60 per cent below to 15 per cent above normal answers most requirements and is conveniently obtained by means of direct-current shunt-wound motors. When alternating current is used, the commutator types are recommended. Recent applications of large fans have used steam engines or turbines, the exhaust of which is used in the heating system. Where water power is available, fans may be operated by water turbines.

(*c*) Fan Location.—It is an advantage to place the fan at the entrance to the heating chamber to reduce the transfer of noise to the duct system. When the fan is located at the exit of the heaters, however, it insures good mixing. If the fan is permitted to discharge directly into the heating or plenum chamber with considerable velocity, much of the velocity head is lost. In this connection care should be used in avoiding all abrupt changes in cross-section of the duct. Exhausters must be installed with the same general precautions as blowers.

(*d*) Noiselessness in Operation.—It is essential that the annoyance of noise should not be transmitted throughout the duct system. Among the more important methods of eliminating noisy systems are low peripheral speed of fans, locating noisy motors in the fan room relieved by using a belt drive with motor in an adjacent room, application of concrete ducts in place of the light fan casings particularly on large outfits, separating engine foundations from the building structure proper, use of cork, wood, etc. as absorption media for the vibration under foundations. At times the installation of fans in wooden housings lined with sound proof materials will be effective. Particular care is needed when locating fans on the roof, especially when a resonant condition may exist.

(7) Duct System.—The proportioning of the duct system is discussed on page 265. In the design two requirements are of importance: low resistance and facilities for cleaning. The first condition requires a careful study of the duct layout. Abrupt changes in cross-section and

[1] "Centrifugal Fans and Pumps and Their Drive," from HUTTIG "Die Zentrifugalventilatoren und Zentrifugalpumpen und ihre Antriebsmaschinen," Otto Spamer, Leipzig, 1919.

sharp turns should be avoided. When transitions are required, they ought to be gradual to preserve streamline flow and as a consequence avoid eddy-current losses. Simplicity in the layout in general is an indication of a well-planned system. For hygienic reasons convenient access should be provided in the design of the ducts. Sheet-steel ducts have to be galvanized with joints smooth on the inside. Flushing ducts with water should be avoided, since the water when not drained properly leads to corrosion. The common belief that high air velocities render the ducts self-cleaning is not always borne out in practice. Ducts in the floor should then be used only when suitable covers are provided for cleaning. In nearly every practical case a solution is possible if architects and engineers will cooperate with the aim of the design in mind.

Ducts should be provided with adjustable dampers to insure a proper distribution of the air in the required amounts to each room (equalizing the flow). In many cases the individual registers of the supply and exhaust ducts are provided with special dampers for the same purpose.

(8) Air Movement within the Room.—Ventilation of a room may take place in one of the three following ways: upward from below, downward from above, and in from above and out from above. With the ever-changing types of rooms that must be accommodated, the standardization of design is impossible. The problem becomes more complicated due to the effect of the heat emitted by persons, and by radiators, the influence of cold outer walls, the transmission of air currents especially when doors or windows are opened, with the many possible variations of pressure differences in the spaces so connected. There are many limitations imposed in designing ventilation systems for large rooms. While the problems are difficult at times, the engineer has an opportunity to display his skill. In general there are certain factors common to all problems.

The ideal ventilation system consists of providing the correct amount of air properly tempered at the origin of the source of contamination. Thus in the case of auditorium ventilation a good solution appears in supplying each person with the required air and moreover suggests that the air supply should be from below. It might be noted that due to the density of the air admitted, it tends to rise and have no disturbing effect. In the auditorium of the Research Laboratory, Charlottenburg, shown in Fig. 145, very good results were obtained by the system described above, and moreover the conditions were maintained satisfactorily for both heating and cooling. Indeed, ten changes of air supply could be introduced per hour without producing discomfort from draft.

It must not be assumed that upward ventilation is the proper solution for all cases. Due to reasons of economy in installation and convenience in construction, particularly for tall buildings, air is supplied from above at several places in the room and withdrawn near the floor level. An unsuitable application for this method is that of ventilating a large area in which considerable smoking is taking place unless provision is made for suitably exhausting the air at well-chosen locations. Many large air inlets in which only very low air velocities are maintained will always be a great help for the designing engineer.

Air heating systems which supply and exhaust the air at or near the floor level are sometimes used where economy in heating is desired. When a system must serve several uses, combinations of duct systems may be justified.

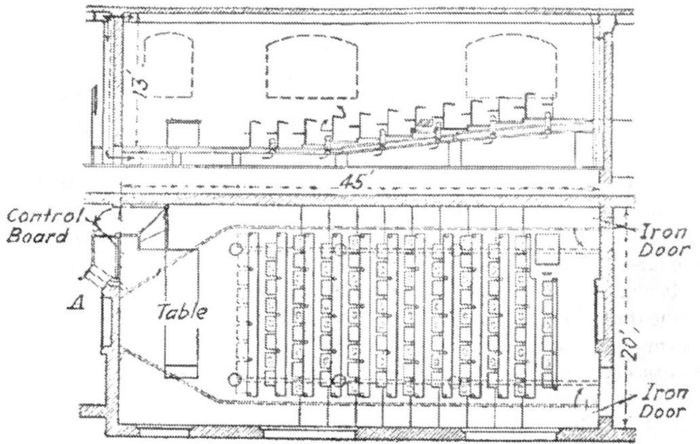

FIG. 145.—Auditorium of Research Laboratory, University of Berlin-Charlottenburg.

b. Exhausting the Air.—If air is supplied to a room of tight construction without provision for exhaust outlets, it will force its way into the walls and ceiling causing them to absorb undesirable odors. In such cases the introduction of the required air supply would be possible only when exhaust ducts are employed. Exhaust ducts may be eliminated only in cases where the leaky building construction is such as to permit the escape of the required air supply.

The location of exhaust openings should be chosen so that short circuiting of the air from the supply ducts is prevented. The velocity of the air at the exhaust registers may be considerably higher than at the supply registers if the occupants are not near such openings. The upper limit of velocity is fixed by the noise produced which becomes objectionable at speeds of about 3 ft. per second.

If the exhaust air passes through a wall (at floor level for instance) Z shaped ducts are frequently used. The transmission of noise is an objection to the design.

Apparatus for warming exhaust air has been discussed on page 137, and exhaust fans were considered on page 147. It still remains to consider methods of duct layout. Individual exhaust ducts to an elevation above the roof will be necessary to prevent a reversal of current within them, as in hospital work, or if sound transmission from combined ducts is to be avoided, as in prisons, for instance. When ducts are closely spaced, short circuiting may be prevented by ending the ducts at different elevations.

In other cases the main exhaust duct may be terminated in the attic space and this in turn be ventilated. This has the advantage of eliminating ill effects due to wind velocity but has the disadvantage of precipitating moisture and consequent injury to woodwork and also the danger due to fire. Because of the fire hazard such termination of ducts is prohibited in some localities. Separate exhaust ducts may be brought together in a fireproofed main duct installed with a short riser and specially ventilated. In any event the reversal of the air current within the duct should be made impossible. In some exhaust systems the air is drawn downward and out through an opening at the base of a shaft, the air movement being insured by the use of a heating medium at the shaft or by means of fans.

FIG. 146.—Control board.

Exhaust ducts like supply ducts are sometimes provided with two sets of dampers; one for a preliminary adjustment of the entire system (distribution equalization), and the second for the desired control of the exhaust from the room. When the wind may effect the operation of the exhaust outlet adversely, a double outlet may be provided, one facing the wind and the other oppositely directed

c Central Control—Every large ventilation system should be provided with apparatus for its control It is best when this control is in some central location and contains the following apparatus:

Remote controls (starters, damper regulation, etc).

Instruments showing operation at remote places (distant thermometers, damper setting indicators, etc.)

Electrical instruments (ammeters, voltmeters, wattmeters, etc)

Mechanical instruments (air meters, pressure recorders, etc)

Main valves for the heating mains of the ventilating system and sometimes the main valves of the heating system

Examples of control boards are shown in Fig 146, for the air-conditioning system of a large office building.

SECTION IV

COOLING OF ROOMS

A. GENERAL

To keep within the scope of this text, the cooling of rooms occupied by people only will be considered This is not alone a question of cooling but also of dehumidification of the air that is desired The highest outside temperatures in New York may be taken as 95° F with a relative humidity ranging between 70 and 90 per cent. Where temperatures have a wide range, they should be taken from the meteorological tables for the locality in question In general inside temperature should not be maintained more than 10 to 12° F below the outside Increase in air motion and a decrease in relative humidity makes for comfortable atmospheric conditions Good distribution of the air supply can be obtained only by using numerous places of air supply and by carefully choosing the known air currents.

B. COOLING METHODS

The simplest method of cooling rooms is by opening windows at night so as to bring the building and contents in contact with the cooler night air. In this way the absorption of heat during the night may offset the rise of temperature during the day so that the effect tends toward a mean temperature below the extreme temperature of the day The influence of this method of cooling is but slight and uncertain It may be augmented by accelerating air motion at night by means of fans and so cooling walls to a greater depth A better arrangement is possible when cellars are available through which the air at night may be passed with the use of fans if necessary

Effectiveness of cooling may be obtained by using cool massive walls, pillars, etc Fans operated during the day circulate the air supply over the cool walls When such cooling of air is employed and the temperature of the air supply is found to be too low, it may be introduced into the rooms prior to occupancy when conditions permit

Cooling Media

1. Indirect Contact with Air.

a Coolers within the Room—Sometimes installations are constructed so that cooling fluids are conducted through the heaters in the room (cold water, for example) Apparatus of this kind is undesirable since the important dehumidification of the air within the room is not effected to sufficient extent and since moisture condenses on the cooling apparatus

b Cooling Units outside the Room—If the cooling units are located outside of the room, the objections due to condensation on the cooling surface is eliminated Sufficient cooling can be attained only by means of large and expensive cooling units The process is assisted when cool spring water, ice, or brine is used

(1) Ordinary Water Supply—The cooling effect depends upon the temperature of the water supply. For example, in Berlin the water temperature ranges from 50 to 55° F and in Vienna from 45 to 47° F. Where the water temperature is as low as the latter, it is possible to cool the air supply to, say, 55° F in order to dehumidify it, if sufficiently large surfaces are available This means that the cooling surface in Berlin must be greater than that in Vienna; for conditions in New York the water temperature is entirely too high in summer to accomplish the desired result

(2) Spring Water—The temperature of spring water varies, and if it exceeds 60° F as it does in Vienna, it becomes ineffective unless previously cooled

(3) Brine—When brine is used, the required cooling surface is decreased Even in this case, however, the room required for coolers and refrigerating apparatus is large Moreover, the cost of installation and the operating expense are high. If the cooling is essential, this type of apparatus may be chosen on economic grounds.

(4) Ice—If ice is used directly for cooling the air, the ice water may be passed through cooling surfaces and so be used as an indirect means of cooling the air supply

(5) Other Means—Artificial cooling media such as ammonia, carbon dioxide, etc are used for cooling the water The water may be filtered and recirculated In certain cases restrictions are placed on the cooling media This was the case in the New York installation previously referred to where ammonia refrigeration in connection with air conditioning was prohibited

2. Direct Contact with Air.

a Ice—Ice may be used as a refrigerant only when it is available at a reasonable price or where the cooling is required at infrequent intervals For gravity circulation the ice is placed above the room to be cooled so that the denser cool air descends Fan operation may be necessary when the ice chamber cannot be located above the room to be cooled When ice is placed in the main supply duct, it is supported on wooden frames constructed so as to provide sufficient refrigerant surface in contact with the air and to permit replenishing of the supply Reheating the air may be accomplished by mixing with uncooled air or by means of special heating units The latter practice while more expensive affords simple and reliable control

b Water—When water is used, it is either sprayed into the air by means of spray nozzles, or the air is made to pass over surfaces over which the water trickles In the latter type stone filters are employed with water pouring over the stones and exposing considerable surface The air supply is then brought into contact with the extended water surface An installation of this kind has given satisfactory service in the auditorium of the Research Laboratory, Charlottenburg, and also in a New York office building.

3. Design of the System.

Computations in connection with the design of cooling systems are given on page 281 *et seq* The installation itself is considered in connection with ducts and cooling surfaces The duct system is to be designed the same as that for any ventilating system and should be hygienic in every respect Pipe coils or extended surface are usually used as cooling surface with a special covering to protect them against sweating Particular attention should be paid to insure noiseless operation of the refrigerating apparatus so that little if any noise be transmitted to the rooms

PART III

THE DESIGN OF HEATING SYSTEMS

SECTION I

HEAT TRANSMISSION

The problem of heat transmission may be divided under two headings. (1) the transient state, i e, the heating-up period, existing while the temperature is being raised; and (2) the steady state, i e., condition prevailing when thermal equilibrium is reached and therefore the temperature is constant.

Since at present it is impossible to determine the transient state the usual practice is to compute the heat transmission during the steady state and include the additional heat transmission required for the transient state by a suitable correction factor The conditions prevailing under the steady state will therefore be considered first.

Since the coefficients of heat transfer for practical constructions are not exact and other assumed data are also only approximate, much labor is saved in computation by rounding figures and using an appropriate factor of safety. This is done in the following text.

A HEAT TRANSMISSION THROUGH MATERIALS FOR THE STEADY STATE

1. General Theory.

Consider the heat transmission through a wall as shown in Fig 147. Assume that the wall is (1) of homogeneous material and (2) at the steady state.

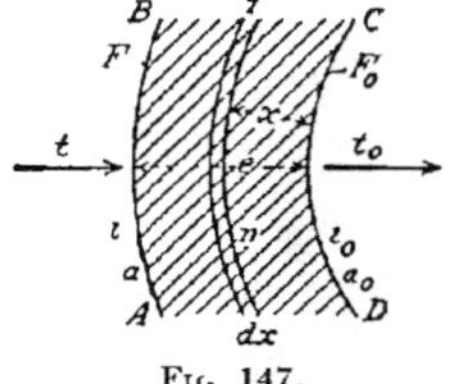

Fig 147.

In addition, let

AB = heat-absorbing surface F in square feet

CD = heat-emitting surface F_0 in square feet

t = air temperature on the side AB (inside temperature) in degrees Fahrenheit

t_0 = air temperature on the side CD (outside temperature) in degrees Fahrenheit.

i and i_0 = surface temperature on the sides AB and CD respectively in degrees Fahrenheit

a = coefficient of absorption in B t u. per square foot per hour per degree Fahrenheit for the inner wall, i e, heat absorbed per square foot of surface per hour for a temperature difference of 1° F. between the surface and the average temperature of the air

a_0 = coefficient of emission in B t u per square foot per hour per degree Fahrenheit, *i e*, heat emitted per square foot per hour for a temperature difference of 1° F between the surface and the average temperature of the air

In the steady state the heat absorbed must be equal to the heat emitted. Expressed mathematically,

$$W = Fa(t - i) = F_0a_0(i_0 - t_0) \tag{1}$$

where W is the heat transfer in B t u per hour. From Eq (1)

$$i = t - \frac{W}{Fa}; \quad i_0 = t_0 + \frac{W}{F_0a_0} \tag{2}$$

The heat W transmitted through the surface f (Fig 147) of a differential thickness dx at a temperature n° F and located at the variable distance x will increase (1) with increasing surface f, (2) with increasing conductivity c, and (3) with increasing temperature difference dn between the two surfaces of the layer f It will decrease with increasing thickness dx of the layer f

The coefficient of conductivity c (or simply the conductivity) is defined as the amount of heat in B t u transmitted hourly through each square foot of the material of a thickness of 1 ft when the temperature difference between the surfaces is 1° F

The proportionality for the steady state is expressed mathematically by

$$W = \frac{fcdn}{dx}, \quad \text{or} \quad dn = \frac{Wdx}{fc} \tag{3}$$

Integrating between the limits $x = 0$ when $n = i_0$ and $x = e$ when $n = i$, there results

$$i - i_0 = \frac{W}{c}\int_0^e \frac{dx}{f} \tag{4}$$

Equation 4 is valid only when W is constant (*i e*, for the steady state) and c is constant (*i e*, for a homogeneous material)

Substituting the values of i and i_0 from Eq 2 in Eq 4, the general equation becomes

$$W = \frac{t - t_0}{\dfrac{1}{Fa} + \dfrac{1}{F_0a_0} + \dfrac{1}{c}\displaystyle\int_0^e \frac{dx}{f}} \tag{5}$$

2. Special Cases.

In heating practice the following special cases are of interest

a Pipe Lines—Consider the heat transmission from the interior to the exterior of the pipe shown in Fig 148 For a length of 1 ft of pipe the surface of the shell f is from Fig 148

$$f = \pi(d + 2x)$$

$$e = \frac{D - d}{2}$$

Substituting these values in Eq. 5,

$$\int_0^e \frac{dx}{f} = \int_0^{\frac{D-d}{2}} \frac{dx}{\pi(d + 2x)} = \frac{1}{2\pi} \log_e \frac{D}{d}$$

from which the value of W becomes

$$W = \frac{\pi(t - t_0)}{\frac{1}{Da_0} + \frac{1}{da} + \frac{1}{2c} \log_e \frac{D}{d}} \tag{6}$$

b. Plane Walls with Parallel Outer Surfaces.—(1) For single-wall construction as shown in Fig. 149 without intervening air space, the conditions are such that

$$F = F_0 = f.$$

Hence

$$\int_0^e \frac{dx}{f} = \frac{e}{f}$$

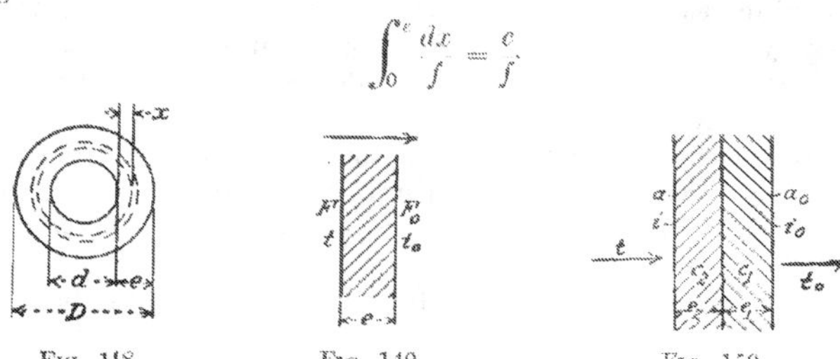

Fig. 148. Fig. 149. Fig. 150.

Substituting in the general Eq. 5,

$$W = \frac{F(t - t_0)}{\frac{1}{a} + \frac{1}{a_0} + \frac{e}{c}}.$$

For convenience let K be defined by the equation

$$\frac{1}{K} = \frac{1}{a} + \frac{1}{a_0} + \frac{e}{c} \tag{7}$$

Hence

$$W = FK(t - t_0). \tag{8}$$

In Eq. 8 K is the coefficient of heat transmission, *i.e.*, the amount of heat in B.t.u. transmitted through one square foot of wall surface per hour per degree Fahrenheit temperature difference between inside and outside air.

(2) Double-wall Construction without Air Space.—If the wall consists of a double layer as shown in Fig. 150 where there is contact between the two walls, the surface temperatures at the contacting surfaces may be considered equal. On this account the values of a at the junction vanish from the equation and Eq. 7 becomes

$$\frac{1}{K} = \frac{1}{a} + \frac{1}{a_0} + \frac{e_1}{c_1} + \frac{e_2}{c_2} \tag{9}$$

where the notation corresponds with that of Fig. 147.

(3) Plane Walls of Multilayer Construction without Air Spaces — For the multilayer walls with contacting surfaces the general equation is therefore

$$\frac{1}{K} = \frac{1}{a} + \frac{1}{a_0} + \frac{e_1}{c_1} + \frac{e_2}{c_2} + \frac{e_3}{c_3} + \qquad + \frac{e_n}{c_n} \tag{10}$$

Professor Knoblauch[1] writes this equation

$$\frac{1}{K} = \frac{1}{a} + \frac{1}{A} + \frac{1}{a_0} \tag{11}$$

where

$$\frac{1}{A} = \frac{e_1}{c_1} + \frac{e_2}{c_2} + \frac{e_3}{c_3} + \quad \cdots + \frac{e_n}{c_n} \tag{12}$$

He defines A as the factor of heat conductivity, i.e., the amount of heat transmitted per hour per square foot of wall surface due to a temperature difference of 1° F between the two outer wall surfaces

From the definition of A there results

$$W = FA(i - i_0) \tag{13}$$

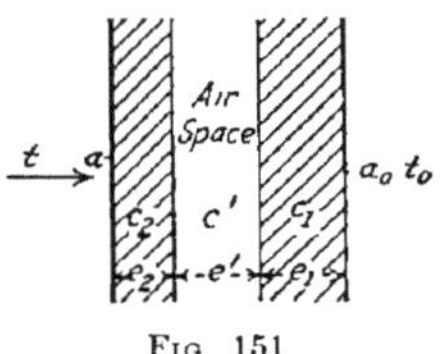

Fig 151

(4) Multilayer Walls with Air Spaces —For multilayer walls in accordance with the ideas of Wierz,[2] Péclet,[3] Nusselt,[4] and Hencky,[5] the conception of an imaginary solid body is introduced instead of the air space, thereby using the law of heat transfer developed by Fourier Taking Fig 151 into consideration, Eq 12 will take the form

$$\frac{1}{A} = \frac{e_1}{c} + \frac{e_2}{c_2} + \quad \cdot \quad \frac{e_n}{c_n} + \frac{e_1'}{c_1'} + \qquad + \frac{c_m'}{c_m'} \tag{14}$$

where

$e_1'e_2'$, etc = thickness of air space in feet.

$c_1'c_2'$, etc = equivalent coefficient of heat conductivity in B.t.u per square feet per hour per degree Fahrenheit per foot of thickness

[1] "Heat Loss of Various Building Constructions," from KNOBLAUCH-HENCKEY, "Zur Berechnung des Warmebedarfs verschiedener Bauweisen," *Gesundh.-Ing*, p 73, 1920

[2] "Elements of Heat Loss Computations, ' from WIERZ, "Die wissenschaftlichen und praktischen Grundlagen der Warmeverlustberechnung in der Heizungstecknik," Berlin, 1921

[3] "Heat and Its Applications," from PÉCLET-HARTMANN, "Wärme und ihre Anwendung," Leipzig, 1860

[4] "Conductivity of Insulating Materials," from NUSSELT, "Die Warmeleitung von Isolierstoffen," Berlin, 1908

[5] "Insulating Effect of Air Spaces in Vertical Walls," from HENCKY, "Über den Warmeschutz von Luftschichten in vertikalen Wanden," Sitzungsber. des Reichsverbandes zur Forderung sparsamer Bauweisen, Berlin, 1919

According to Wierz the values of $\frac{e'}{c'}$ may be taken from the following Table 1 with sufficient accuracy for heating purposes.

TABLE 1.—CONTAINING THE VALUES OF $\frac{e'}{c'}$ FOR DIFFERENT VALUES OF e' AND DIFFERENT TEMPERATURE FACTORS c

Temperature factor c	Air space e' in inches										
	½	¾	1	1½	2	2½	3	3½	4	4½	5
$c = 6.0$	0.76	0.80	0.80	0.85	0.90	0.90	0.94	0.96	0.98	0.99	1.00
$c = 5.5$	0.79	0.82	0.85	0.89	0.94	0.97	1.00	1.04	1.06	1.07	1.08
$c = 5.0$	0.84	0.87	0.90	0.95	1.00	1.04	1.08	1.11	1.14	1.16	1.17

The temperature factor c which is considered later (see p. 164), in general, may be taken as 5 for windows and 5.50 for walls. In order to show the insulating effect of an air space, the following Table 2 due to Wierz shows the corresponding coefficients of air conductivity for various thicknesses e' and their component and the contribution of each element to the total conductivity. In Table 2 the radiation exchange coefficient C (see p. 165) is taken as 0.15.

TABLE 2

	Air space e' in inches							
	½	¾	1	1½	2	3	4	5
Equivalent coefficient of heat conductivity c'	0.06	0.08	0.09	0.13	0.16	0.23	0.29	0.36
Proportion due to radiation	0.035	0.05	0.06	0.09	0.12	0.18	0.24	0.31
Proportion due to conduction	0.015	0.015	0.015	0.015	0.015	0.015	0.015	0.015
Proportion due to convection	0.010	0.015	0.015	0.025	0.025	0.035	0.035	0.035

Table 2 shows that: (1) with increasing thickness of air space e' the equivalent conductivity coefficient c' increases rapidly;

(2) radiation contributes the major part of c', and the influence by conduction and convection diminishes rapidly.

(3) to obtain good insulation by means of air spaces the following considerations are important:

(*a*) Radiation should be decreased, but this in general is impossible.

(*b*) Other materials should be substituted for the air space in which values of c are smaller than the values of c' in Table 2.

(*c*) The air space should be subdivided by means of thin walls parallel to the limiting surfaces, *e.g.*, double windows, etc. A subdivision of an air space 4 in. thick into two spaces 2 in. thick will reduce, according to

Table 2, the value of c' from 0 29 to 0 16. The dividing wall in the air space was not taken into consideration.

(*d*) A subdivision of the air space in a vertical direction to the limiting surfaces will cause no appreciable change, since the influence of the convection is small On the contrary it will cause an increase in the heat transfer through the connecting sections

(5) Special Values (*a*) Coefficient of Heat Absorption —According to Wierz[1] and based on the work of Dulong and Petit,[2] Stefan,[3] Lorenz,[4] and Péclet[5] the following equation may be used in heating practice

$$a = \frac{C\left[\left(\frac{T_2}{100}\right)^4 - \left(\frac{T_1}{100}\right)^4\right]}{t - i} + 0\ 097L(t - i)^{0\ 25} \qquad (15)$$

Hencky[6] substitutes

$$c = \frac{\left(\frac{T_2}{100}\right)^4 - \left(\frac{T_1}{100}\right)^4}{t - i}$$

and calls especial attention to the influence of this temperature factor c at low temperatures The notation is as follows:

a = coefficient of heat absorption in B t u per hour per degree Fahrenheit per square foot.

C = radiation exchange coefficient measured in B t u per square foot per hour per (degree Fahrenheit)^4

T_2 and T_1 = absolute temperatures of the two counterradiating surfaces in degrees Fahrenheit In this case $T_1 = i + 460$ and $T_2 = t + 460$ approximately

$t - i$ = excess temperature of air over the surface temperature of the wall in degrees Fahrenheit

L = a constant with values ranging from 4 to 6

[1] "Elements of Heat Loss Determination," from WIERZ, "Die wissenschaftlichen und praktischen Grundlagen der Warmeverlustberechnung in der Heizungstechnik," Berlin, 1921

[2] "Temperature Measurements and Laws of Heat Transmission," from DULONG and PETIT, "Des recherches sur la mesure des températures et sur les lois de la communication de la chaleur," Ann Chem Phys , vol 7, Paris, 1817

[3] "Relation between Temperature and Radiation," from STEFAN, "Über die Beziehung zwischen der Warmestrahlung und der Temperatur," Sitzungsber der math-naturwiss Klassse der Kais Akademie der Wissenschaften , Bd 79, Vienna, 1879

[4] "Conductivity of Metals for Heat and Electricity," from LORENZ, "Über das Leitungsvermogen der Metalle fur Warme und Elektrizitat," Ann Phys Chem , Leipzig, 1881

[5] "Heat and Its Applications," from PECLET-HARTMANN, "Warme und ihre Anwendung," Leipzig, 1860

[6] "Heat Loss through Walls," from HENCKY, "Die Warmeverluste durch ebene Wande," R Oldenbourg, Munich and Berlin, 1921

Substituting the value of c in Eq (15) there results

$$a = Cc + 0\ 097L(t - i)^{0\ 25}$$

C is obtained from the following equation

$$\frac{1}{C} = \frac{1}{C_1} + \frac{1}{C_2} - \frac{1}{C'} \tag{16}$$

in which C = radiation exchange coefficient in B t u per square feet per hour per (degree Fahrenheit)[4]

C_1 and C_2 = radiation constants in B t u per square feet per (degree Fahrenheit)[4] of the two surfaces exchanging radiation

C' = radiation constant of a black body (= 0 162)

The value of L according to Ser[1] lies between 6 and 4 as follows:

L = 6 for free air in natural motion (wind, etc)
L = 5 for still free air, for room air being rapidly cooled, for air layers with strong current
L = 4 for enclosed air (room air) when temperature difference $(t - i)$ > 10° F

With $L = 4$, Eq (15) coincides exactly with the observations of Nusselt[2] and Hencky.[3] For temperature differences $t - i$ smaller than 10° F, the equation

$$a = 5\ 7C + 0\ 61 + 0\ 01(t - i) \tag{17}$$

due to Nusselt may be applied

In general work a simplification may be made by using an average value of the radiation constants for building materials, namely, $C_1 = C_2 = 0\ 155$ When making this assumption, the error introduced in Eq (16) is very small and the radiation exchange coefficient becomes $C = 0\ 15$ For practical purposes the values of a may be taken from Table 3 This table is based on an air temperature of 70° F when it is assumed that $C = 0\ 15$.

[1] "Treatise on Industrial Physics," from SER, "Traité de Physique Industrielle," Paris, 1888

[2] "Heat Conduction in Insulated Material," from NUSSELT, "Die Warmeverluste von Isolierstoffen," T Springer, Berlin, 1908

[3] "Heat Loss through Walls," from HENCKY, "Die Warmeverluste durch ebene Wande," R Oldenbourg, Berlin, 1921.

TABLE 3.—HEAT ABSORPTION COEFFICIENT, a, IN B.T.U. PER SQUARE FOOT PER HOUR PER DEGREES FAHRENHEIT FOR THE INNER SURFACES OF OUTER WALLS AND WINDOWS WITH $C = 0.15$ AND AN AIR TEMPERATURE OF 70° F.

Temperature difference $(t - i)$ between room air and inside surface temperature of wall in degrees Fahrenheit (excess temperature)	Heat absorption coefficient a for		
	$L = 4$	$L = 5$	$L = 6$
2	1.5		
5	1.5 } due to Nusseit		
8	1.5		
12	1.6	1.8	2.0
14	1.6	1.8	2.0
16	1.6	1.8	2.0
18	1.7	1.9	2.1
20	1.7	1.9	2.1
22	1.7	1.9	2.1
24	1.7	1.9	2.1
26	1.7	1.9	2.1
28	1.7	1.9	2.2
30	1.7	2.0	2.2
33	1.8	2.0	2.2
36	1.8	2.0	2.2
40	1.8	2.0	2.3

The excess temperature $t - i$ may be estimated or it may be taken from Table 4 according to Rietschel.

TABLE 4.—VALUES OF $t - i$ ACCORDING TO RIETSCHEL

For Brick Walls	Degrees Fahrenheit
4 in. thick	15
8½ in. thick	13
13 in. thick	11
17½ in. thick	9
22 in. thick	7
26½ in. thick	5
31 in. thick	3
35½ in. thick	2
40 in. thick	1
Over 40 in. thick	0
Window, single	35
Window, double	20
Doors, outside	4
Ceilings, with filling	2
Inside walls	0

(*b*) Coefficient of Heat Emission.—Wierz[1] proposes the equation

$$a_0 = Cc + 0.19\sqrt{v} \tag{18}$$

where

a_0 = heat emission in B.t.u. per square foot per hour per ° F.
v = wind velocity in feet per minute.

The other terms have been defined previously. Values computed by aid of Eq. (18) are in substantial agreement with observations of Recknagel[2] and Nusselt.[3] Substituting $C \cong 0.15$ and using an average temperature factor $c = 5.3$, the values of a_0 are given in Table 5.

TABLE 5.—COEFFICIENT OF HEAT EMISSION a_0 IN B.T.U. PER SQUARE FOOT PER HOUR PER DEGREE FAHRENHEIT FOR VARIOUS WIND VELOCITIES v

v in feet per minute	50	75	100	150	200	300	400	500	600	800	1,000	1,200	1,400
a_0	2.0	2.5	2.5	3.0	3.5	4.0	4.5	5.0	5.5	6.0	7.0	7.5	8.0

As Wierz shows, it is incorrect to assume a high wind velocity when computing the values of a_0, since abnormal conditions of air motion must be allowed for by additional exposure factors. For most cases the velocity v may be assumed as 100 ft. per minute.

(*c*) Coefficient of Heat Conductivity.—The coefficient of heat conductivity for a number of building and insulating materials under varying conditions was determined by Knoblauch and his assistants[4] and are included in Tables IV*a* and IV*b* at the end of this text.

[1] "Elements of Heat Loss Determination," from WIERZ, "Die Wissenschaftlichen und praktischen Grundlagen der Wärmeverlustberechnung in der Heizungs-technik," Berlin, 1921.

[2] Pocket Handbook for Heating Engineers," RECKNAGEL, "Kalender für Gesundheitstechniker," R. Oldenbourg, Berlin.

[3] "Heat Conductivity of Insulating Materials," from NUSSELT, "Die Wärmeleitfähigkeit von Wärmeisolierstoffen," Berlin, 1908.

[4] "The Heat Conductivity of Insulation Material," from KNOBLAUCH, RAISCH, and REIHER, "Die Wärmeleitzahl von Bau- und Isolier-stoffen und die Wärmedurchlässigkeit neuer Bauweisen," Munich, 1920; "A Method of Determining the Heat Conductivity," from POENSGEN, "Ein technisches Verfahren zur Ermittlung der Wärmeleitfähigkeit plattenförmiger Stoffe," Berlin, 1912; "The Heat Emission of Heated Bodies," from WAMSLER, "Die Wärmeabgabe geheizter Körper an Luft," Munich, 1909; "The Heat Conductivity of Insulation and Building Material," from GRÖBER, "Die Wärmeleitfähigkeit von Isolier- und Baustoffen," Berlin, 1911; "Calculation of the Heat Loss of Different Buildings," from KNOBLAUCH-HENKY, Berechuung des Wärmebedarfes verschiedener Bauweisen," Munich, 1920; "Insulation of Cooling Chambers," from HENKY, "Untersuchungen zur Isolation von Kühlräumen," Munich, 1915.

(*d*) Coefficient of Heat Transmission K.—The determination of the values of K will be made by a direct computation and verified by a check computation. For this purpose Eqs. (1) and (13) are useful.

$$\left.\begin{aligned} \frac{W}{a} &= t - i. \\ \frac{W}{a_0} &= i_0 - t_0. \end{aligned}\right\} \quad (1)$$

$$\frac{W}{A} = i - i_0 \quad (13)$$

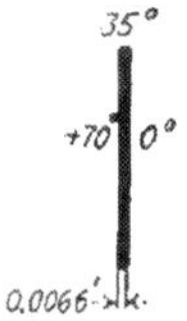

A few examples will show the method to be followed in a computation.

Fig. 152.

Example 1.

For a single window (Fig. 152), assume the following:

Glass thickness $e = 0.0066$ ft.

$c = 0.538 \cong 0.54$ (from Table IV*a*).

Wind velocity $v = 100$ ft. per minute.

Outside temperature $= 0°$ F.

Inside temperature $= 70°$ F.

Direct Computation: According to Table 4 (p. 166) for a single window: $t - i = 35°$ F. From Table 3 (p. 166) for $t - i = 35°$ F., $L = 5$, room air being cooled rapidly, and the value of $a = 2.0$. From Table 5 (p. 167) for $v = 100$ ft. per minute, $a_0 = 2.5$. In addition $\frac{e}{c} = \frac{0.0066}{0.54} = 0.012$. Substituting these data in Eq. (7),

$$\frac{1}{K} = \frac{1}{2.0} + 0.012 + \frac{1}{2.5} = 0.912$$

from which $K = \frac{1}{0.912} = 1.1$ B.t.u. per square foot per hour, per degree Fahrenheit.

In a checking computation for 1 sq. ft. of surface, according to Eq. (8) there results

$$\begin{aligned} W &= 1 \times 1.1 \times (70 - 0). \\ &= 77 \text{ B.t.u. per square foot.} \end{aligned}$$

From Eq. (1)

$$t - i = \frac{W}{a} = \frac{77}{2.0} = 38.5° \text{ F.}$$

$$i_0 - t_0 = \frac{W}{a_0} = \frac{77}{2.5} = 30.8° \text{ F.}$$

From Eq. (13)

$$i - i_0 = \frac{W}{A} = 77 \times 0.012 = 0.9° \text{ F.}$$

The value of $t - i$ according to Rietschel was 35° F. whereas the checking computation resulted in the value of 38.5° F. Table 3 (p. 166) shows that a is the same for both values of $t - i$. Consequently the value of K is 1.1 B.t.u. for practical purposes.

Example 2.

For a double window (Fig. 153), assume the following:

$e_1 = e_2 = 0.0066$ ft.

$e' = 4$ in.

$v = 100$ ft. per minute

$t_0 = 0°$ F.

$t = 70°$ F.

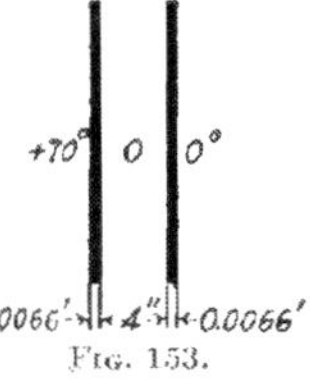

Fig. 153.

Direct Computation: From Table 4 (p. 166) $t - i = 20°$ F.

$L = 4$ (p. 165) for enclosed room air.

$a = 1.7$ (Table 3, p. 166).

$a_0 = 2.5$ (Table 5, p. 167) for $v = 100$ ft. per minute.

$\frac{e'}{c'} = 1.14$ (Table 1, p. 163, where $c = 5.0$).

The values of $\frac{e_1}{c_1} = \frac{e_2}{c_2}$ may be neglected; hence

$$\frac{1}{K} = \frac{1}{1.7} + 1.14 + \frac{1}{2.5} = 2.13$$

from which $K = \frac{1}{2.13} = 0.47$ B.t.u. per square foot per hour per degree Fahrenheit.

As a check there results for 1 sq. ft. of surface:

$$W = 0.47 \times 70 = 32.9 \text{ B.t.u. per hour.}$$

$$t - i = \frac{W}{a} = \frac{32.9}{1.7} = 19.4° \text{ F.}$$

This is near enough to the value of $t - i = 20°$ F. obtained previously.

Example 3.

Assume an outer wall of brick 18 in. thick plastered to a thickness of 0.75 in. (0.063 ft.) on both sides as shown in Fig. 154.

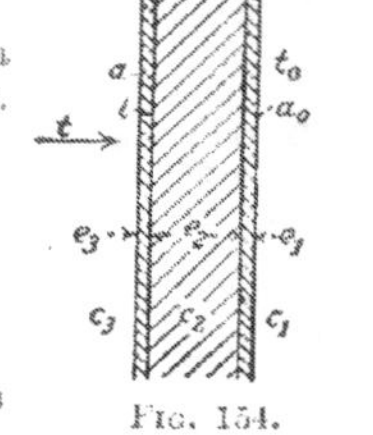

Fig. 154.

Given $c_1 = c_2 = c$ = 0.5 as an average value.

$e_1 = e_3$ = 0.75 in.

e_2 = 18 in.

$t - i$ (estimated) = 9° F. (Table 4, p. 166).

v = 100 ft. per minute.

Solution: From Table 3 (p. 166) and Table 5 (p. 167) it is found that for $L = 4$, $a = 1.5$, and $a_0 = 2.5$; hence

$$\frac{1}{K} = \frac{1}{1.5} + \frac{0.063}{0.5} + \frac{1.5}{0.5} + \frac{0.063}{0.5} + \frac{1}{2.5}.$$

$$= 0.67 + 0.126 + 3 + 0.126 + 0.4 = 4.3.$$

Therefore

$K = 0.23$ B.t.u. per square foot per hour per degree Fahrenheit. *Ans.*

Check: $W = 0.23 \times 70 = 16.1$ B.t.u. per hour for unit of surface.

$$t - i = \frac{16.1}{1.5} = 10.7° \text{ F.}$$

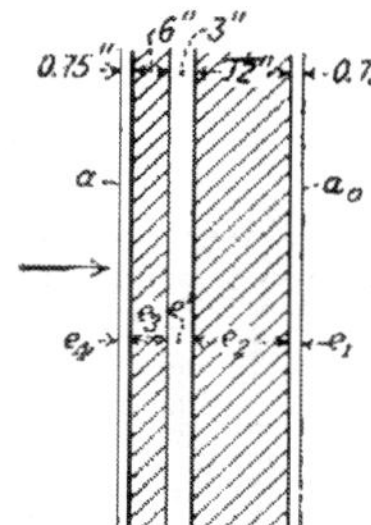

Fig. 155.

Thus the checking solution is 10.7° instead of 9° as shown in the computation. From Table 3 for 10.7° F. temperature difference between room air and inside temperature of wall surface, $a = 1.6$ instead of 1.5 as shown in the computation. The value of K determined with either value of a is substantially the same.

Example 4.

Use the same data as example 3 except that the wall construction shall be 6 and 12 in. thick with a 3-in. air space between as shown in Fig. 155. Assume the following:

$t - i = 8°$ F. (estimated from Table 4, p. 166).

$a = 1.5$ for $L = 4$ (Table 3, p. 166).

$c = 5.5$; $\frac{e'}{c'} = 1.0$ (Table 1); $a_0 = 2.5$ as before.

Therefore

$$\frac{1}{K} = \frac{1}{1.5} + \frac{0.063}{0.5} + \frac{0.5}{0.5} + 1 + \frac{1}{0.5} + \frac{0.063}{0.5} + \frac{1}{2.5}$$

$$= 0.67 + 0.126 + 1.0 + 1.0 + 2.0 + 0.126 + 0.4 = 5.3$$

$$K = 0.19 \text{ B.t.u. per square foot per hour per degree Fahrenheit}$$

Check $W = 0.19 \times 70 = 13.3$ B.t.u. per hour per unit of surface

Therefore

$$t - i = \frac{13.3}{1.5} = 8.9° \text{ F.}$$

The slight difference against 8° as assumed causes no change in the values of K

Example 5.

Use the data of example 4 with the exception that the air space is filled with blast-furnace slag ($c = 0.064$ from Table IVb)

$$\frac{1}{K} = \frac{1}{1.5} + \frac{0.063}{0.5} + \frac{0.5}{0.5} + \frac{0.25}{0.064} + \frac{1}{0.5} + \frac{0.063}{0.5} + \frac{1}{2.5}$$

$$= 0.67 + 0.126 + 1 + 3.9 + 2 + 0.126 + 0.4$$

$$\frac{1}{K} = 8.2 \quad K = 0.12 \text{ B.t.u. per square foot per hour per degree Fahrenheit}$$

Check:

$$W = 0.12 \times 70 = 8.4 \text{ B.t.u. per hour for unit of surface}$$

$$t - i = \frac{8.4}{1.5} = 5.6° \text{ F. instead of } 8° \text{ F. which causes no change in } K$$

Conclusions Derived from Examples

1 The heat transfer through a single window is more than twice that of a double window

2 A brick wall construction (18 in thick) having a 3-in air space decreases the heat transfer from 0.23 for a solid wall to 0.19 for the hollow construction. This represents a decrease in heat loss of about 17 per cent

3 Filling the air space with blast-furnace slag results in a coefficient of heat transmission $K = 0.12$, or a decrease of about 48 per cent over that of a solid wall This assumption is true only when the blast-furnace slag does not absorb moisture Recent heating practice avoids filling air spaces owing to the possibility of damp walls, pollution by vermin, etc

REMARK When figuring these examples with a wind velocity of 15 miles per hour (1,320 ft per minute), in accordance with American heating practice, higher K values than those given in American literature are obtained

Additional examples of the application of principles to heating problems are discussed in the publication of Wierz, frequently referred to in this text Attention is directed by him to several factors which influence the heat transfer coefficient K, such as the effect of rain or snow and wind on windows; the heat required to melt snow on skylights, etc

(*e*) Coefficient of Heat Requirement —Hencky[1] combines the heat losses previously discussed with those caused by air changes and as a result develops the coefficient of heat requirement. For this purpose, however, several assumptions must be made in reference to the formulas and to their application which in the present state of the art are open to

[1] "Heat Losses through Plane Walls," from HENCKY, "Die Warmeverluste durch ebene Wande," Springer, Berlin 1921

objections With the foregoing in mind, it is advisable to treat the influence of air changes separately

(*f*) Rietschel's Heat-loss Calculation —In 1890 Rietschel introduced a simple formula for determining the heat loss based on the theory of Péclet, and arrived at the expression

$$W = FK(t - t_0). \tag{19}$$

A sufficient number of values of K were published at that time This method was further amplified by the Verband der Zentralheizungs-Industrie and the values of K increased in proportion. K values used in American heating practice are given in Table VI In addition Table VII contains the so-called "Safety Factors" for the heat loss calculation.

B HEAT LOSS PRIOR TO ATTAINING STEADY STATE

The amount of heat required to attain a steady state cannot as yet be determined satisfactorily For plane walls the following empirical formulas may be used

1. Average-sized Rooms.

The additional allowance Z for heating up according to Rietschel is given below

For rooms which are heated daily and where operation is discontinued at night,

$$Z = \frac{0\ 063W_1(n - 1)}{z} \tag{20}$$

For rooms not heated daily,

$$Z = \frac{0\ 1W(8 + Z)}{z}, \tag{21}$$

where z = duration of heating-up period in hours.

Z = allowance in B t u for heating up

W_1 = B t u loss per hour through outer walls and windows during the steady state (note that the ceiling under an unheated attic must also be considered as an outer wall)

W = total heat loss per hour of the room during the steady state.

n = number of hours for which operation is discontinued

For instance in the case of class rooms the heat is turned on at 5 a m and the final temperature is reached at 8 a m Hence z = 3 hr., again if heating is continued to 5 p m then n = 12 hr

Owing to the expense of installation, additions greater than 30 per cent are not usual. In cases of this kind the duration of the heating-up period z is increased For halls which on account of their small size cannot be placed under the heading of the next subchapter, the heating-up period z may be from 5 to 6 hrs. If in a building several rooms are used at times when others are unheated, such rooms should have not only an additional allowance but also a independent control of the heating system One way is to provide a separate system of mains, the control valves for which are in the boiler room.

Based on the foregoing formulas, additional allowances have been determined, which are given in Table VII

2. Large Rooms Infrequently Heated and Occupied for Short Periods Only.

In rooms seldom used or used only for short periods, such as halls, churches, etc , it is inadvisable to attempt reaching the steady state By means of rapid heating it is possible to bring the air temperature up to requirements before the walls attain a temperature normal under the circumstances Upon discontinuing operation the cooling of the room is effected more rapidly

Under these circumstances no great losses can take place At most only a part of the wall thickness absorbs heat, and the only heat loss of any moment is that which is transmitted by the windows The determination of the heat loss in cases of this kind is based upon empirical data There are two special cases.

Case 1—When the heating surfaces are favorably distributed proposes the formula

$$W = \frac{FK(t - t_0)}{2} + F_1\left(8\ 5 + \frac{t - t_1}{z}\right) \qquad (22a)$$

Case 2—When the heating surfaces are unfavorably situated, the proposed formula is

$$W = \frac{FK(t - t_0)}{2} + F_1\left[15 + \frac{2(t - t_1)}{z}\right] \qquad (22b)$$

In these equations

W = heating requirements in B t u per hour
F = glass surface in square feet
F_1 = superficial area of all walls, ceilings, floors, pillars, etc , in square feet
K = transmission coefficient for glass (for single windows of large dimensions Rietschel allows 1 1 B t u per square foot per hour per degrees Fahrenheit)
t = required inside temperature in degrees Fahrenheit
t_1 = initial room temperature when heating up occurs (assume at approximately 32° F)
t_0 = lowest outside temperature
z = heating-up period in hours

For rooms exceeding 40 ft in height 5 per cent should be added to above equations for each additional 3 ft in height The equations take into consideration the infiltration of air and assume good building construction and good condition of building (absence of broken panes, holes, or cracks in walls or ceilings, etc)

With shorter heating-up periods, the heat absorption of the walls diminishes while air circulation increases It is advisable for purposes of calculation to assume a heating-up period that is not too long (*e g*, for churches approximately 5 to 6 hrs) In actual practice, however, this time should be exceeded by a few hours

3. Temperature Variation throughout the Room.

Assumed inside and outside temperatures have been discussed (p 3) The inside temperature, however, varies with increasing height above the floor Experiments show that for all practical purposes the increase of temperature up to a height of 10 ft may be neglected For greater heights the temperature is approximated given by the formula

$$t' = t[1 + 0\ 017(h - 10)], \text{ when } t' \gtreqless 1\ 3t \qquad (23)$$

In this expression

t = temperature in degrees Fahrenheit at eye level

t' = temperature in degrees Fahrenheit at a distance h ft. above floor

In Eq. (23) it is assumed that the heating system operates at normal load without the heat added by excessive illumination For outside temperatures over 50° F with the usual illumination of electric arc or metal filament lamps the temperature t' is

$$t' = t[1 + 0\ 005(h - 10)] \qquad (24)$$

The lowest outside temperatures to be assumed has been given on page 3 The temperature of unheated rooms under varying conditions may be assumed as follows·

For unheated and closed rooms, situated between two heated rooms,

$$t_0 = 40° \text{ F}$$

For unheated and closed rooms in which a single wall is in contact with a heated room (a cellar room for example),

$$t = 32° \text{ F}$$

For unheated rooms which are open occasionally to the outside (garages, vestibules, corridors, etc),

$$t_0 = 20° \text{ F}$$

For unheated attics with cement roofs,

$$t_0 = 20° \text{ F}$$

For unheated attics with slate or tile roofs,

$$t_0 = 10° \text{ F}$$

Occasionally it may be required to determine the temperature of an unheated room For example it may be desired to heat a room only when the heat emission of the neighboring rooms is insufficient to insure a certain minimum temperature (*e.g.*, the freezing point). In Fig. 156 rooms A are heated to a temperature $t°$. Assume further that the existing outside temperature is $t_0°$. Suppose that B is an unheated room the temperature of which, t_x, is to be determined. The heat emission from rooms A to B in the steady state is $\Sigma FK(t - t_x)$ while that emitted from room B under steady-state conditions is $\Sigma F_0K_0(t_x - t_0)$. Since this equality must be maintained at all times

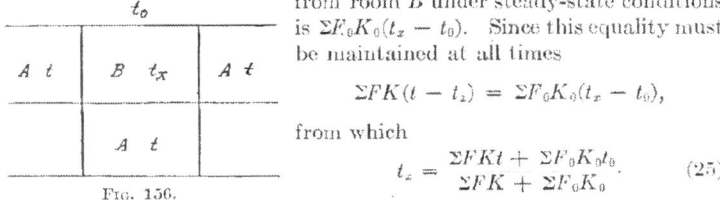

$$\Sigma FK(t - t_x) = \Sigma F_0K_0(t_x - t_0),$$

from which

$$t_x = \frac{\Sigma FKt + \Sigma F_0K_0t_0}{\Sigma FK + \Sigma F_0K_0}. \qquad (25)$$

Fig. 156.

C. DETAILED COMPUTATION OF HEAT LOSS ACCORDING TO RIETSCHEL

Example 6.

Let it be required to find (Fig. 157) the temperature t_x of the unheated room II and the heat loss from room I.

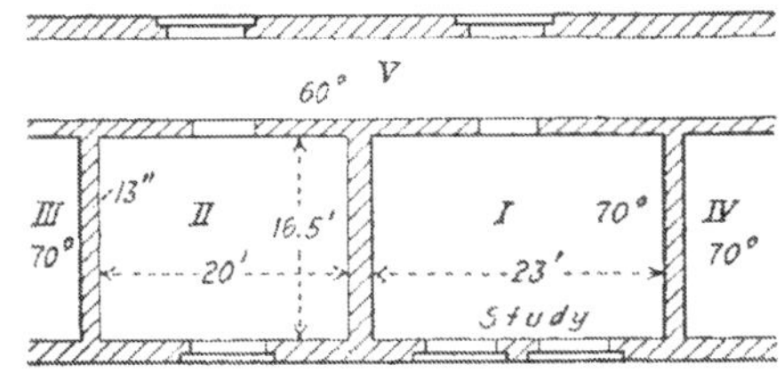

Fig. 157.

Assume that the heat transfer coefficients for several materials from Table VI are:

Double windows (D. W.)[1]	$K = 0.6$
Outer wall (O. W.) of brick 18 in. thick with furring lath and plaster	$K = 0.2$
Inner wall (I. W.) stud partition, lath and plaster both sides	$K = 0.34$
Inner door (I. D.)[2]	$K = 0.45$
Floor (F) of wood, filling, air space, lath and plaster	$K = 0.06$
Ceiling (C) insulex insulation lath and plaster under joists	$K = 0.08$
Height of room	14 ft.
Window sizes	6 by 8 ft.
Door sizes	5 by 9 ft.
Lowest outside temperature	0° F.
Temperature of attic	10° F.
Temperature of rooms I, III, and IV	70° F.
Temperature of room V	60° F.

The attic is assumed to have a slate roof. The direction of exposure and other data are taken from Fig. 157.

[1] Single windows will be designated by S. W.

[2] Outer doors will be designated by O. D.

Since the height of the rooms exceeds 10 ft., the average room temperature must be determined. For this purpose the ceiling temperature is first ascertained. Thus for rooms I, III, and IV, from Eq. (23),

$$t' = 70[1 + 0.017(14 - 10)]$$
$$= 74.8° \text{ F}$$
$$\text{Average temperature} = \frac{70 + 74.8}{2}$$
$$= 72° \text{ approximately}$$

For room V,

$$t' = 60[1 + 0.017(14 - 10)]$$
$$= 64° \text{ F}$$
$$\text{Average temperature} = \frac{60 + 64}{2}$$
$$= 62° \text{ F}$$

For room II

$$t' = t_x[1 + 0.017(14 - 10)]$$
$$= 1.07t_x$$
$$\text{Average temperature} = \frac{t_x + 1.07t_x}{2}$$
$$= 1.04t_x \text{ approximately.}$$

Solution. Determination of temperature from Eq. (25).
Room II receives

From room I through inner wall	$= 16.5 \times 14 \times 0.34(72 - 1.04t_x)$	$= 5{,}655 - 81.7t_x$
From room III through inner wall	$= 16.5 \times 14 \times 0.34(72 - 1.04t_x)$	$= 5{,}655 - 81.7t_x$
From room V through inner wall	$= (20 \times 14 - 5 \times 9) \times 0.34(62 - 1.04t_x°)$	$= 4{,}954 - 83.1t_x$
From room V through I. D.	$= 5 \times 9 \times 0.45(60 - t_x)$	$= 1{,}215 - 20.3t_x$
From underneath room II	$= 16.5 \times 20 \times 0.06(74.8 - t_x)$	$= 1{,}481 - 19.8t_x$
Total		$18{,}960 - 286.6t_x$

Room loses

Through windows $6 \times 8 \times 0.6 \times 1.04t_x$	$= 30\ t_x$
Addition of 15 per cent on account of northern exposure and 10 per cent on account of sudden change of temperature	$= 7.5t_x$
Through outer wall $(20 \times 14 - 6 \times 8) \times 0.2 \times 1.04t_x$	$= 48.3t_x$
Addition of 15 per cent on account of exposure and 10 per cent on account of sudden temperature change	$= 12.1t_x$
Through ceiling $16.5 \times 20 \times 0.08 \times (1.07t_x - 10)$	$= 28.3t_x - 264$
Total	$= 126.2t_x - 264$

Heat received = heat lost

$$18{,}960 - 286.6t_x = 126.2t_x - 264$$

for which

$$t_x = \frac{19{,}224}{412.8} = 47° \text{ F approximately}$$

Ceiling temperature $= 1.07 \times 47 \cong 50°$ F.

Average temperature $= \dfrac{47 + 50}{2} = 48.5°$ F

The heat loss from room I is computed in tabular form as follows

Room						Cooling surface									Temperature, degrees Fahrenheit			Coefficient of heat transmission	B.t.u.	Additional quantity for direction of exposure, B.t.u.	Additional quantity for sudden temperature change, B.t.u.	Total B.t.u.		Total B.t.u. for continuous operation	Interruption of operations		Remarks
Number	Classification	Length, feet	Width, feet	Height, feet	Volume, cubic feet	Classification	Direction of exposure	Length, feet	Height or width, feet	Area, square feet	Number	To subtract, square feet	For computation, square feet	Thickness of wall, inches	Inside	Outside	Difference					Emission	Received		Addition[a]	Total B.t.u.	
1	Study	28	16.5	14	5,313	D. W.	N	8	6	48	2	..	96	..	72	0	72	0.6	4,147	622	415	5,184	...	......	653	.....	In operation from 8 o'clock in the morning to 10 at night, 3 hr. heating up. $n = 7$ $z = 3$
						O. W.	N	23	14	322	1	96	226	18	72	0	72	0.2	3,254	488	325	4,067	...	......	512		
						I. D.	...	5	9	45	1	..	45	..	70	60	10	0.45	203	....	....	203					
						I. W.	...	23	14	322	1	45	277	..	72	62	10	0.34	942	....	....	942					
						I. W.	...	16.5	14	231	1	..	231	..	72	48.5	23.5	0.34	1,924	....	....	1,846					
						I. W.	...	16.5	14	231	1	..	231	..	72	72	0	0.31	0	....	....	0					
						C.	...	16.5	23	379.5	1	..	379.5	..	74.8	10	64.8	0.08	1,967	...	....	1,967	...	......	248		
						F.	...	16.5	23	379.5	1	..	379.5	..	70	74.8	−4.8	0.06	−109	....	....	.....	109				
																						14,209	109	14,100	1,413	15,513	

[a] These additions are calculated by Eq. (20).

The detailed computation given involves considerable labor. On this account approximate methods are frequently used. In several American tests[1] similar methods are used which check up fairly well with the method developed in the foregoing. It is hoped, however, that the Rietschel method which rests on fundamentals will be useful in unusual cases, *e.g.*, air spaces, hollow walls, etc.

D. DETERMINATION OF HEAT LOSS FROM CUBICAL CONTENTS

In some cases the heating requirements must be estimated without detailed information of the building construction. Uber,[2] with the aid of statistics, determined the heat loss per cubic foot of contents and compiled these data according to the nature and size of building. The following table is abridged from his publication:

Total cubical contents in cubic feet	Per cent of entire volume	Total volume to be heated in cubic feet	Average B.t.u. per cubic foot per hour	Total heat loss in B.t.u. per hour
Courthouses:				
150,000 to 250,000	50	75,000 to 125,000	3.4	250,000 to 425,000
250,000 to 350,000	60	150,000 to 200,000	3.0	450,000 to 600,000
350,000 to 1,500,000	70	250,000 to 1,000,000	2.80	700,000 to 2,800,000
1,500,000	75	....	2.2	
Prisons:				
150,000 to 350,000	50	75,000 to 175,000	3.6	270,000 to 630,000
350,000 to 700,000	65	230,000 to 450,000	3.0	700,000 to 1,350,000
700,000 to 1,500,000	75	500,000 to 1,100,000	2.5	1,250,000 to 2,750,000
Hospitals in prisons:				
50,000 to 150,000	55	30,000 to 80,000	7.0	200,000 to 550,000
Office buildings:				
250,000 to 1,500,000	70	180,000 to 1,000,000	3.1	550,000 to 3,100,000
1,500,000 to 4,000,000	60	900,000 to 2,500,000	2.2	2,000,000 to 5,500,000
High schools:				
350,000 to 700,000	65	230,000 to 450,000	3.4	780,000 to 1,500,000
Seminaries:				
500,000	65	325,000	2.7	880,000
1,000,000	65	650,000	2.5	16,000,000

[1] "Code of Minimum Requirements for the Heating and Ventilation of Buildings, 1927;" American Society of Heating and Ventilating Engineers;" "Engineering Standards, 1924," Heating and Piping Contractors National Association; "Ideal Heating Practice, 1921," American Radiator Company."

[2] "Design and Operation of Central Heating Systems," from Uber, "Bau-und Betriebstechnisches für Zentralheizungen," Berlin, 1916.

SECTION II

COMPUTATION OF HEATING SURFACE

A. GENERAL THEORY

The general theory[1] of heat flow applied to heating surfaces may be subdivided as follows (1) both fluids circulating (double-current apparatus), (2) one fluid circulating (single-current apparatus), (3) and no fluids circulating (eddying-current apparatus) [2] A fluid is considered circulating when flow takes place in a defined direction. Non-flowing fluids are not necessarily in a state of rest, but on the contrary may have a strong eddying motion while maintaining a uniform temperature

1. Double-current Apparatus.

Assume that the heating fluid is below the surface F (Fig 158) and that the cooling liquid is above it

While the fluid below the heating surface is being cooled from a temperature T_1 to a temperature T_2, the fluid above the surface is being heated from a temperature t_1 to t_2. In the steady state (when thermal equilibrium is reached) the temperatures t and T at any point f of the surface F are dependent only on the position of the surface f If f denotes the functional relation, it may be expressed mathematically by the relation

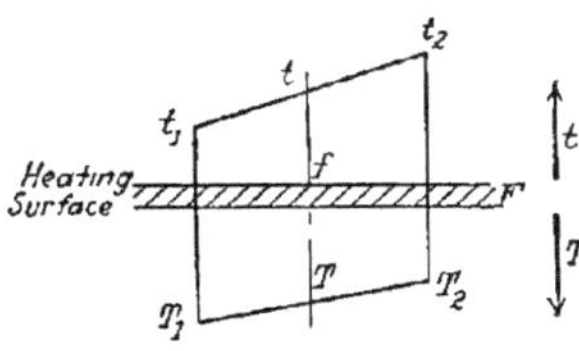

FIG 158 —Double current apparatus

$$t = f(p).$$

2. Single-current Apparatus.

In Fig 159, let the fluid above the heating surface (or cooling surface) be at a temperature t_1 at a time z_1, the temperature being uniform because of a strong eddying motion At the same time z_1, let the lower fluid (the cooling water) enter at a constant temperature T_0 and be heated to T_1 At a later time z_2, the upper fluid will have cooled to a uniformly distributed temperature t_2, while the lower fluid will have reached a temperature T_2. At a still later time z_3, let the new temperatures in the two fluids

[1] "Principles of Heating Apparatus," from BERLOWITZ, "Das System der Warmeapparate," Berlin, 1910

[2] "Design and Calculation of Heating and Cooling Coils," from GROBER, 'Die Berechnung von Heiz- und Kuhlrohren," Munich, 1920, "Fundamental Laws of Heat Conduction and Heat Transmission," from GROBER, "Die Grundgesetze der Warmeleitung und des Warmeuberganges," Berlin, 1921

be t_2 for the upper and T'_3 for the lower. It is to be noted that in general the temperatures t and T at any position f on the surface F depend not only on the position of f but also on the time z. In functional notation this may be expressed

$$t = f(f, z).$$

3. Eddying-current Apparatus.

With reference to the diagram in Fig. 160, let both currents have an eddying motion and uniform temperature distribution at all times. It

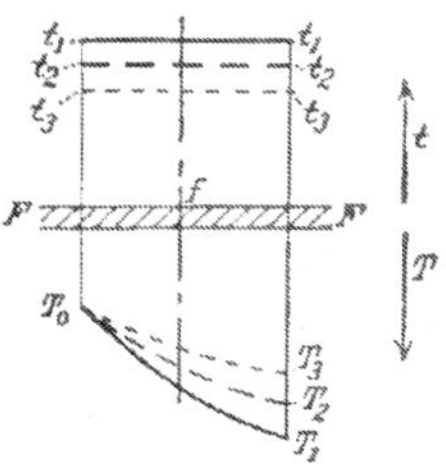

FIG. 159.—Single current apparatus.

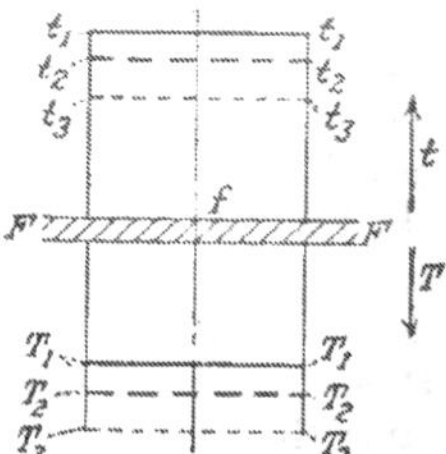

FIG. 160.—Eddying current apparatus.

will be seen that the temperatures in this case are independent of the position of f and dependent only upon the time. Consequently

$$t = f(z).$$

In heating practice as treated in this text, only double-current apparatus will be considered. The design may employ the countercurrent principle shown in Fig. 161, the parallel current shown in Fig. 162, or the transverse current shown in Fig. 163. The heat transmission in transverse current apparatus has been theoretically developed by Nusselt,[1]

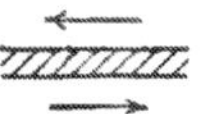

FIG. 161.—Countercurrent.

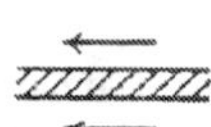

FIG. 162.—Parallel current.

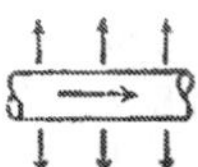

FIG. 163.—Transverse current.

who found that the transfer effect lies between the results obtained for countercurrent and parallel-current apparatus.

B. COUNTERCURRENT AND PARALLEL-CURRENT APPARATUS

The general equation (see Fig. 164) for heat transfer may be written

$$dW = K(u' - u_1)dF \tag{26}$$

[1] "Heat Transmission in Pipe Systems," from NUSSELT, "Der Wärmeübergang in Rohrleitungen," Berlin, 1900.

where

dW = heat transfer through surface dF in B t u

K = coefficient of heat transmission in B t u per square foot per degree Fahrenheit per hour.

u' = temperature of heat emitting fluid in degrees Fahrenheit

u_1 = temperature of heat absorbing fluid in degrees Fahrenheit

In the steady state the heat transfer through the surface dF (i e , the value dW) is to the total heat transfer of the surface (i e , W) as the temperature change over the surface dF (i e , du) is to the total temperature change over the surface F (i e , $t_2 - t_1$) This statement is mathematically expressed as

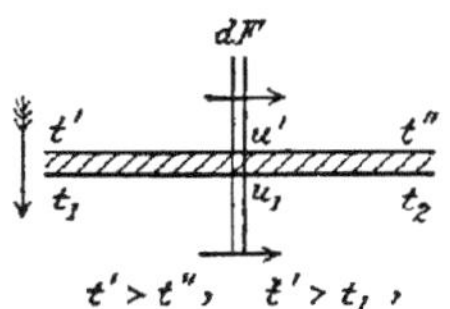

FIG 161

$$\frac{dW}{W} = \frac{du_1}{t_2 - t_1} \tag{27}$$

Substituting the value of dW of Eq (27) for the corresponding value in Eq (26), there results.

$$dF = \frac{Wdu}{K(u' - u_1)(t_2 - t_1)} \tag{28}$$

Equation (28) contains two variables u' and u_1 A relationship between them may be found as follows· In the steady state the decrease in temperature of the fluid emitting heat to the surface dF is to the temperature increase of the fluid absorbing heat to the surface dF as the total temperature decrease of the one fluid is to the total temperature decrease of the other, i e ,

$$\frac{t' - u'}{u_1 - t_1} = \frac{t' - t''}{t_2 - t_1} \tag{29}$$

Adding +1 to both sides, there results

$$\frac{u_1 - t_1 + t' - u'}{u_1 - t_1} = \frac{t_2 - t_1 + t' - t''}{t_2 - t_1}.$$

Let $t' - t_1 = \Delta_1$ and $t'' - t_2 = \Delta_2$.

Then

$$\frac{u_1 - u' + \Delta_1}{u_1 - t_1} = \frac{\Delta_1 - \Delta_2}{t_2 - t_1}$$

whence

$$u' - u_1 = \frac{u_1(\Delta_2 - \Delta_1) + \Delta_1 t_2 - \Delta_2 t_1}{t_2 - t_1}$$

Substituting this in Eq (28) and integrating between the limits,

$$F = 0, \text{ then } u_1 = t_1$$
$$F = F, \text{ then } u_1 = t_2$$

After some transformation

$$F = \int_{t_1}^{t_2} \frac{Wdu_1}{K[u_1(\Delta_2 - \Delta_1) + \Delta_1 t_2 - \Delta_2 t_1]} \tag{30}$$

In the above W is assumed constant, a condition true for the steady state. To obtain simple formulas for use in practice, it is further assumed that the value of K may be considered constant. In reality the value of K depends upon (1) the shape and area of surface of the radiator, (2) the state of flow of the two fluids, (3) their absolute temperatures, densities, and viscosity, and (4) the temperature differences.

To carry out the above simplification and attain a reasonable degree of accuracy for the formula, the value of K is determined by experiment for each particular case.

If N is used to designate the denominator of Eq. (30), on differentiation

$$dN = K(\Delta_2 - \Delta_1)du_1$$

where Δ_1, Δ_2, t_1, and t_2 are held constant. On multiplication of this differential by $\frac{W}{K(\Delta_2 - \Delta_1)}$, it will be observed that the product is identical with the numerator of Eq. (30), and hence the integral is the $\log_e$ form. Therefore

$$F = \frac{W}{K(\Delta_2 - \Delta_1)} \log_e \frac{t_2(\Delta_2 - \Delta_1) + \Delta_1 t_2 - \Delta_2 t_1}{t_1(\Delta_2 - \Delta_1) + \Delta_1 t_2 - \Delta_2 t_1},$$

$$= \frac{W}{K(\Delta_2 - \Delta_1)} \log_e \frac{\Delta_2(t_2 - t_1)}{\Delta_1(t_2 - t_1)},$$

or finally

$$F = \frac{W}{K(\Delta_2 - \Delta_1)} \log_e \frac{\Delta_2}{\Delta_1}$$

On substitution of the values of Δ_2 and Δ_1, then

$$F = \frac{W}{K[(t' - t_1) - (t'' - t_2)]} \log_e \frac{t' - t_1}{t'' - t_2} \text{ (for parallel currents)} \quad (31)$$

and similarly (see Fig. 164a)

$$F = \frac{W}{K[(t' - t_2) - (t'' - t_1)]} \log_e \frac{t' - t_2}{t'' - t_1} \text{ (for countercurrents)} \quad (32)$$

In both cases it will be observed that (1) the denominator shows the difference of the temperature variations at the beginning and at the end of the heating surface, and (2) the numerator shows the logarithm to the base e of this difference.

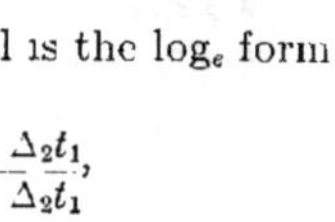

$t' \longrightarrow t''$

$t_2 \longleftarrow t_1$

$t' > t''$. $t_1 > t_2$,
$t' > t_2$, $t'' > t_1$.

Fig. 164a.—Countercurrent.

Developing $\log_e x$ into a series, there results

$$\log_e x = 2\left[\frac{x - 1}{x + 1} + \frac{1}{3}\left(\frac{x - 1}{x + 1}\right)^3 + \cdots\right]$$

If

$$x = \frac{t' - t_1}{t'' - t_2}$$

and since for all practical purposes terms beyond the first may be omitted, Eq. (31) becomes

$$F = \frac{W}{K\left(\frac{t' + t''}{2} - \frac{t_1 + t_2}{2}\right)}. \tag{33}$$

For countercurrents Eq (32) will also reduce to Eq (33) when only the first term of the expression is used In this case when the temperature differences are so small that the first term alone of the above series is sufficient, the surface F required for a given heat transfer is the same for both parallel-current and for countercurrent apparatus.

C. APPLICATION OF EQUATIONS TO THE USUAL FORMS OF HEATING SURFACES

1. Hot-water Radiation for Gravity and Forced-circulation Systems. In Eq (33) let

t' = flow temperature t_f of water in degrees Fahrenheit.
t'' = return temperature t_r of water in degrees Fahrenheit.
t_1 = entering temperature of air in degrees Fahrenheit
t_2 = exit temperature of air in degrees Fahrenheit

With radiators in general it is inaccurate to assume that a directed motion of the air exists In addition it is inconvenient to use the temperatures of the (entering and exit) air For this reason the temperature t at the center of the heated space measured at eye level[1] (5 ft) above the floor is used in place of the values t_1 and t_2 With this in view the hot-water heating surface F_w in square feet may be found from the formula

$$F_w = \frac{W}{K\left(\frac{t_f + t_t}{2} - t\right)}. \tag{34}$$

When making tests to find the value of K, it is essential that Eq (34) shall be used in computing it.

2. Steam Radiation.

The surface F_d required for steam radiators is computed in much the same way as the surface for hot-water radiation except that in place of $\frac{t_f + t_r}{2}$ the average steam temperature t_d is used, and this may be assumed constant Hence the formula may be written

$$F_d = \frac{W}{K(t_d - t)}. \tag{35}$$

[1] In the opinion of the author it would be better to use a lower point e g , knee height, 1½ ft from floor See also BRABBÉE, "The Heating Effect of Radiators," *Jour* A S H & V E , 1925, 1926, 1927

The values of K for water and steam radiators most usually found in practice have been determined[1] experimentally. They are compiled in Tables VIII and IX and all other details concerning radiators The location of radiators within the room have been considered on page 45 and following.

3. Heat-transfer Apparatus Having Rapid and Directed Circulation.

In the design of tubular boilers, countercurrent apparatus, etc , Eqs (31) and (32) should be used for determining the required heating surface

[1] *Bulls* 1, 3, 4, and 8 of the Research Laboratory, Charlottenburg (*Mitteilungen der Versuchsanstalt*), R Oldenbourg, Munich

SECTION III

PIPING EQUATIONS

A INTRODUCTION

Research conducted in 1910[1] on the frictional resistance of pipe and fittings resulted in data which made it possible to devise simplified methods of computing hot-water systems. Additional studies resulted in a similar method for computing duct sizes in ventilating and warm-air heating systems.[2] Later work made it possible to place the treatment of high- and low-pressure heating systems on a similar basis.[3] This method of computation has been proved extensively in practice and with excellent results. Moreover, the method is appreciated by teachers and students since the design of all piping used in heating and ventilating practice is based on the same general method. In what follows, the theoretical development has been carried out only as far as necessary to insure completeness in presentation. The original sources from which the equations are derived are given in the footnotes. The equations also hold for compressed air, water, gas, etc.

In customary practice, the solution of any given problem may be arrived at (1) making a rapid computation for estimating the cost of installation (bids), or (2) by a check computation for subsequent installation in greater detail.

B GENERAL THEORY

The fundamental relation which must be satisfied in any system involving the flow of fluids is

$$H = \Sigma(lR + Z), \tag{36}$$

[1] "Friction and Local Resistances in Hot-water Heating Systems," from BRABBÉE, "Reibungswiderstände in Warmwasserheizungen," and "Einzelwiderstände in Warmwasserheizungen," Bulletins 14 and 15, Research Laboratory, Charlottenburg. Pub., R. Oldenbourg, Munich 1913.

[2] "Simplified Methods for Determining Ventilation and Air Heating Systems," from BRABBÉE-BRADTKE, "Vereinfachtes Zeichnerisches und rechnerisches Verfahren zur Bestimmung der Rohrleitungen von Luftungs- und Luftheizanlagen," R. Oldenbourg, Munich, 1915.

[3] "Simplified Method of Determining Steam Systems," from BRABBÉE-WIERZ, "Vereinfachtes Zeichnerisches oder rechnerisches Verfahren zur Bestimmung der Durchmesser von Dampfleitungen," R. Oldenbourg, Munich, 1915.

where

H = pressure head in inches of water column[1] at 62° F
R = frictional resistance (pipe friction) 1 ft of pipe in in. w c.
l = length of pipe in feet
Z = frictional resistance (local resistance) of valves, ells, tees, bends, etc in in w c.

Equation (36) states that the pressure head must balance the sum of the pipe and fitting resistances

Pressure Head

The pressure head in all heating systems is stated (1) as a difference in weight of two fluid columns, or (2) as a given value

As an example under the first heading, consider a section of a gravity hot-water heating system In Fig. 165 water of density s'' (in pounds

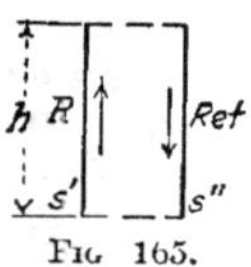

Fig 165.

per cubic foot) flows downward in the return pipe, causing the water in the supply main of density s' (also in pounds per cubic foot) to be forced upward The pressure head is accordingly

$$H = 0\ 192h(s'' - s') \qquad (37)$$

where

h = vertical height of the counteracting columns in feet
H = pressure head in inches of water column at 62° F.

The constant 0 192 is a conversion factor to express H in the units chosen

As a given value, e g , in a steam system operated at a pressure of 2 lb per square inch, the head pressure is simply H = 2 lb per square inch = 55 4 in w c

Pressure head may also be expressed as a combination of items 1 and 2 For example in the case of forced-circulation heating systems wherein a pump operates with a known pressure P and further, by virtue of gravity, a pressure head of H is available, the total pressure head is

$$T = P + H$$

[1] In the future inches of water column will be expressed as in w c

PIPE FRICTION

Tests at the Research Laboratory, Charlottenburg,[1] show that the law of friction may be expressed with sufficient accuracy for heating and ventilation practice as

$$R = \frac{p_2 - p_1}{l} = a\frac{v^n}{d^m}, \tag{38}$$

where

R = pressure drop per foot of run
p_2 = initial pressure
p_1 = terminal pressure
l = length of run in feet
a = a constant
v = average velocity[2] in cross-section of pipe in feet per second
d = inside diameter of pipe in inches
$n, m,$ = numerical exponents

RESISTANCE OF FITTINGS (LOCAL RESISTANCES)

The resistance of fittings such as ells, tees, valves, etc may be expressed in the form

$$Z = \Sigma\zeta\frac{0.192sv^2}{2g} \tag{39}$$

where Z = pressure drop in in. w c.
ζ = coefficient of resistance
s = density in pounds per cubic feet
v = average velocity in cross-section in feet per second.
g = acceleration of gravity in feet per second per second

Attention is directed to the fact that tests show that the previously accepted values of the resistance through fittings are in error For instance the usual allowance was to estimate the resistance at 15 to 25 per cent of the total resistance of the piping systems The facts show that the allowance should be 50 per cent for ordinary hot-water and steam heating systems with pipe sizes from ½ to 12 in with an increase of this percentage for pipe sizes beyond those given For air ducts having an inside diameter of about 20 in the local resistance is 80 per cent of the total, increasing to 90 per cent when the diameter reached about 40 in

For the steady state under constant-flow conditions the following equation is useful:

$$Q = 25\frac{\pi d^2}{4}vs, \tag{40}$$

[1] See footnote on p 184

[2] By average velocity is meant that average which if uniformly maintained over the cross-section would give the same flow

in which

Q = quantity of the fluid flowing in pounds per hour and in which the other symbols have the previous meaning

C SUMMARY OF FORMULAS

The group of equations may be assembled as follows·

$$\left.\begin{aligned} H &= 0\ 192h(s'' - s') \\ R &= \frac{p_2 - p_1}{l} = \frac{av^n}{d_m} \\ Z &= 0\ 192\Sigma\zeta\frac{sv^2}{2g} \\ Q &= 25\frac{\pi d^2}{4}vs \end{aligned}\right\} \qquad (41)$$

These will be used subsequently in forming tables A more detailed explanation of their use will follow when discussing the different heating systems

SECTION IV

DESIGN OF VARIOUS TYPES OF HEATING SYSTEMS

A LOCAL HEATING PLANTS

Open Fireplaces

Due to the many and intricate factors which enter into this form of heating, its computation, is as yet not possible For practical purposes, however, the heating efficiency of the fireplace may be assumed to range from 5 to 10 per cent

Conduit Heating

Accurate determination of the required heating surface for conduit systems is not possible Satisfactory results may be obtained if the heating surface is proportioned for the following heat emission.

For brick ducts,	300–400 B t u per square feet
For ribbed cast-iron pipe,	400–450 B t u. per square feet
For plain cast-iron pipe,	550 B t u per square feet

Example 7.

Assume a church of brick construction Let the ceiling height be 50 ft , total window area, 1,200 sq ft , combined area of wall, floor, ceiling, and pillars, 30,000 sq ft It is required to design a conduit heating system using plain cast-iron pipe ducts Let it be assumed further that

Lowest outside temperature	0° F
Required inside temperature	60° F
Inside temperature at beginning of heating-up period	32° F
Duration of heating-up period	6 hr
Maximum length of cast-iron pipe ducts	100 ft
Heat emission from plain cast-iron pipe ducts	550 B t u per square foot per hour
Coefficient of heat transmission K for single church windows	1 1

Solution: Since the heating surfaces are not necessarily in the best location for favorable heat emission, Eq (22b) applies Substituting, therefore, the values from above,

$$W = \frac{1,200 \times 1.1(60 - 0)}{2} + 30,000\left[15 + \frac{2(60 - 32)}{6}\right]$$

$$\cong 770,000 \text{ B t u}$$

Since the ceiling height exceeds 40 ft (p 172), the correction to be applied is $\frac{(50 - 40)}{3} \times 5 \cong 16.7$ per cent Hence, there is to be added $770\,000 \times 167 \cong$ 129,000 B t u , making the total heat requirement $770,000 + 129,000 = 899,000$ B t u This will require $\frac{899,000}{550} \cong 1,640$ sq ft of duct surface since each square foot of cast-iron pipe emits 550 B t u per square foot per hour Choosing ducts of rectangular cross-section having inside dimensions of 0 8 by 1 ft which have a heating surface of 3 6 sq ft per lineal foot, $\frac{1,640}{3\ 6} \cong 455$ lin ft of duct will be required With maximum duct length of 100 ft there will be required $\frac{455}{100} \cong 5$ ducts approximately

Stove Heating

1. Iron Stoves

The calculation of the required sizes of iron stoves should be based on the equation

$$F = \frac{W}{K}$$

where F = stove heating surface in square feet
W = heat loss from room in B t u per hour
K = heat emission in B t u per square feet of stove heating surface per hour

Assuming usual room temperatures, the values of K are generally taken as follows

For plain stoves

- K = 550–750 B t u per square feet per hour for constant operation
- K = up to 950 B t u per square foot per hour for intermittent operation

For stoves with outside ribs

- $K = 350-500$ B t u per square feet per hour for constant operation
- K = up to 750 B t u per square feet per hour for intermittent operation

The heat loss W may be computed as outlined on page 174 The method, however, is seldom employed since the empirical methods used in practice when based on good judgment give satisfactory results In these methods it is customary to base the amount of heating surface on the cubical contents of the room or on the number and size of the front windows of the room to be heated

Some objections to the empirical methods have been overcome by Wierz[1] who has compiled Tables XXIII*a* and *b* (p 323) from which the necessary amount of heating surface may be taken in proportion to the length of the outside wall of the room to be heated For cold inner walls, for walls heated from one side only, and for the additional heat loss due to wind and exposure, the correcting factors noted below the tables must be taken into consideration The heating surface of the stove is to

[1] "Elements of Heating," from Wierz, "Heizungstechnische Grundlagen" 'The Iron Stove (Der eiserne Zimmerofen)," issued by the Association of German Iron Stove Manufacturers, R Oldenbourg, Munich and Berlin, 1923

include top and bottom of the combustion chamber, all surfaces heated by radiation (ash pit included) and such surfaces as are in contact with incandescent fuel or in contact with the flue gas, and also such ornamentation designed which increase the heating surface (no legs) From the foregoing it would therefore be reasonable to include the area of the flue connection from the stove to the chimney as effective heating surface

Example 8

Assume a corner room on the ground floor of a building exposed on all sides Let its dimensions be 11 ft 6 in by 15 ft and 11 ft 6 in high Assume single windows

Solution The length of the outer wall is 11 5 + 15 = 26 5 ft From Table XXIII*a* the required heating surface should be 13 8 sq ft Due to the exposure 15 per cent additional should be allowed, or the equivalent of 13 8 × 15 ≅ 2 1 sq ft, making a total heating surface of 13 8 + 2 1 = 15 9 sq ft A stove of 16 sq ft of surface will answer the requirements.

2. Tile Stove.

Since tile stoves are rarely used in the United States, no attempt will be made to go into detail as to the necessary sizes. The computation, however, is similar to that for iron stoves

3. Oil Stoves.

The heat emitted from an oil stove is dependent upon the heat value of the fuel For stoves connected with a chimney by means of a pipe, the emission is approximately 70 per cent of the heat in the fuel When the products of combustion are discharged into the room, the heat emission is identical with the heat value of the fuel used The latter types are objectionable from a hygienic viewpoint.

4 Gas Stoves.

As a rule the calculation of the capacity of gas stoves is based upon the cubical contents of the room to be heated More favorable results, however, would be obtained if the capacity were based on the actual heat loss Unlike iron stoves, gas stoves must be properly proportioned, since they cannot be forced

Allowing for a heat loss of from 10 to 15 per cent through the chimney, approximately 400 to 500 B t u per cubic foot of gas is available for heating These values depend upon the heat value of the gas used and will vary widely in different localities

5. Electrical Heaters.

The heat emission W (in B t u) of electrical heaters is computed from the known equation.

$$W = 3.41 I^2 R,$$

where

I = current in amperes

R = resistance of heater in ohms

The capacity of electrical heaters should be determined from the heat losses. If it is possible to distribute electric heaters to the places of greater heat losses, e g , windows, cold walls and entries, comfortable conditions can be obtained with room temperatures considerably lower than the usual figure of 70° F

B. HOT-WATER HEATING

Required Boiler Heating Surface

1. Heating-up Period.

There is a certain interval between setting a heating plant into operation and the time when the steady state is reached This interval is known as the heating-up period An accurate determination of the heat required during this period offers some difficulty Experience shows, however, that the necessary heating surface of the boiler may be computed by the following formula with satisfactory results:[1]

$$F_1 = \frac{1.1\{z(W + Z) + (A + 0\ 12B)(t_1 - t_2)\}}{zK_1}, \qquad (42)$$

in which

F_1 = total boiler heating surface in square feet.[2]
W = steady-state heat loss from building in B t u per hour.
Z = additional allowance for heat loss of building in transient state.
z = heating-up period in hours
A = water content of system in pounds.
B = weight of iron in system in pounds
t_1 = average temperature of boiler flow and return in the steady state in degrees Fahrenheit
t_2 = temperature to which the system cools between intervals of operation (for daily operation it may be taken as 80° F)
K_1 = permissible boiler load in B t u per hour during heating-up period This value is published by the boiler manufacturer together with a guaranteed efficiency for the given load
1 1 = a factor of 10 per cent to allow for the heat loss of the system (piping, etc)

According to tests conducted on European boilers K_1 should not exceed 4,000 B t u per square foot In the absence of more accurate data, for purposes of computation the water content and the weight of iron may be assumed as follows:

For each square foot of radiator surface, 2 lb of water
For each square foot of radiator surface, 7 5 lb of iron.

Water content of piping for each 100 B t u transmitted 0 4 lb.

[1] In the heating-up period neither Z nor W is constant Thus the above formula must be considered as only approximate

[2] Total boiler heating surface is the total surface touched by fire or combustion gases

2. Steady State.

The required boiler heating surface in the steady state may be proportioned according to the formula

$$F_2 = \frac{1.1W}{K_2}, \tag{43}$$

where

1.1 = additional allowance for heat loss of system (piping, etc.)
F_2 = total boiler heating surface in square feet [1]
W = heat emitted by system in B.t.u. per hour during the steady state (= heat loss from building)
K_2 = permissible boiler load per square foot of heating surface during the steady state = 3,000 B.t.u. per hour

From the foregoing it may be seen that the heat load on the boiler during the heating-up period is larger than the load in the steady state. As a consequence, the former requirement will necessitate the larger heating surface. The conditions under which a boiler must operate deserve consideration. For example, suppose that a hot-water heating system is to operate so that when an outside temperature of 32° F. or more prevails, the heating system operates intermittently, and when outside temperatures below 32° F. are encountered, the system operates continuously. A case of this kind will require a computation for both the heating-up and the steady-state conditions, the larger value for the required boiler heating surface will then be used.

Example 9

Let it be required to find the boiler heating surface needed for a house for which the following data are assumed:

Heat required by house during steady state	2,000,000 B.t.u. per hour at 0° F.
Additional B.t.u. for heating up Z (see p. 171)	300,000 B.t.u. per hour at 0° F.
Heating-up period (z)	3 hr.
Water content of system	25,000 lb.
Total weight of iron in system	70,000 lb.
Flow temperature of boiler	t_f = 180° F.
Temperature difference between flow and return	$t_f - t_r$ = 30° F.
K_1 (for heating-up period)	4,000 B.t.u. per square foot per hour
K_2 (for continuous operation)	3,000 B.t.u. per square foot per hour

Assume further that for temperatures 32° F. and upwards intermittent operation will be used and for temperatures below 32° F. continuous operation is to be maintained.

Solution: 1. *Heating-up Period.*—In order to use Eq. 42 which covers the heating up period, it is necessary to compute the heat loss of the building for an outside temperature of 32° F. It may be approached in the following manner:

Temperature difference for 70° F. inside and 32° F. outside	= 38° F.
Temperature difference for 70° F. inside and 0° F. outside	= 70° F.

[1] See footnote, p. 191.

The heat loss according to specifications is computed on a 70° F temperature basis (70° F inside and 0° F outside) Strictly speaking the heat loss of a building is not directly proportional to the temperature difference between inside and outside, but for the purpose of this problem the assumption is safe Consequently for an outside temperature of 32° F the heat loss is

$$W_{32°} = \frac{2,000\,000}{70} \times 38 \cong 1,090,000 \text{ B t u per hour approximately.}$$

Also

$$Z_{32°} = \frac{300,000}{70} \times 38 = 163,000 \text{ B t u per hour approximately.}$$

The average boiler temperature for 0° F outside is

$$t_m = \frac{t_f + t_r}{2} = \frac{180 + 150}{2} = 165° \text{ F}$$

Assuming values of t_{f_1} and t_{r_1} to be 120 and 100° respectively for outside temperatures of 32° F the average boiler temperature under these conditions becomes

$$t_{m_1} = \frac{120 + 100}{2} = 110° \text{ F.}$$

Substituting these values in Eq 42, the required boiler heating surface in square feet is

$$F_1 = \frac{1\ 1\{3(1,090,000 + 163,000) + (25,000 + 0\ 12 \times 70,000)(110 - 80)\}}{3 \times 4,000}$$

$$\cong 436 \text{ sq ft}$$

2 *Steady State*—Substituting the corresponding values in Eq (43), the boiler heating surface is

$$F_2 = \frac{1\ 1 \times 2,000,000}{3,000} \cong 733 \text{ sq ft}$$

From manufacturers' catalogues a boiler is selected the heating surface of which corresponds to the larger of the two values, namely, to 733 sq ft of heating surface This method of figuring boilers per square foot of heating surface is a fundamental deviation from the American practice to figure boilers according to their output, given in square feet of steam rating

Radiators Surface Required

The required amount of hot-water radiation is obtained from Eq. (34) *i e*,

$$F = \frac{W}{K\left(\frac{t_f + t_r}{2} - t\right)}$$

p 182 in which,

W = the heat loss of the room in B t u. per hour

t_f = the temperature of the inflowing water in degrees Fahrenheit

t_r = the temperature of the water leaving the radiator in degrees Fahrenheit

t = the temperature of the room in degrees Fahrenheit

K = the coefficient of heat transmission in B t u per square foot per hour per degrees Fahrenheit

From tests of radiators of varying designs and sizes a series of values for K are found which may be applied to most existing practical conditions The more important of these values are compiled in Table VIII (p 298).

Example 10.

The heatloss of room I (Fig. 157, p. 174) ≅ 15,500, B.t.u. per hour for an outside temperature of 0° F. Assume further:

Flow temperature of water	180° F.
Room temperature	70° F.
Temperature drop in radiator	30° F.

The radiator is to be installed under a window and is to have no enclosure. It is to be supported at a height 4 in. above the floor by means of brackets. The distance from floor to sill is to be 30 in.

Solution:

$$t_f = 180°.$$
$$t_r = 150°.$$
$$\frac{t_f + t_r}{2} - t = 95°.$$

From Table VIII it will be seen that values of K have been compiled for varying radiator heights and temperature differences. Let it be assumed (Fig. 166) that it shall be a two-column radiator leaving at least 2 in. between top of radiator and the sill. The height of the radiator will then be 30 − 4 − 2 = 24 in. for which K = 1.4.[1] Substituting these values in Eq. 34,

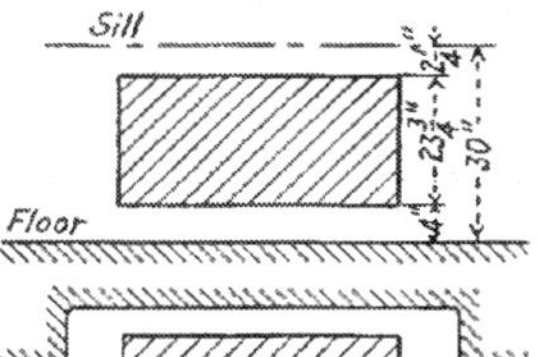

FIG. 166.—Radiator in window recess.

$$F = \frac{15{,}500}{1.4 \times 95} \cong 117 \text{ sq. ft.}$$

Since a heating surface of 117 sq. ft. is excessive for a single radiator of the two-column type, it is better to divide this surface into two radiators each having 58.5 sq. ft. From a manufacturer's catalogue it will be found that a two column peerless radiator, *e.g.* the Ideal Fitter (American Radiator Co.), corresponding to the requirements has the following dimensions:

Total height, 26 in.
Height of legs, 2¼ in.
Therefore height of sections, 23¾ in.
Heating surface per section, 2⅔ sq. ft.
Width of section, 7⅜ in.

The heating surface will therefore consist of two 22-section two-column radiators.

Determination of Pipe Sizes

1. Gravity Hot-water Heating. Two-pipe System.

a. Heat Loss of Piping Neglected. (1) Circulating Pressure Head.—As previously mentioned on page 185, the circulating pressure head is determined by means of Eq. 37 in inches of water column pressure.

$$H = 0.192h(s'' - s').$$

Application of this formula to the many types of hot-water heating systems will be illustrated in later examples.

[1] It is assumed that the radiator legs are approximately 2 in. high.

(2) Pipe and Fitting Frictions.—Frictional resistance of pipes (more briefly, pipe friction) is due to the friction between the flowing water and the inner surface of the pipe. The resistance due to fittings (fitting resistance) is the resistance in elbows, tees, crosses, valves, radiators, boilers, etc. Numerous tests at the Research Laboratory, Charlottenburg, have made possible a mathematical expression of the magnitude of these resistances.[1] The results for frictional resistances for water flowing through iron pipes are expressed in a formula similar to Eq. (38) (p. (186) which now takes the form

$$R = 0.058\frac{v^{1.85}}{d^{1.26}} \text{ for ordinary pipe} \tag{44}$$

$$R = 0.076\frac{v^{1.86}}{d^{1.37}} \text{ for flanged pipe} \tag{45}$$

where

R = frictional resistance per foot in inches of water column (62° F.)
v = average water velocity in feet per second
d = inside diameter of pipe in inches

In both equations the average water temperature is assumed to be 165° F.[2]

The results of the tests for the resistances of fittings are expressed by the formula

$$Z = \Sigma\zeta \frac{0.192\ s \cdot v^2}{2g} \tag{39}$$

where

Z = pressure drop in inches of water column
ζ = coefficient of resistance determined by test.
v = average velocity of water in feet per second.
g = acceleration of gravity (= 32.2).
s = density of the water in pounds per cubic foot

The coefficient of resistance ζ is independent of the temperature of the water; ζ has been determined by the tests mentioned and will be discussed later.

For water heating systems, Eq. (40) (p. 186) takes the following form.

$$v = \frac{W}{Ad^2(t' - t'')} \tag{46}$$

[1] See footnote 1, p. 184.

[2] For average water temperatures of 180 to 140° F. the values given by the above formulas may be used without corrections. If lower average water temperatures are used, the values of R must be multiplied by a factor as follows

1.05 for average water temperature of 120° F

1.10 for average water temperature of 100° F

where

v = velocity of flow in feet per second
W = quantity of heat in B t u per hour transmitted through the pipe to supply the heat requirement of the radiation connected thereto
A = a factor which for low-pressure hot-water heating systems is taken as 1,200
d = inside diameter of pipe in inches
$t' - t''$ = temperature drop through radiator on which the proportioning of the system is based

To summarize, the following group of equations will apply to the design of any hot-water heating system

$$\left.\begin{aligned} H &= 0\,192h(s'' - s') \\ R &= \begin{cases} 0\,058\,\dfrac{v^{1\,85}}{d^{1\,26}} \text{ for ordinary pipe} \\ 0\,076\,\dfrac{v^{1\,86}}{d^{1\,37}} \text{ for flanged pipe} \end{cases} \\ Z &= 0\,192\Sigma\zeta\frac{v^2}{2g}s \\ v &= \frac{W}{Ad^2(t' - t'')}. \end{aligned}\right\} \quad (47)$$

At first sight it might appear difficult to apply the equations in practice As a matter of fact by means of charts and tables to be now described, the design of hot-water systems is extremely simple

Chart 1 at the end of the book consists of two divisions. fitting resistances and piping resistances Consulting the section on pipe friction it will be noticed that the chart contains.

(1) Frictional loss R per lineal foot of pipe in units of thousandths of an inch of water column

(2) Pipe diameters ranging from ½ to 12 in

(3) Quantity of heat to be transmitted per hour (under line I)

(4) Corresponding average water velocity in the pipe in feet per second (under line II)

The section on fitting resistances includes

(1) Average velocity of water in the pipe in feet per second

(2) Fitting resistances in inches of water column for values of ζ ranging from 1 to 15

(3) Coefficients of resistance for the more important fittings, valves, etc

The values of ζ for tees are valid only so long as no large velocity changes occur in these fittings For all usual cases this assumption is sufficiently accurate. In exceptional cases Tables 19 and 20 of *Bulletin*

15 of the Research, Laboratory Berlin-Charlottenburg, should be consulted Crosses may be regarded as double tees necessitating a doubling of the straightway and branch values of ζ. Distribution or collecting headers such as manifolds may be divided into elbow elements With reference to the resistance through boilers, see footnote [1]

The influence of viscosity of hot water flowing in pipes having a rough inner surface has not yet been determined with certainty. Neither do the equations given here and in *Bulletins* 14 and 15 of the Research Laboratory present final data The equations given here have been determined from tests, and while empirical, nevertheless they include these influences to a certain extent and are sufficiently accurate for practical purposes Indeed thousands of heating systems, large and small, have been designed by the methods here given with satisfactory results as to operation and economy.

(3) Method of Using Chart 1.—Chart 1 is compiled for a temperature drop of 30° F through the system If for any reason a different temperature drop is selected, for example, t° F , it will be necessary to convert the quantity of heat to be transmitted to a drop corresponding to 30° before using Chart 1 This is done in the following manner: Divide the quantity of heat transmitted by the various pipe sections,[2] by t° and multiply by 30

(*a*) Preliminary Piping Layout for Estimating Purposes —A heating system of several radiators may be subdivided into elementary water circuits in which the circuit may be traced from the boiler through the piping and radiator back to the boiler In such systems certain radiators are more favorably located than others with respect to circulation through them In design it is common practice to start with the least-favored radiator. This radiator in general is the one which is vertically nearest the boiler and horizontally the remotest

The circulating head for any radiator is found according to methods described on page 185. Reference to Table XVI will show the head required to overcome the frictional resistances of the fittings on the line for the circuit under consideration. The difference between the available head and the head required to overcome the fitting friction will be the head available for overcoming pipe friction. If this difference is divided by the length of the pipe circuit, the result will be the friction loss R per foot of pipe To comply with the units used in Chart 1 the value of R so found must be multiplied by 1,000 and is then ascertained in the first column. Proceeding horizontally to the quantity of heat to

[1] "Piping Systems in Heating and Ventilation," from KUTHE, "Die Rohrnetze in der Heiz- und Luftungstechnik," *Gesundh -Ing* , Berlin, 1916

[2] Each circuit is subdivided into pipe sections By the term is meant a length of pipe in which the diameter, heat transmitted, and the heat loss (when heat loss is considered) remains unchanged.

be transmitted for the pipe section in question, the desired pipe diameter will be found at the top of the column As will be shown by later examples the remaining circuits are proportioned by similar methods

(*b*) Check Computation for Construction Purposes —In making the check computation, it is important to note three factors, namely, the pipe size as determined in the preliminary estimate, the quantity of heat to be transmitted by the circuit in question and finally the circulating pressure head The main problem is to insure a circulating pressure head at least equivalent to the sum of the resistances of the pipes and fittings. The need appears therefore for the value of the velocity v, the friction loss R per lineal foot of pipe, and the sum of the coefficients $\Sigma\zeta$ of the fitting resistances installed in the circuit Turning again to Chart 1, proceed vertically along the column corresponding to the assumed pipe diameter until the heat transmission of the pipe is reached Directly below this will be found the corresponding velocity of the water Proceeding horizontally to the extreme left the friction-loss column will be found. It should be remembered that these values of R are given in units of thousandths of an inch of water column for one foot of pipe.

R is to be multiplied with l, which is the length (in feet) of the pipe section in question The next step is to determine the sum of the resistances for tees, valves, and other pipe fittings installed on the pipe section under consideration To do this the small table to be found directly underneath the fitting resistance table of Chart 1 will be needed. With the values of ζ so found attention is then directed to the second division of Chart 1 In this table the water velocities have been combined with the resistance coefficients ζ in a manner to allow immediate reading of the total head necessary to overcome fitting resistances Finally the sum of the values $lR + Z$ computed for the complete circuit should be equal or somewhat less than the available circulating pressure head If the conditions are not attained at first trial, then one or more of the pipe sections should be changed until the conditions are met. Examples to follow will illustrate the method

(*c*) *Influence of Heat Loss from Mains* —In hot-water systems with basement distributing mains it is found that the heat loss from the mains causes a slight decrease in the circulating head This is not very essential, and in view of several other minor influences, not usually included in the computations, the heat loss of those mains may be neglected

On the other hand for overhead distribution the circulating forces are appreciably augmented, and the effect should be included in the design. This influence will now be discussed

(1) Preliminary Design for Estimating Purposes —In a manner similar to the previous discussion the circulating head is first determined without regard to the heat losses. For the usual types of installations

the increase in the circulating head caused by the cooling effect of the piping is given in Table XIV at the end of this book The total head available to cause circulation is therefore equal to the calculated head plus the additional head as taken from Table XIV. The remainder of the computation is identical with that indicated under subdivision *a* on page 197 relative to preliminary layout It is necessary to mention, however, that when considering the heat losses of the piping, due attention must be paid to the enlargement of the radiator heating surfaces (see Table XIV*b*) necessitated by the ensuing decrease of the radiator temperature drop

(2) Check Computation for Construction Purposes —For the check computation the cooling effect of each pipe section must be considered, also the circulating head which is caused thereby, and the heating system proportioned accordingly To determine the temperature drop through a pipe section, the following equation may be used·

$$\delta = \frac{lk(1-\eta)(t_f-t_z)}{l'Q}, \tag{48}$$

in which

δ = temperature drop in degrees Fahrenheit of the water flowing in the pipe section

$\frac{lk}{l'}$ = quantity of heat emitted by the uninsulated pipe section for 1° temperature difference between the averages of the water and surrounding air temperatures. Values of $\frac{k}{l'}$ may be taken from Table I at the end of this book

η = efficiency of the insulation expressed as a ratio

t_f = temperature of the water entering the pipe section in degrees Fahrenheit

t_r = temperature of the water leaving the pipe section in degrees Fahrenheit

t_z = temperature of the surrounding air in degrees Fahrenheit. For an uninsulated pipe running through a room, it will be the room temperature For an insulated pipe run in chases in the wall, it is assumed as 100° F. For an uninsulated pipe running through chases in the wall, it is assumed as 115° F.

Q = quantity of water flowing through the pipe section in pounds per hour

It would be more accurate in the above equation to replace the value of t_f by the expression $\frac{t_f + t_r}{2}$ which is the average water temperature The complication, however, is not justified by the change, and for this reason Eq 48 as written will be used.

The computations may best be carried out by some form of tabulation for clearness. The following is an example:

Number	Q	l	d	$\frac{k}{l'}$	$\frac{lk}{l'}$	$1 - \eta$	t_f	t_s	$t_f - t_s$	δ	t_r	t_m	h'

In the above tabular form $t_m = \frac{t_f + t_r}{2}$. Also h' = circulating head for the circuit in question, and is obtained by multiplying the vertical height of the circuit by 0.192 $(s_m - s')$ where s_m and s' are the average densities of the water in the return and flow risers, respectively.

For risers of usual lengths properly insulated, the heat loss is comparatively small and may be neglected. In the exceptional cases of long risers the temperature drop is determined as previously shown.

As a particular case consider Fig. 167 in which several return risers connect with a return main. Assume that the water from the first return riser enters the main return at A with a temperature of 140° F. at a rate of 18 cu. ft. per hour. In flowing from A to B the water will lose heat, and this will be indicated by a drop in temperature. According to Eq. 48, assuming the surrounding air temperature to be 50° F., the efficiency of the insulation to be 60 per cent (*i.e.*, 0.60 for use in equation, 48) the temperature drop will be about 1.2° F. At this place an additional 18 cu. ft. or 18 × 61.37 = 1,105 lb. of water enter from the return riser II at 140° F. to mix with that flowing from A. The temperature of the mixture will be less than 140 and greater than 138.8° F. The temperature of the mixture is found as follows: Assume the temperature of the mixture to be x° F. Then water flowing from A will be heated to this temperature necessitating 1,105 × $(x - 138.8)$ B.t.u. This heat will be taken from the warmer water of the second return riser, thereby causing its temperature to drop $(140 - x)$° F. In other words the heat given up by the water of the return riser a B must equal the heat absorbed by the water at this place from the riser I. Expressing this in the form of an equation

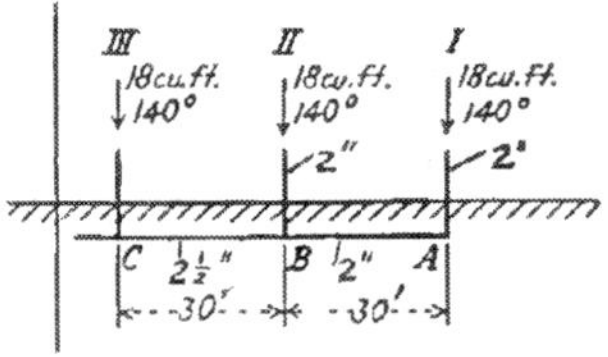

FIG. 167.

$$1{,}105 \times (x - 138.8) = 1{,}105 \times (140 - x). \qquad (49)$$

Cancelling the 1,105 from both sides and transposing,

$$2x = 140 + 138.8.$$
$$x = 139.4° \text{ F.}$$

In flowing from B to C the water cools to a temperature of about 138.7°, and on mixing with the water from the third riser, it becomes reheated to a temperature of about 139° F. In actual practice conditions are still more favorable than those chosen, for the following reasons:

(1) The temperature at the ceiling of the cellar is usually higher than 50° F.

(2) The insulation of the return main will have a higher efficiency than 60 per cent.

(3) The return risers nearer the boiler deliver water at a higher temperature than assumed in the above calculation.

With the slight vertical height of the return main in mind and considering the above, it is obvious that for most practical cases the heat loss from the main return may be neglected. For unusual cases the heat loss may be considered by means of Eq. (48).

Having determined the circulating head by means of the method described above, it remains to compare this head with the one found in the assumption and to determine whether the condition

$$H = \Sigma(lR + Z)$$

is valid for the pipe sizes assumed. Should this not be the case alterations of the diameters in one or more of the circuits may eliminate slight departures. Data furnished by flow and return temperatures also give the basis upon which the heating surface of the radiators is proportioned. These matters will receive additional attention in examples to follow.

2. One-pipe Gravity Hot-water Heating System.

a. Heat Losses Not Included. (1) Circulating Head. (a) Temperature Computations.—In Fig. 168 let t' represent the flow temperature and t'' the return temperature of a one-pipe hot-water heating system. On the assumption that there is no heat loss from the piping, the temperature of the water in sections 4, 5, and 6 will be the same as the boiler flow temperature. Similarly $t_3 = t''$ and $t_8 = t_9$. In order to determine the temperatures t_7 and t_{10}, the temperature drop of the radiators must be assumed. These may not be chosen at random but on the contrary must satisfy a definite relationship to the temperature drop of the entire system $(t' - t'')$, to the total heat emission ΣW of all the radiators connected to the same main, and W_h the largest heat output of any radiator on the riser. If ΔH represents the temperature drop through a radiator, then ΔH is chosen so that

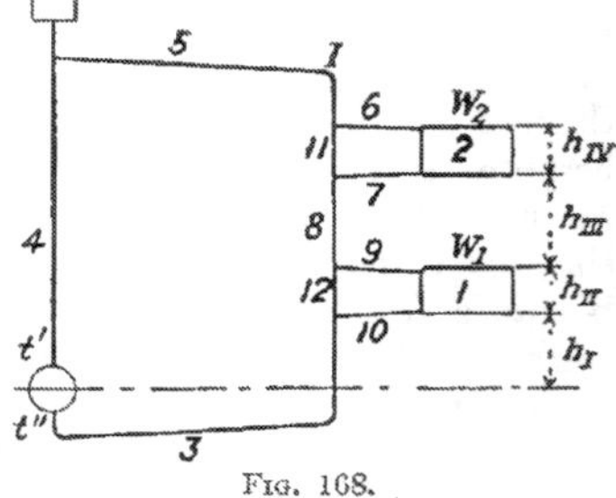

FIG. 168.

$$\Delta H > \frac{W_h}{\Sigma W}(t' - t''). \quad (50)$$

After deciding upon the temperature drop in accordance with the above, the temperatures t_7 and t_{10} are found by subtraction from t_6 and t_9 respectively.

It is still necessary to compute the temperature t_8 For this purpose use is made of the relation

$$W_1 + W_2 = \Sigma W = Q(t' - t'')$$

where Q represents the amount of water flowing in the system in pounds per hour Similarly the heat output of radiator 2 is

$$W_2 = Q(t' - t_8)$$

and for radiator 1

$$W_1 = Q(t_8 - t'').$$

Consequently

$$t_8 = t'' + \frac{W_1}{Q}, \quad \text{or} \quad t_8 = t' - \frac{W_2}{Q}. \tag{51}$$

In this manner all temperatures throughout the system are determined

(b) Circulating Head for the Circuit of Riser I —In conformity with previous discussions the circulating head H in inches of water column pressure is

$$H = 0.192\{h_{\text{I}}(s'' - s') + h_{\text{II}}(s_1 - s') + h_{\text{III}}(s_8 - s') + h_{\text{IV}}(s_2 - s')\} \tag{52}$$

in which

s'' = density of water in pounds per cubic foot at t''° F.

s_1 = density of water in pounds per cubic foot at t_1° F, i e, the average temperature of water in radiator 1 (see Fig. 168).

s_8 = density of water in pounds per cubic foot at t_8° F, i e., the average temperature of the water in section 8 of the piping system.

(c) Circulating Head for the Circuit of Radiator 1 —By the circuit at radiator 1 is understood flow in sections 9, 10, and 12 (Fig 168) of the piping system. Due to the loss of heat from the radiator, it produces a circulation as a result of the increase in density of the water within the radiator There is also a short circuit of water through section 12 of the piping system, the frictional resistance of which acts as an additional circulating force for the radiator circuit and must be added to its pressure head Expressing this in formula,

$$H_{\text{radiator circuit}} = 0\ 192h_{\text{II}}(s_1 - s_{12}) + (lR + Z)_{12} \tag{53}$$

(2) Preliminary and Check Computation of Piping —The available pressure head is determined by the method indicated above The preliminary and check computations of the pipe sizes may then be carried out as outlined previously (p 197)

b *Piping Computations When Heat Losses Are Included* —The method of designing one-pipe systems is similar to that for two-pipe systems In using Table XIV attention is directed to the fact that the radiators are replaced by the risers, therefore the values taken from the table must correspond to the radiator at or nearest the center of the vertical distance between the boiler and the highest radiator It has been determined in tests that the influences in one-pipe heating systems are such that only one-half of the tabular values need to be taken in computation

3. Loft Heating Systems.

When a single story is to be heated, the circulating head is due solely to the heat losses from the piping and the radiators. The computations will be indicated on page 222 with the exception, however, that Table XV is to be used in assuming the circulating head The usual method is to be followed for the preliminary and checking computation of the piping

4. Greenhouse Heating.

The mode of procedure in the design of greenhouse heating systems is similar in every respect to corresponding hot-water heating systems for other purposes Mention should be made of the fact that the coefficient of resistance of the pipe coils must not be taken as equivalent to that of radiators The resistance of pipe coils should be computed by means of the formula previously discussed:

$$H = \Sigma(lR + Z) \tag{54}$$

5. Accelerated Circulation Hot-water Systems.

Special systems of hot-water heating in which circulation is augmented[1] by any means are designed much the same way as ordinary systems. The principal difference, however, lies in determining the circulating head These data are usually furnished by the manufacturers of such systems. Should Chart 1 at the end of the book prove inadequate, Chart 2 may be used

6. Pump Heating Systems.

In systems using pumps to accelerate the flow, the circulating head has components, namely, (1) circulating head due to the pump pressure and (2) gravity head of the system The total head H expressed in inches of water column pressure is

$$H = H_p + H_g \tag{55}$$

where H_p and H_g are the heads due to the pumps and to gravity

In some cases the head due to gravity may be neglected when it is small in comparison to the pumping head.

For all systems, however, the resistance due to pipe and fittings must be equal to or less than the available circulating head, or

$$H_p \text{ or } H = \Sigma(lR + Z). \tag{56}$$

It is to be noted that the circuit to each radiator includes the pump and its connections The influence of fitting resistances is important in connection with pump systems. Whereas the fitting resistance of long

[1] See also Pump Heating Systems

mains amounts to from 10 to 20 per cent of the entire resistance of the system, according to Table XVI, from 70 to 90 per cent of the circulating head is used to overcome the fitting resistances at the pumping station. It is therefore of utmost importance that the resistances at the pumping station be reduced as much as possible. This may be done by avoiding abrupt changes in cross-sections, using gate valves instead of ordinary globe valves, etc. Owing to the high water velocities in pumping systems, Chart 1 may prove inadequate, in which case Chart 2 furnishes additional data.

7. Numerical Examples.

a. Two-pipe Hot-water Heating System. (1) Without Regard to Heat Losses.

Example, 11.

Determine the pipe sizes for a basement main hot-water heating system. The general layout is that sketched in Fig. 169. Assume boiler flow temperature to be 180° F. and the temperature drop of the radiators to be 30° F. The lengths of the various pipe sections may be taken from the Summary A on page 208.

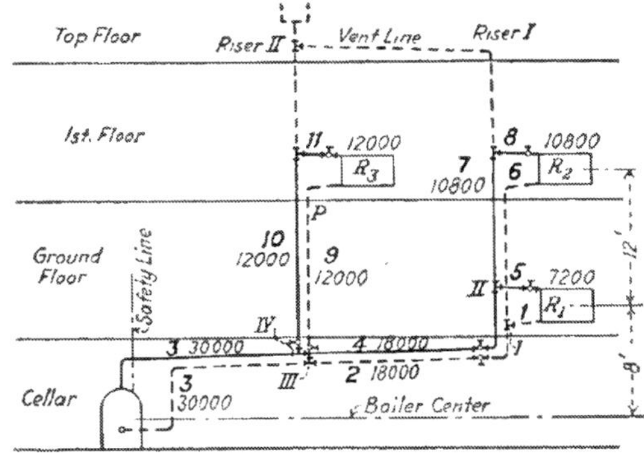

FIG. 169.

Computation: 1. *Assumption of Pipe Sizes. a.* Least-favored Circuit.—As previously noted, the computation is first made for the least-favored radiator. By this is meant the radiator nearest to the boiler in a vertical direction and the remotest from the boiler in a horizontal direction. In the example chosen this is obviously radiator 1, consisting of the circuit 1, 2, 3, 4, and 5. The available circulating head (pressure head) for the circuit is found from Eq. 37, namely,

$$H = 0.192h(s'' - s') \text{ in. w.c.,}$$

in which for the example at hand

h = vertical height from center of boiler to center of radiator in feet = 8 ft.
s' = density of water in flow riser at 180° F. (= 60.58 from Table XIII).
s'' = density of water in return riser at 180 − 30 = 150° F. (= 61.19 from Table XIII).
0.192 = factor to convert pounds per square foot into inches of water column.

Hence on substitution in above equation,

$$H = 0.192 \times 8 \times (61.19 - 60.58) = 8 \times 0.117$$
$$= 0.94 \text{ in. w.c.}$$

Allowing 50 per cent for fitting resistances (Table XVI)	= 0.47	in. w.c.
Balance available for pipe friction	= 0.47	in. w.c.
Length of circuit 1, 2, 3, 4, and 5	= 115	ft.
Friction loss per foot of pipe $= \frac{0.47}{115}$	= 0.00409 in. w.c.	
or in $\frac{1}{1,000}$ in. w.c.	= 4.09	

Consult the table to the right in Chart 1, find in the column to the extreme left the nearest value to 4.09, and proceed horizontally to the column giving the heat transfer of the pipe in question. At the head of this column will be found the required pipe diameter. For example, from Fig. 169, it will be seen that 18,000 B.t.u. per hour must be supplied by pipe section 2. In the table the nearest value to 4.09 is 4.0 (*i.e.*, in thousandths of an inch of water column). Proceeding horizontally, the table shows that 15,600 B.t.u. are transmitted by a 1¼-in. pipe with a velocity of 0.28 ft. per second and a friction loss per lineal foot of 0.004 in. w.c. For a 1½-in. pipe, however, with the same friction loss per lineal foot, 25,100 B.t.u. is transmitted with a water velocity of 0.3 ft. per second. The required pipe size of section 2 is somewhat larger than 1¼ and smaller than 1½ in. Since commercial piping is to be had only in standard sizes, one of the above diameters must be chosen. If 18,000 B.t.u. (or next higher tabular value) is to be transmitted by both 1¼- and 1½-in. pipes, the following result is attained.

Pipe Size	Friction Loss R	Velocity v
1¼ in.	0.006 in. w.c.	0.36 ft. per second
1½ in.	0.0023 in. w.c.	0.24 ft. per second

As will be seen the friction loss for a 1¼-in. pipe is higher than permissible and that for the 1½-in. pipe is lower than that required in the computation. Therefore the 1½-in. pipe is to be chosen and the reserve may be used to offset departures in selecting the pipe sizes for the remaining sections. In a similar manner the pipe sizes, friction losses, and velocities are found for the remaining radiators. These data are compiled in Summary A on page 208.

b. Circuit for Radiator 2.—Neglecting the heat loss from the piping, there are then the same flow and return temperatures as in the circuit for radiator 1. The available circulating head for 20-ft. elevation is

$H = 20 \times 0.117$	= 2.34 in. w.c.
Allowing 50 per cent for fitting resistances (see Table XVI) .	= 1.17 in. w.c.
Remaining head available for pipe friction	= 1.17 in. w.c.

The circuits for radiators 1 and 2 have sections 2, 3, and 4 in common. These have been considered in connection with the circuit of radiator 1. As found there, the pipe friction was 0.00409 in. w.c. per lineal foot of pipe so that the total head for 105 ft. = 105 × 0.00409 = 0.429 in. w.c., the remaining head is 1.17 − 0.429 = 0.741 in. w.c. for sections 6, 7, and 8.

The length of sections 6, 7, and 8 is 34 ft., so that the frictional resistance per lineal foot of pipe is $\frac{0.741}{34} = 0.0218$ in. w.c.

Second Method.—The same result may be obtained in a different way. Assume a plane passed horizontally through the center of the radiator 1. Consider this as a zero plane and let friction heads be considered positive when circulation is directed toward this plane and negative when directed away from the plane. In approaching the zero plane from junction II, Fig. 169, pipe section 5 must be traversed necessitating a circulating head of 5 × 0.00409 = 0.02045 in. w.c. From the zero plane, flow takes place through pipe section 1 requiring a circulating head of −5 × 0.00409 = −0.02045 in. w.c. The difference of the circulating heads between junction points I and II will result not only for the available head for sections 1 and 5 but also for sections 6, 7, and 8. Therefore

Available head at junction I and II	0.0409 in. w.c.
Additional head due to elevation of radiator 2 above radiator 1 = 12 × 0.117	= 1.404 in. w.c.
Allowing 50 per cent (Table XVI) for fitting resistances	= 0.702 in. w.c.
Remaining head	= 0.702 in. w.c.
Total circulating head = 0.702 + 0.0409	= 0.743 in. w.c.
Total length of circuit 6, 7, and 8	= 34 ft.
Friction loss per foot of pipe	= 0.0218 in. w.c.

The next step is to tabulate the corresponding friction losses, velocities, and fitting resistances in Summary A.

c. Circuit of Radiator 3.—Since this radiator has the same height above the center of the boiler as radiator 2 and since also the same temperature differences prevail, therefore,

Circulating head for radiator 3	= 2.34 in. w.c.
Allowing 50 per cent for fitting resistances	= 1.17 in. w.c.
Remaining head available for pipe friction	= 1.17 in. w.c.
Radiator 3 has pipe section 3 in common with radiators 1 and 2, and since its length is 47 ft., it requires a head of 47 × 0.00409	= 0.192 in. w.c.
Hence remaining head available for sections 9, 10 and 11	= 0.978 in. w.c.
Consequently the friction loss per foot of pipe $\frac{0.978}{42}$	= 0.0233 in. w.c.

As in the previous case the pipe sizes are chosen from Chart 1.

Second Method.—If the friction head required by pipe sections 2 or 4 be added the head previously determined at junction point II, the sum will equal the available circulating head at the junction point III (see Fig. 169).

Therefore head[1] at junction point III = −0.02045 − (29 × 0.00409)	= −0.139 in. w.c.
Head at junction point IV	= +0.139 in. w.c.
Available head is therefore 0.139 − (−0.139)	= +0.278 in. w.c.
Head due to height of radiator 3 above radiator 1 = 12 × 0.117	= 1.404 in. w.c.
Deducting 50 per cent for fitting resistances	= 0.702 in. w.c.
Available head remaining	= 0.702 in. w.c.
Additional head from junction points III and IV........	= 0.278 in. w.c.
Friction head available for pipe sections 9, 10, and 11 . .	= 0.980 in. w.c.
Length of pipe sections = 42 ft.; hence friction loss per foot of pipe	0.0233 in. w.c.

[1] In the above the positive sign is chosen when the junction point lies before the zero plane of the radiator as determined by the direction of flow.

This completes the computations required for the assumption of the pipe sizes In general the computations above would suffice for estimating purposes For actual installation, however, it is desirable to use a check computation which will be discussed in what follows.

Check Computations for Piping: It will be necessary to determine the magnitude of the fitting resistances in each of the pipe sections The more important values of ζ for pipe fittings used in heating practice are compiled in Chart 1

For each section of piping the values of ζ are determined and tabulated in Summary A (p 208) For the present purpose it will be assumed that all radiator connections are as follows

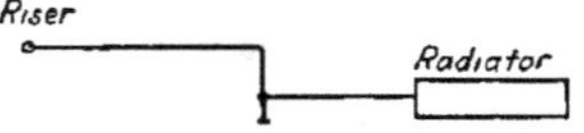

It is customary to assign one-half of the radiator resistance each to the flow and to the return. For pipe section 1 the resistances according to the sketch are

Half radiator resistance	ζ = 1 5
2 1-in elbows	= 3 0
1 tee branch	= 1 5
	$\Sigma\zeta$ = 6 0

Similarly for pipe section 2

1 1½-in elbow	ζ = 1 0
1 1½-in riser valve	= 8 0
1 tee straightway	= 1 0
	$\Sigma\zeta$ = 10 0

Having determined the sum of the resistances of the fittings for each pipe section, the values of Z may be determined with the aid of Chart 1 It now remains to be seen if the values of $[lR + Z]$ as given in Summary A coincide sufficiently with the total circulating head as found in the preliminary computation Should this not be the case, one or more of the pipe sizes of the sections should be altered until the condition is met.

SUMMARY A

Number of pipe section	Quantity of heat, B.t.u.	Length l in feet	Assumed d in inches	Computation — Original values: R per ft., in. w.c.	v, ft. per sec.	lR, in. w.c.	$\Sigma\zeta$	Z, in. w.c.	Corrected values: d, in.	R per ft., in. w.c.	v, ft. per sec.	lR, in. w.c.	$\Sigma\zeta$	Z, in. w.c.	Difference: lR $g-m$	Z $i-o$
(a)	(b)	(c)	(d)	(e)	(f)	(g)	(h)	(i)	(j)	(k)	(l)	(m)	(n)	(o)	(p)	(q)

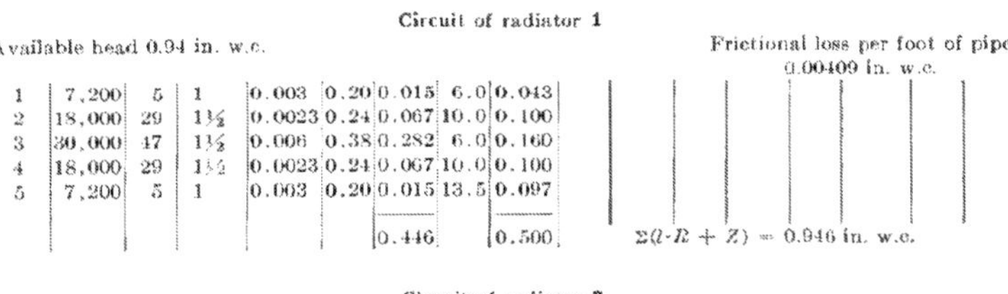

Circuit of radiator 1

Available head 0.94 in. w.c.

Frictional loss per foot of pipe 0.00409 in. w.c.

(a)	(b)	(c)	(d)	(e)	(f)	(g)	(h)	(i)	(j)	(k)	(l)	(m)	(n)	(o)	(p)	(q)
1	7,200	5	1	0.003	0.20	0.015	6.0	0.043								
2	18,000	29	1½	0.0023	0.24	0.067	10.0	0.100								
3	30,000	47	1½	0.006	0.38	0.282	6.0	0.160								
4	18,000	29	1½	0.0023	0.24	0.067	10.0	0.100								
5	7,200	5	1	0.003	0.20	0.015	13.5	0.097								
						0.446		0.500								

$\Sigma(l \cdot R + Z) = 0.946$ in. w.c.

Circuit of radiator 2

Available head 2.34 in. w.c.

Frictional loss per foot of pipe 0.0218 in. w.c.

(a)	(b)	(c)	(d)	(e)	(f)	(g)	(h)	(i)	(j)	(k)	(l)	(m)	(n)	(o)	(p)	(q)
6	10,800	17	1	0.007	0.33	0.119	7.0	0.140								
7	10,800	12	¾	0.026	0.55	0.312	1.0	0.055								
8	10,800	5	¾	0.026	0.55	0.130	14.0	0.770								
						0.561		0.965								

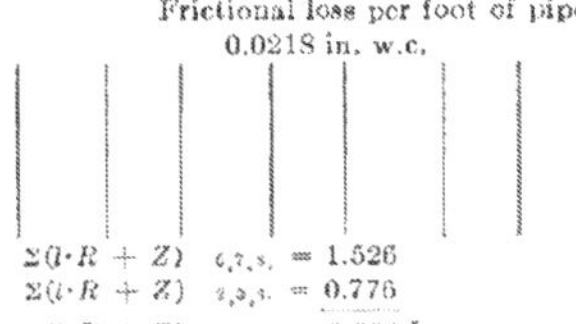

$\Sigma(l \cdot R + Z)_{6,7,8} = 1.526$

$\Sigma(l \cdot R + Z)_{2,3,4} = 0.776$

$\Sigma(l \cdot R + Z) = 2.302$ in. w.c.

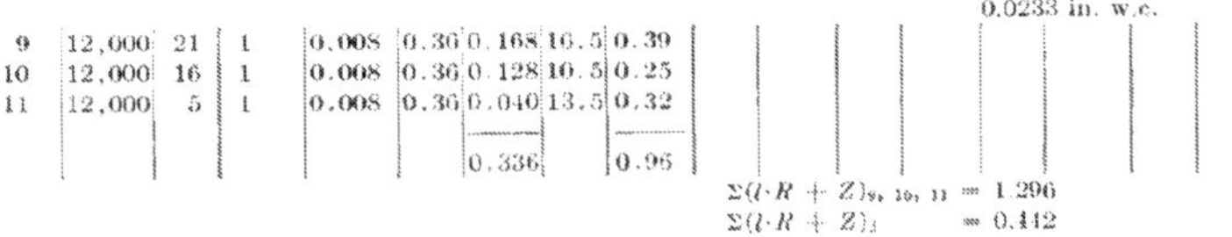

Circuit of radiator 3

Available head 2.34 in. w.c.

Frictional loss per foot of pipe 0.0233 in. w.c.

(a)	(b)	(c)	(d)	(e)	(f)	(g)	(h)	(i)	(j)	(k)	(l)	(m)	(n)	(o)	(p)	(q)
9	12,000	21	1	0.008	0.36	0.168	16.5	0.39								
10	12,000	16	1	0.008	0.36	0.128	10.5	0.25								
11	12,000	5	1	0.008	0.36	0.040	13.5	0.32								
						0.336		0.96								

$\Sigma(l \cdot R + Z)_{9,10,11} = 1.296$

$\Sigma(l \cdot R + Z)_3 = 0.442$

1.738 in. w.c. $\cong$ 1.74.

It will be noticed that in the case of the current circuit of radiator 1 the available head is less than 1 per cent smaller than the value of $\Sigma(lR + Z)$ which is sufficiently accurate. In all other cases the available head is larger than $\Sigma(lR + Z)$; therefore no changes in pipe sizes are necessary.

Example 12.

Assume a piping layout as shown in Fig. 170 having a basement distribution system. Let the heat loss be negligible. Assume that the flow temperature of the water leaving the boiler is 180° F. and that the temperature drop through the radiators is 30° F. Additional data are to be taken from the summary, and the radiator connections are in accordance with the illustration on page 207. Determine the pipe sizes.

1. *Preliminary Computations of the Piping.* *a.* The Least-favored Circuit.

The length of pipe sections 1, 2, 3, 4, 5, 6, 7, 8, 9, 10, 11, 12, 13 and 14 . . . . = 226 ft.

The circulating pressure head for this circuit is $0.192 \times 10(s_{180} - s_{150}) = 0.192 \times 10 \times 0.61 = 10 \times 0.117$ . . . . = 1.17 in. w.c.

Deducting 50 per cent (Table XVI) for resistances of fittings . . . . = 0.585 in. w.c.

Available head for pipe friction . . . . = 0.585 in. w.c.

Frictional loss per foot of pipe $= \frac{0.585}{226}$ . . . . $\cong$ 0.00259 in. w.c.

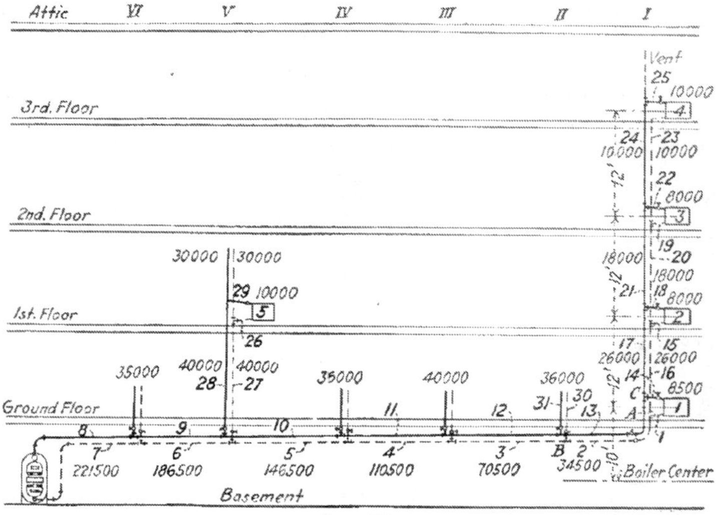

Fig. 170.

From these data the pipe diameters of the circuit may be approximated from Chart 1 and entered into Summary B.

b. Circuit of Radiator 2.

Available circulating head of circuit $= 22 \times 0.117$ . . . . $\cong$ 2.57 in. w.c.

Deducting 50 per cent (Table XVI) for resistance of fittings . . . . $\cong$ 1.285 in. w.c.

Available head for pipe friction . . . . $\cong$ 1.29 in. w.c.

Radiator 2, however, has pipe sections 2, 3, 4, 5, 6, 7, 8, 9, 10, 11, 12, and 13 in common with radiator 1 which require a head of 216×0.00259 . . . . $\cong$ 0.559 in. w.c.

Available head for pipe sections 15, 16, 17, and 18 . . . . $\cong$ 0.73 in. w.c.

Length of pipe sections 15, 16, 17, and 18 . . . . = 30 ft.

Frictional resistance per foot of pipe $= \frac{0.73}{30}$ . . . . = 0.0243 in. w.c.

c. Circuit of Radiator 3.

Circulating head for the circuit $= 34 \times 0.117$ . . . . = 3.98 in. w.c.

Deducting 50 per cent for resistance of fittings . . . . = 1.99 in. w.c.

Available head for pipe friction . . . . = 1.99 in. w.c.

This circuit has the pipe sections 2 to 13, 16, and 17 in common with the circuits previously considered The head necessary to overcome the frictional resistance of these pipe sections is as follows

Pipe section 2 to 13	= 0 559	in w c
Pipe section 16 and 17 = 24 × 0 0243	= 0 583	in w.c
	1 142	in. w c.
Head available for pipe sections 19, 20, 21, and 22 = 1 99 − 1 14	= 0 85	in w c
Frictional resistance per foot of pipe = $\frac{0\ 85}{30}$	= 0 0283	in w c

Proceeding in a similar way for radiators 4 and 5, the following results are obtained

Radiator	Circulating Head	Frictional Loss per Foot of Pipe
4	5 38 in w c	0 029 in w c
5	2 57 in w c	0 0297 in w c

The pipe diameters may now be approximated as outlined previously, and entered in the summary B

Checking Computation of the Piping

RESISTANCES OF THE FITTINGS

Pipe section 1, d = 1¼ in		Pipe section 8, d = 4 in.	
Half radiator	ζ = 1 5	Balance of boiler	ζ = 1 0
2 long-sweep ells	= 2 0	Long-sweep ell	= 0 5
Tee (branch)	= 1 5	Tee (straightway) .	= 1 0
	$\Sigma\zeta$ = 5 0		$\Sigma\zeta$ = 2 5
Pipe section 2, d = 2 in		Pipe section 9, 10, 11, and 12	
Riser valve	ζ = 7 0	Tee (straightway) .	$\Sigma\zeta$ = 1 0
Tee (straightway)	= 1 0	Pipe section 13, d = 2″	
Long-sweep ell	= 0 5	Tee (straightway) .	ζ = 1 0
		Riser valve	= 7 0
	$\Sigma\zeta$ = 8 5	Long-sweep ell	= 0 5
Pipe section 3, d = 2½ in			
Tee (straight-way)	ζ = 1 0		$\Sigma\zeta$ = 8 5
Pipe sections 4, 5, and 6		Pipe section 14, d = 1¼ in	
Tee (straighway)	$\Sigma\zeta$ = 1 0	Tee (branch) .	ζ = 1 5
Pipe section 7, d = 4 in		Angle valve .	= 9 0
2 long-sweep ells	ζ = 1 0	Long-sweep ell	= 1 0
Part of boiler	= 1 5	Half radiator	= 1 5
	$\Sigma\zeta$ = 2 5		$\Sigma\zeta$ = 13 0

In a similar manner the resistances of the other fittings for circuits are found and recorded in Summary B When ascertaining the value $\Sigma(lR + Z)$ for each current circuit, it will be found that of the 29 pipe sections, only sections 5, 16, and 19 will require corrections The total cost of the installation as approximated from the preliminary computations differs only slightly from the actual cost of the installation as determined from the corrected (final) computations

SUMMARY B

Number of pipe section	B.t.u. required	Length, feet	Diameter, inches	Computation												
				Original values					Corrected values						Difference	
				R per ft.	v	lR	Σζ	Z	d, in.	R per ft.	v	lR	Σζ	Z	lR	Z
(a)	(b)	(c)	(d)	(e)	(f)	(g)	(h)	(i)	(j)	(k)	(l)	(m)	(n)	(o)	(p)	(q)
Current circuit radiator 1																
Available pressure head, 1.17 w.c.										Frictional loss per foot of pipe, 0.00259						
1	8,500	5	1¼	0.0014	0.16	0.007	5.0	0.023								
2	34,500	20	2	0.0018	0.26	0.036	8.5	0.104								
3	70,500	12	2½	0.0023	0.33	0.028	1.0	0.020								
4	110,500	16	3	0.0020	0.36	0.032	1.0	0.024								
5	146,500	24	3	0.0035	0.50	0.084	1.0	0.045	3½	0.0016	0.33	0.038	1.0	0.02	0.046 +	0.025
6	186,500	12	3½	0.0026	0.45	0.031	1.0	0.037								
7	221,500	26	4	0.0018	0.40	0.047	2.5	0.073							0.071	
8	221,500	22	4	0.0018	0.40	0.040	2.5	0.073								
9	186,500	12	3½	0.0026	0.45	0.031	1.0	0.037								
10	146,500	24	3	0.0035	0.50	0.084	1.0	0.045								
11	110,500	16	3	0.0020	0.36	0.032	1.0	0.024								
12	70,500	12	2½	0.0023	0.33	0.028	1.0	0.020								
13	34,500	20	2	0.0018	0.26	0.036	8.5	0.104								
14	8,500	5	1¼	0.0014	0.16	0.007	13.0	0.060								
$\Sigma(l \cdot R + Z)^{1,14} =$ 0.523 + 0.689 = 1.21 in. w.c. − 0.071																
$\Sigma(l \cdot R + Z)^{1,14} =$ 1.14 in. w.c. which corresponds sufficiently with head available																
Circuit radiator 2																
Available head, 2.57 in. w.c.										Frictional loss per foot = 0.0243						
15	8,000	3	¾	0.016	0.4	0.048	6.0	0.170								
16	26,000	12	1¼	0.012	0.5	0.144	1.0	0.045	1	0.035	0.8	0.420	1.0	0.12	+0.276	0.075
17	26,000	12	1¼	0.012	0.5	0.144	1.0	0.045								
18	8,000	3	¾	0.016	0.4	0.048	13.5	0.415							0.35	
$\Sigma(lR + Z)_{15}^{18} =$ 0.384 + 0.675 = 1.06																
$\Sigma(lR + Z)_{2}^{13}$ = 1.04																
2.10 in. w.c. + 0.35 = 2.45 in. w.c.																
Circuit radiator 3																
Available head, 3.98 in. w.c.										Frictional loss per foot, 0.0283 in. w.c.						
19	8,000	3	¾	0.016	0.4	0.048	6.0	0.170	½	0.12	0.9	0.36	6.0	0.88	+0.312	+0.71
20	18,000	12	1	0.016	0.5	0.192	1.0	0.045								
21	18,000	12	1	0.016	0.5	0.192	1.0	0.045							1.02	
22	8,000	3	¾	0.016	0.4	0.048	13.5	0.415								
$\Sigma(lR + Z)_{19}^{22}$ = 0.480 + 0.675 = 1.16																
$\Sigma(lR + Z)_{2}^{13}$ = 1.04																
$\Sigma(lR + Z)_{16}^{17}$ = 0.73																
2.93 + 1.02 from pipe section change = 3.95 in. w.c.																
Current circuit radiator 4																
Available head, 5.38 in. w.c.										Frictional loss per foot = 0.029 in. w.c.						
23	10,000	15	¾	0.023	0.5	0.345	7.0	0.320								
24	10,000	12	¾	0.023	0.5	0.276	1.0	0.045								
25	10,000	3	¾	0.023	0.5	0.069	13.5	0.610								
$\Sigma(l \cdot R + Z)_{23,25}$ = 0.690 + 0.975 = 1.67 in. w.c.																
$\Sigma(l \cdot R + Z)_{2,13}$ = 1.04																
$\Sigma(l \cdot R + Z)_{16,17}$ = 0.73																
$\Sigma(l \cdot R + Z)_{20,21}$ = 0.47																
$\Sigma(l.R + Z)$ circuit 4 = 3.91 in. w.c. against 5.38 in. w.c. no further change possible.																
Current circuit radiator 5																
Available head, 2.57 in. w.c.										Frictional loss per foot = 0.0297 in. w.c.						
26	10,000	3	¾	0.023	0.5	0.069	6.0	0.27								
27	40,000	14	1½	0.010	0.5	0.140	9.5	0.43								
28	40,000	17	1½	0.010	0.5	0.170	9.5	0.43								
29	10,000	3	¾	0.023	0.5	0.069	13.5	0.61								
$\Sigma(lR + Z)_{26}^{29}$ = 0.448 + 1.74 = 2.19 in. w.c.																
$\Sigma(lR + Z)_{6, 7, 8, \text{ and } 9}$ = 0.37																
Total $\Sigma(lR + Z)$ for current circuit = 2.56 in. w.c. against 2.57 in. w.c. available.																

Example 13.

Assume an overhead distribution system of hot-water heating which is to include the heat losses of the piping. Let the flow temperature at the boiler be 175° F. with a temperature drop of 30° in the radiators. Assume the riser installed in a chase in the wall where the temperature is about 100° F. so that the heat loss from the riser may be neglected because of its insulation efficiency of 85 per cent. Let the distribution mains have an insulation efficiency of 80 per cent and be run in the attic where the temperature is 32° F. In addition let the down-feed risers be located in chases having an air temperature of 100° F. and let the riser insulation efficiency be 60 per cent. The radiators are to be connected as shown on (p. 207). Additional data are to be taken from Fig. 171.

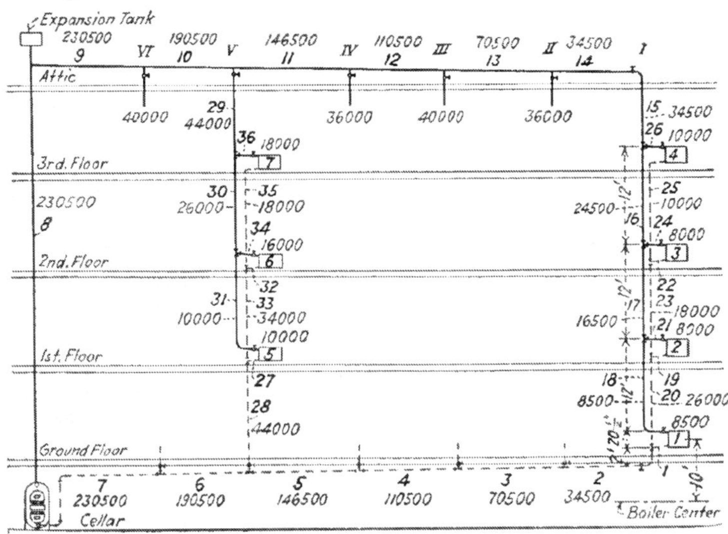

Fig. 171.

1. *Preliminary Design of Piping and Radiators for Estimating Purposes.*—As usual the design is begun by considering the least-favored current circuit, *i.e.*, that of radiator 1 in this case. The circuit consists of pipe sections from 1 to 18.

a. Radiator 1:

s_1 = density of water at 175° F. = 60.68 lb. per cubic foot
s_2 = density of water at 145° F. = 61.28 lb. per cubic foot

$s_2 - s_1$ = 0.6 lb. per cubic foot

The circulating pressure head per foot of pipe is

$$0.192(s_2 - s_1) = 0.192 \times 0.6 = 0.115 \text{ in. w.c.}$$

The head for 10 ft. of pipe exclusive of the heat losses is therefore $10 \times 0.115 = 1.15$ in. w.c. The additional head (Table XIV[1]) for 195° F. flow temperature = 1.0 in. w.c. and for 175° F. = 1.0 − 30 per cent = 0.7 in. w.c. Consequently

[1] The horizontal dimension of the system is about 100 ft.

Total head = 1 15 + 0 7 ..	1 85	in w c
Less 50 per cent for resistances of fittings (Table XVI)	= 0 925	in w c
Available for pipe friction circuit 1 to 18	= 0 925	in w c.
Length of pipe section 1 to 18	= 317 5	ft
Friction per foot of pipe $= \frac{0.925}{317.5}$	= 0 00291	in w c

With the aid of Chart 1 the pipe sizes may be approximated and entered in Summary C

b Radiator 2

Available head for 22 ft of pipe excluding heat loss as for radiator 1 = 22 × 0 115	= 2 53	in w c
Additional head due to heat loss (Table XIV)	= 0 7	in w c
Total head	3 23	in w c
Less 50 per cent for loss through fittings	= 1 61	in. w c
Available head for pipe friction	= 1 62	in w c
Pipe sections 2 to 17 in common with radiator 1 require a head of 295 5 × 0 00291	= 0 86	in w c
Remaining head for pipe sections 19, 20, and 21	= 0 76	in w c
Length of pipe sections 19, 20, and 21	= 22 ft	
Frictional resistance per foot of circuit $= \frac{0.76}{22}$	=0 0345	in w c

In a similar manner the pipe sizes for the remaining circuits are approximated, and with the use of Chart 1 the values are then tabulated in Summary C

c Additional Allowance to the Heating Surface for Estimating the Costs—It is assumed that the heating surface has been determined on a basis which excludes the heat loss of the piping For cost estimation Table XIVB contains additional allowances for correcting the heating surface previously found with respect to the heat emission from the piping For instance in the case of radiator 1 the table shows that its heating surface should be increased 5 per cent

2 *Checking Computation of the Piping and Radiation Surface* *a* Pipe Sizes for Final Design —The approximate pipe sizes have been determined and are entered in Summary C They are to be used in computing the actual circulating heads for which purpose Eq (48) may be used The results of the computations are tabulated in Summary D The notation used was discussed on p 210

The last column of Summary D represents the head due to the pipe sections under consideration For instance in pipe section 15 it is ascertained as follows

Average temperature in pipe section $= \frac{173.3 + 173}{2} \cong 173.2°$ F.

Compared with the corresponding head in the riser

$$H = 0.192h(s' - s'')$$

Length h of pipe section 15 = 10 ft

s' = density[1] at 173 2° F = 60 73 lb per cubic foot

s'' = density[1] at 175 0° F = 60 68 lb per cubic foot

$s' - s''$ = 0 05 lb per cubic foot

$$H = 0.192 \times 10 \times 0.05 = 0.096 \text{ in w c}$$

The circulating heads for the remaining pipe sections are found in a similar manner. In Summary D the actual head for the individual current circuits are found as follows

[1] For practical purposes the densities when accurate within 0 5° F are sufficient (see Table XIII)

b. Radiator 1:

$\Sigma h = h_{15} + h_{16} + h_{17} + h_{18} +$ radiator $1 + h_2 + h_7.$

$= 0.096 + 0.115 + 0.138 + 0.184 + 0.134 + 0.257 + 0.925 \cong 1.85$ in. w.c.

It should be noted that since $\Sigma(lR + Z)^{18} = 1.71$ in. w.c., no change in pipe sizes is necessary.

c. Radiator 2:

$\Sigma h = h_{15} + h_{16} + h_{17} +$ radiator $2 + h_{20} + h_2 + h_7.$

$= 0.096 + 0.115 + 0.138 + 0.124 + 1.52 + 0.257 + 0.925 \cong 3.18$ in. w.c.

This result is in sufficient agreement with the value 2.72 previously found to require no change in the pipe sizes. The remaining circulating heads may be found by the method indicated above.

In determining the percentage increase of the heating surface, when including the heat emission of the piping, it will be observed from Table XVIII that the values of K increase with increasing difference $(t_m - t_r)$. It will be noted from Table XVIII, that the values are tabulated for intervals of 20° F. From the trend shown in the piping computations, it will be seen that the decrease in temperatures is usually small. The values of K may, as a consequence, be considered constant.

The increase of heating surface for two cases where the heat emission is involved follow:

$W = F_1K(t_m - t_r)$ for radiators excluding heat loss of piping.

$W = F_2K(t_{m_1} - t_r)$ for radiators including heat loss of piping.

$$F_2 = \frac{F(t_m - t_r)}{t_{m_1} - t_r}.$$

In accordance with Table XIVB the heating surface should be increased by 5 per cent. A more exact computation as indicated results in 5 per cent, thus checking with the approximation.

Summary C

Number of pipe section	B.t.u. transmitted	Length feet	Diameter, inches	Computation											Difference	
				Original values					Corrected values							
				R per foot	v	lR	$\Sigma\zeta$	Z	d	R per foot	v	lR	$\Sigma\zeta$	Z	lR	Z
(a)	(b)	(c)	(d)	(e)	(f)	(g)	(h)	(i)	(j)	(k)	(l)	(m)	(n)	(o)	(p)	(q)
Circuit of radiator 1. Approximate head = 1.85 in. w.c. Pressure drop per foot = 0.00291 in. w.c. Actual pressure head = 1.85 in. w.c.																
1	8,500	5	1	0.0040	0.24	0.0200	5.0	0.052								
2	34,500	18	2	0.0018	0.26	0.0324	9.0	0.110								
3	70,500	17	2½	0.0023	0.33	0.0391	1.0	0.020								
4	110,500	16	3	0.0023	0.38	0.0368	1.0	0.026								
5	146,500	22	3	0.0035	0.50	0.0770	1.0	0.045								
6	190,500	13	3½	0.0026	0.45	0.0338	1.0	0.037								
7	230,500	21	4	0.0030	0.45	0.0420	2.5	0.091								
8	230,500	54.5	4	0.0020	0.45	0.1090	2.5	0.091								
9	230,500	16	4	0.0020	0.45	0.0320	1.0	0.037								
10	190,500	13	3½	0.0026	0.45	0.0338	1.0	0.037								
11	146,500	22	3	0.0035	0.50	0.0770	1.0	0.045								
12	110,500	16	3	0.0023	0.38	0.0368	1.0	0.026								
13	70,500	17	2½	0.0023	0.33	0.0391	1.0	0.020								
14	34,500	16	2	0.0018	0.26	0.0288	8.0	0.098								
15	34,500	10	2	0.0018	0.26	0.0180	1.0	0.012								
16	24,500	12	1½	0.0040	0.30	0.0480	1.0	0.016								
17	16,500	12	1¼	0.0020	0.22	0.0240	1.0	0.009								
18	8,500	17	1	0.0040	0.24	0.0680	13.0	0.140								

$\Sigma(l \cdot R + Z)_1^{18} = 0.7956 + 0.912 \cong 1.71$ in. w.c.

SUMMARY C.—(*Continued*)

Radiator 2

Approximate head = 3.23 in. w.c. | Pressure drop per foot = 0.0345 in. w.c. | Actual head = 3.18 in. w.c.

19	8,000	5	3/4	0.016	0.4	0.080	6.0	0.17
20	26,000	12	1	0.035	0.8	0.420	1.0	0.12
21	8,000	5	3/4	0.016	0.4	0.080	13.5	0.42

$\Sigma(l \cdot R + Z)_{19}^{21} = 0.580 + 0.71 = 1.29$ in. w.c.
$\Sigma(lR + Z)_{2}^{17}$ from radiator 1 $= 1.43$

2.72 in. w.c.

Radiator 3

Approximate head = 4.61 in. w.c. | Pressure drop per foot = 0.0485 in. w.c. | Actual head = 4.53 in. w.c.

22	8,000	5	3/4	0.016	0.4	0.080	6.0	0.17
23	18,000	12	3/4	0.070	0.9	0.840	1.0	0.15
24	8,000	5	3/4	0.016	0.4	0.080	13.5	0.42

$\Sigma(l \cdot R + Z)_{22}^{24} = 1.000 + 0.74 = 1.740$ in. w.c.
$\Sigma(lR + Z)_{2}^{16} =$ from radiator 1 $= 1.397$
$\Sigma(l \cdot R + Z)_{20} =$ from radiator 2 $= 0.540$

3.677 = 3.68 in. w.c.

Pipe sections 22 or 23 cannot be changed.

Radiator 4

Approximate head = 5.99 | Pressure loss per foot = 0.055 in. w.c. per foot | Actual head = 5.89

25	10,000	17	3/4	0.023	0.5	0.391	7.5	0.34
26	10,000	5	3/4	0.023	0.5	0.115	13.5	0.61

$\Sigma(lR + Z)_{25}^{26} = 0.506 + 0.95 = 1.456$
$\Sigma(lR + Z)_{2}^{15} = 1.33$
$\Sigma(lR + Z)_{23} = 0.99$
$\Sigma(lR + Z)_{20} = 0.54$

Total head ∴ = 4.316 in. w.c. = 4.32

Radiator 5

Approximate head = 2.81 in. w.c. | Pressure loss per foot = 0.0177 in. w.c. | Actual head = 2.91 in.

27	10,000	5	1	0.006	0.30	0.030	5.0	0.082
28	44,000	15	1 1/2	0.012	0.55	0.180	9.5	0.520
29	44,000	11	1 1/2	0.012	0.55	0.132	9.5	0.520
30	26,000	12	1 1/4	0.012	0.50	0.144	1.0	0.045
31	10,000	17	1	0.006	0.30	0.102	14.0	0.230

$\Sigma(lR + Z)_{27}^{31} = 0.588 + 1.307 = 1.90$ in. w.c.
$\Sigma(lR + Z)_{6,7,8,9,10} = 0.54$

2.53 in. w.c.

Circuit of radiator 6

Approximate head = 4.19 in. w.c. | Frictional resistance per foot = 0.06 in. w.c.

32	16,000	3	3/4[a]
33	34,000	12	1
34	16,000	3	3/4

Circuit of radiator 7

Approximate head = 5.57 in. w.c. | Frictional resistance per foot = 0.0702 in. w.c.

35	18,000	15	3/4[a]
36	18,000	3	3/4

[a] Taken from Chart 2.

The circuits of radiators 6 and 7 are dimensioned merely for the cost approximation.

SUMMARY D

Number of pipe section	Pounds per hour	l, feet	d, inches	$\frac{k}{P}$	$1 - \eta$	t_f	t room	δ, degrees Fahrenheit[a]	t_r, degrees Fahrenheit	h
Distributing main										
9	7,683	16	4	2.2	0.2	175	32	~0.1	174.9	
10	6,350	13	3½	2.3	0.2	174.9	32	~0.1	174.8	
11	4,883	22	3	2.0	0.2	174.8	32	~0.3	174.5	
12	3,683	16	3	2.0	0.2	174.5	32	~0.2	174.3	
13	2,350	17	2½	1.7	0.2	174.3	32	~0.4	173.9	
14	1,150	16	2	1.4	0.2	173.9	32	~0.6	173.3	
Downfeed riser I										
15	1,150	10	2	1.2	0.4	173.3	100	~0.3	173	0.096[1]
16	817	12	1½	0.95	0.4	173	100	~0.4	172.6	0.115
17	550	12	1½	0.95	0.4	172.6	100	~0.6	172	0.138
18	283	17	1	0.80	0.4	172	100	~1.4	170.6	0.184[c]
25	333	17	¾	0.60	0.4	143	100	~0.5	142.5	1.47
The 333 lb. of water flowing through pipe section 25 having a terminal temperature of 142.5° mix with 267 lb. of water from radiator 3, having a temperature of 142.6° F. Temperature of mixture according to Eq. (49) = 142.5°.										
23	600	12	¾	0.6	0.4	142.5	100	~0.2	142.3	1.50
The 600 lb. of water flowing through pipe section 23 mixes with the return water from radiator 2. Temperature of mixture = 142.2° F.										
20	867	12	1	0.76	0.4	142.2	100	~0.2	142	1.52
The 867 lb. of water from pipe section 20 mixes with 283 lb. from radiator 1. Temperature of mixture = 141.7° F.										
2	1,150	2[b]	2	1.1	0.4	141.7	100	0.0	141.7	0.257
Since the heat losses from the return main are negligible, 141.7° F. represents the temperature of return at the boiler.										
7	.....	7.2	...	....	...	141.7	...	.....	141.7	0.925
radiator 1	.....	1.7	...	....	...	170.6	...	.....	140.6	0.134
radiator 2	.....	1.7	...	....	...	172	...	.....	142	0.124
radiator 3	.....	1.7	...	....	...	172.6	...	.....	142.6	0.121
radiator 4	.....	1.7	...	....	...	173	...	.....	143	0.118
Downfeed riser V										
29	1,467	11	1½	0.95	0.4	174.8	100	0.2	174.6	0.021
30	867	12	1¼	0.83	0.4	174.6	100	0.3	174.3	0.023
31	333	17	1	0.80	0.4	174.3	100	0.1	174.2	0.046
35	600	15	¾	0.60	0.4	144.6	100	0.3	144.3	1.410
From radiator 6,534 lb. of water mixes with water flowing through pipe section 35. Temperature of mixture = 144.3° F.										
33	1,133	12	1	0.76	0.4	144.3	100	0.1	144.2	1.428
From radiator 5 333 lb. of water mixes with water flowing through pipe section 33. Temperature of mixture = 144.2° F.										
28	1,467	15	1½	0.90	0.4	144.2	100	0.2	144	1.785
radiator 5	.....	1.7	...	....	...	174.2	...	.....	144.2	0.108
radiator 6	.....	1.7	...	....	...	174.3	...	.....	144.3	0.104
radiator 7	.....	1.7	...	....	...	174.6	...	.....	144.6	0.104

[1] See page 200.

[a] Accuracy to one decimal sufficient.

[b] Since the heat loss of the return main may be neglected, only 2 ft. of pipe section 2 is calculated.

[c] It will be noted that under column l the entire length of the pipe section has been given. While this is necessary for determining the temperature drop, only the vertical height of the pipe is used in calculating the pressure head, *e.g.* pipe section 18, 25 vertical height = 12′. For pipe section 7 the vertical height is given in Fig. 171.

b. One-pipe Hot-water Heating System.

Example 14.

In the example to follow it will be assumed that the heat emission from the piping is included. Let also the temperature t' of the water leaving the boiler be 185° F., the temperature drop in all risers to be 30° F., and the temperature drop Δh of all radiators on the same riser be 20° F.[1] The heat loss of the riser and main return is to be neglected. The temperature of the attic is assumed at 32° F. and room temperatures 70° F. Let the efficiency of insulation in the overhead distribution main be 80 per cent. The risers running through rooms are not to be insulated. The additional details are to be taken from Fig. 172.

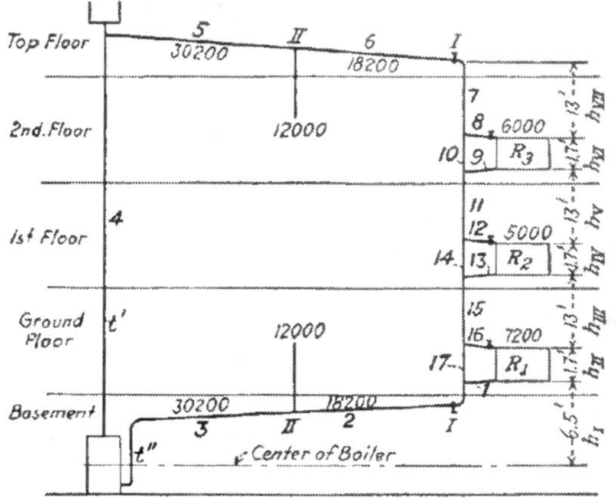

FIG. 172.

1. *Preliminary Computation of Piping and Radiation.* *a.* Least-favored Current Circuit. (1) Circulating Pressure Head. (*a*) Computation of Temperatures Exclusive of Piping Heat Losses.

$$t' = 185° \text{ F.} = t_4 = t_5 = t_6 = t_7 = t_8 = t_{10}.$$

$$Q = \frac{\Sigma W}{t' - t''} = \frac{6{,}000 + 5{,}000 + 7{,}200}{30} = \frac{18{,}200}{30} \cong 607.$$

$$t_9 = t_8 - 20 = 185 - 20 = 165° \text{ F.}$$

$$t'' = t_3 = t_2.$$

$$t_{11} = (\text{See Eq. (51)})\ t' - \frac{W_3}{Q} = 185 - \frac{6{,}000}{607} \cong 175° \text{ F.}$$

$$t_{12} = 175° \text{ F.} = t_{14}.$$

$$t_{13} = 175 - 20 = 155° \text{ F.}$$

$$t_{15} = (\text{See Eq. (51)})\ t_{11} - \frac{W_2}{Q} = 175 - \frac{5{,}000}{607} \cong 167° \text{ F.}^2$$

$$t_{16} = 167° \text{ F.} = t_{17}$$

$$t_1 = 167 - 20 = 147° \text{ F.}$$

Average temperature in R_3 = 175° F.
Average temperature in R_2 = 165° F.
Average temperature in R_1 = 157° F.

[1] $\Delta h \geq \frac{7{,}200}{18{,}200} \times 30 \geq 11.9°$ F. See Eq. (50), p. 201.

[2] Accuracy within 0.5° F. is taken as sufficient for practical purpose.

(*b*) Computation of the Circulating Pressure Head.

$$H = 0.192\{h_I(s'' - s') + h_{II}(s_{RI} - s') + h_{III}(s_{15} - s') + h_{IV}(s_{R_{II}} - s') + h_V(s_{II} - s') + h_{VI}(s_{R_{III}} - s') + h_{VII}(s' - s')\}$$

$$= 0.192\{6.5(s_{155} - s_{185}) + 1.7(s_{157} - s_{185}) + 13(s_{167} - s_{185}) + 1.7(s_{165} - s_{185}) + 13(s_{175} - s_{185}) + 1.7(s_{175} - s_{185}) + 13(s_{185} - s_{185})\}$$

$$= 0.192(6.5 \times 0.64 + 1.7 \times 0.60 + 13 \times 0.39 + 1.7 \times 0.43 + 13 \times 0.22 + 1.7 \times 0.22 + 0)$$

$$= 0.192(4.16 + 1.02 + 5.07 + 0.73 + 2.86 + 0.37)$$

$$= 0.192 \times 14.21 = 2.73 \text{ in. w.c.}$$

(2) Approximation of the Current Circuit in Downfeed Riser I.

Circulating pressure head exclusive of heat losses		2.73 in. w.c.
Considering the heat losses according to Table XIV, Downfeed risers uninsulated for three floors		
Horizontal length of system		up to 75 ft
Height of radiator above boiler 22.05 ft. center to center		
Upfeed and downfeed main 40 ft. apart		
Additional circulating pressure head (Table XIV)	= 1.00 in. w.c.	
Taking one-half of this value, since it is a one-pipe system	= 0.50	
Reduction of 15 per cent because of 185° F. flow temperature . .	= 0.075	
Actual allowance for heat loss is therefore	= 0.43 in. w.c.	0.43
Total circulating pressure head		3.16
Allowing 50 per cent for resistance of fittings		1.58
Circulating head available for pipe friction		1.58 in. w.c.
Length of pipe sections 1, 2, 3, 4, 5, 6, 7, 8, 9, 11, 12, 13, 15, and 16	= 200 ft	
Friction loss per foot of pipe $= \dfrac{1.58}{200}$	= 0.0079 in. w.c.	

The pipe sizes may now be determined from Chart 1 and entered in Summary E.

In this example it will be noted that temperature drops of from 20 to 30° F. are dealt with. Chart 1 has been computed on the basis of a 30° temperature drop. To use this chart the quantities must be converted to this drop. For example, pipe section 8 transmits 6,000 B.t.u. with a temperature drop of 20° F. For a 1° drop the transmission is $\dfrac{6{,}000}{20}$ B.t.u. and for 30° it is $30 \times \dfrac{6{,}000}{20} = 9{,}000$ B.t.u. Chart 1 may then be used as in previous cases.

b. Short-circuiting Pipe Sections 10, 14, and 17.

d_{10} assumed one pipe size smaller than d_8 or d_9	$d_{10} = \frac{3}{4}$ in.
d_{14} assumed one pipe size smaller than d_{12} or d_{13}	$d_{14} = \frac{3}{4}$ in.
d_{17} assumed one pipe size smaller than d_{16} or d_1.	$d_{17} = \frac{3}{4}$ in.

c. Approximation of the Radiator Sizes.—Assume that 20-in. high two-column radiation is to be used. To illustrate, radiator 2 will be discussed. Exclusive of the heat emission the average temperature of the radiator temperature (see p. 217) is

$$\frac{t_f + t_r}{2} = 165° \text{ F.}$$

The required surface F is therefore

$$F = \frac{W}{K\left(\frac{t_a + t_e}{2} - t_r\right)} = \frac{5{,}000}{1.4(165 - 70)}$$

$$= \frac{5{,}000}{1.4 \times 95} = 37.6 \text{ sq. ft.}$$

The values of K are taken from Table VIII.

From Table XIVB an additional allowance of 10 per cent is found to compensate for the heat loss for two-pipe systems. For one-pipe systems only one-half of this should be taken, i.e., 5 per cent of 37.6 ≅ 1.9 sq. ft. The total surface is therefore 39.5 sq. ft.

2. Check Computation

RESISTANCES OF THE FITTINGS

Pipe section 1, d = 1 in.

Tee (branch)	ζ = 1.5
Half radiator	= 1.5
	$\Sigma\zeta$ = 3.0

Pipe section 2, d = 1¼ in.

Long-sweep ell	ζ = 1.0
1 riser valve	= 9.0
Tee (straightway)	= 1.0
	$\Sigma\zeta$ = 11.0

Pipe section 3, d = 1½ in.

3 long-sweep ells,	$\Sigma\zeta$ = 1.5
Part of boiler	= 1.5
	$\Sigma\zeta$ = 3.0

Pipe section 4, d = 1½ in.

Remainder of boiler	ζ = 1.0
Long-sweep ell	= 0.5
	$\Sigma\zeta$ = 1.5

Pipe section 5, d = 1½ in.

Tee (branch)	ζ = 1.5

Pipe section 6, d = 1¼ in.

Tee (straightway)	ζ = 1.0
Valve	= 9.0
Long-sweep ell,	= 1.0
	$\Sigma\zeta$ = 11.0

Pipe section 7, d = 1¼ in. $\Sigma\zeta$ = 0

Pipe section 8, d = 1 in.

Tee (branch)	ζ = 1.5
Valve	= 2.0
Half radiator	= 1.5
	$\Sigma\zeta$ = 5.0

Pipe section 9, d = 1 in.

Half radiator	ζ = 1.5
Tee (branch)	= 1.5
	$\Sigma\zeta$ = 3.0

Pipe section 11, d = 1¼ in. $\Sigma\zeta$ = 0

Pipe section 12, d = 1 in.
Same as pipe section 8

Pipe section 13, d = 1 in.
Same as pipe section 9

Pipe section 15, $\Sigma\zeta$ = 0

Pipe section 16, same as pipe section 8

Pipe section 10,
2 tees (straightway) $\Sigma\zeta$ = 2.0

Pipe section 14,
2 tees (straightway) $\Sigma\zeta$ = 2.0

Pipe section 17,
2 tees (straightway) $\Sigma\zeta$ = 2.0

SUMMARY E

Number of pipe section	Quantity of heat in B.t.u. for temperature drop of			Length of pipe section, feet	Diameter d, inches	Computation											Difference	
						Original values					Corrected values							
	30° F.	20° F.	1° F.			R per foot	v, feet per sec.	lR	Σζ	Z	d, in.	R per foot	v, feet per sec.	lR	Σζ	Z	lR (n-h)	Z (p-j)
	(a)	(b)	(c)	(d)	(e)	(f)	(g)	(h)	(i)	(j)	(k)	(l)	(m)	(n)	(o)	(p)	(q)	(r)
	Current circuit drop main 1																	
	Assumed pressure head 3.16 in. w.c.					R per foot = 0.0079 in. w.c.					Actual pressure head = 3.84							
1	10,800	7,200	360	3	1	0.0070	0.33	0.021	3.0	0.059								
2	18,200	.....	...	25	1	0.0180	0.55	0.450	11.0	0.600								
3	30,200	.....	...	23	1¼	0.0140	0.55	0.322	3.0	0.160								
4	30,200	.....	...	55	1½	0.0060	0.38	0.330	1.5	0.039	1¼	0.014	0.55	0.77	1.5	0.083	0.44	0.044
5	30,200	.....	...	16	1½	0.0060	0.38	0.096	1.5	0.039								
6	18,200	.....	...	24	1¼	0.0060	0.36	0.144	11.0	0.260								
7	18,200	.....	...	13	1¼	0.0060	0.36	0.078										
8	9,000	6,000	300	3	1	0.0045	0.26	0.014	5.0	0.061								
9	9,000	6,000	300	3	1	0.0045	0.26	0.014	3.0	0.037								
11	18,200	.....	...	13	1¼	0.0060	0.36	0.078										
12	7,500	5,000	250	3	1	0.0035	0.22	0.011	5.0	0.044								
13	7,500	5,000	250	3	1	0.0035	0.22	0.011	3.0	0.026								
15	18,200	.....	...	13	1¼	0.0060	0.36	0.078										
16	10,800	7,200	360	3	1	0.0070	0.33	0.021	5.0	0.099								
								1.67		1.42						Σ(lR + Z) = 0.48		
				$\Sigma(lR + Z)_1^{16}$ = 1.67 + 1.42 = 3.09 in. w.c. 3.09 + 0.48 = 3.57 in. w.c.														
10	9,200	.....	...	1.7	¾	0.020	0.50	0.034	2.0	0.091								
14	10,700	.....	...	1.7	¾	0.026	0.55	0.044	2.0	0.110								
17	7,400	.....	...	1.7	¾	0.014	0.38	0.024	2.0	0.052	½	0.1	0.8	0.17	2.0	0.23	[a]	

[a] See Chart 2.

Checking Computation of Piping and Radiators. a. Current Circuit of Downfeed Riser I.—(1) Computation of Temperatures Including Heat Loss of the Piping (see Eq. (48)).

Number	Q	l	d, inches	$\frac{k}{l'}$	$\frac{l \cdot k}{l'}$	$1 - \eta$	t_f	t_{room}	$t_f - t_{room}$	∂	t_r	t_m	h'*
5	1,007	16	1½	1.2	19.2	0.2	185	32	153	0.6	184.4		
6	607	24	1¼	1.0	24	0.2	184.4	32	152.4	1.2	183.2		
7	607	13	1¼	0.96	12.5	1.0	183.2	70	113.2	2.3	180.9	182	
8	300	3	1	0.86	2.58	1.0	180.9	70	110.9	0.9	180		
9	300	3	1	0.80	2.40	1.0	160	70	90	0.7	159.3		
11	607	13	1¼	0.87	11.3	1.0	171	70	101	1.9	169.1	170	
12	250	3	1	0.83	2.49	1.0	160.1	70	90.1	1.0	168.1		
13	250	3	1	0.80	2.40	1.0	148.1	70	78.1	0.7	147.4		
15	607	13	1¼	0.87	11.3	1.0	160.9	70	90.9	1.7	159.2	160	
16	360	3	1	0.80	2.40	1.0	159.2	70	89.2	0.6	158.6		
1	360	3	1	0.76	2.28	1.0	138.6	70	68.6	0.4	138.2		

$t_9 = t_8$ − temperature drop in radiator 3 = 180 − 20 = 160° F.

$$t_{11} = t_7 - \frac{W_3}{Q} = 180.9 - \frac{6{,}000}{607} = 171° \text{ F.}†$$

$t_{13} = t_{12}$ − temperature drop radiator 2 = 168.1 − 20 = 148.1° F.

$$t_{15} = t_{11} - \frac{W†}{Q} = 169.1 - \frac{5{,}000}{607} = 160.9° \text{ F.}$$

$t_1 = t_{16} - 20 = 138.6°$ F.

$$t_2 = t_{15} - \frac{W_1}{Q} = 159.2 - \frac{7{,}200}{607} = 147.3° \text{ F.}$$

* The computation of the pressure heads h' is carried out under subdivision 2.

† It would be more accurate to add the heat loss of the pipe sections 8, 9 and 12, 13, and 16, 1 to the heat transmitted W_3, W_2, and W_1 respectively. The influence for short radiator connections, however, is negligible. Note that the computed pressure head, due to the above, will remain less than the actual pressure head.

Since the heat loss from the main return may be neglected, the temperature of the returns to the boiler is 147.3° F.

$$\text{Average temperature in radiator } R_3 = \frac{180 + 160}{2} = 170° \text{ F.}$$

$$\text{Average temperature in radiator } R_2 = \frac{168.1 + 148.1}{2} = 158.1° \text{ F.}$$

$$\text{Average temperature in radiator } R_1 = \frac{158.6 + 138.6}{2} = 148.6° \text{ F.}$$

(2) Determination of the Circulating Head.—With the temperatures of each pipe section determined, it is now possible to find the effective pressure head and also the heat loss from the piping.

Temperature to be used in connection with height h_I = 147.3° F.
Temperature to be used in connection with height h_{II} = 148.6° F.
Temperature to be used in connection with height h_{III} = 160.0° F.
Temperature to be used in connection with height h_{IV} = 158.1° F.
Temperature to be used in connection with height h_V = 170.0° F.
Temperature to be used in connection with height h_{VI} = 170.0° F.
Temperature to be used in connection with height h_{VII} = 182.0° F.

$$H = 0.192\{6.5(s_{147.3} - s_{185}) + 1.7(s_{148.6} - s_{185}) + 13(s_{160} - s_{185}) + 1.7(s_{158.1} - s_{185}) + 13(s_{170} - s_{185}) + 1.7(s_{170} - s_{185}) + 13(s_{182} - s_{185})\}.$$

In practice when the densities are accurate to within 0.5° F they may be used with satisfactory results. With the use of Table XIII the following results

$$H = 0.192(6.5 \times 0.78 + 1.7 \times 0.76 + 13 \times 0.53 + 1.7 \times 0.58 + 13 \times 0.33 + 1.7 \times 0.33 + 13 \times 0.07)$$
$$= 0.192(5.07 + 1.29 + 6.89 + 0.99 + 4.29 + 0.56 + 0.91)$$
$$= 0.192 \times 20 = 3.84 \text{ in. w. c.}$$

The pressure head found by the preliminary computation was 3.16 in. w. c. and $\Sigma(lR + Z)_1^{16} = 3.09$ which is sufficiently close to 3.16 in. w. c. The checking computation, however, shows that that there is a discrepancy of $3.84 - 3.09 = 0.75$ in. w. c. This may be corrected by decreasing the diameter of pipe section 4, in which case $\Sigma(lR + Z)_1^{16} = 3.57$ in. w. c., which agrees fairly well with 3.84 in. w. c. computed above.

b. Short-circuit Pipe Sections 10, 12, and 17.

For pipe section 10

From Eq. (53)

$$H = 0.192h_{VI}(s_{R_3} - s_{10}) + (lR + Z)_{10} = 0.192 \times 1.7 \times 0.24 + 0.125.$$
$$= 0.078 + 0.125 = 0.203 \text{ or } \cong 0.20 \text{ in. w. c.}$$

Again

$$0.20 \geqq (lR + Z)_{8,9}$$
$$\geqq 0.13$$

The pipe section under discussion therefore needs no correction.

For pipe section 14

$$H = 0.192h_{IV}(s_{R_2} - s_{14}) + (lR + Z)_{14} = 0.192 \times 1.7(s_{158.1} - s_{169})$$
$$= 0.192 \times 1.7 \times 0.23 + 0.154 = 0.075 + 0.154 = 0.229$$
$$\cong 0.23 \text{ in. w. c.}$$
$$0.23 \geqq (lR + Z)_{12,13}$$
$$\geqq 0.09$$

Pipe section 14 remains unchanged.

For pipe section 17

$$H = 0.192h_{II}(s_{R_1} - s_{17}) + (lR + Z)_{17} = 0.192 \times 1.7(s_{148.6} - s_{169}) + 0.076$$
$$= 0.069 + 0.076 = 0.145 \cong 0.15$$

Also

$$0.15 \geqq \Sigma(lR + Z)_{1,16} = 0.2$$

Pipe section 17 must therefore be changed to ½ in. so that $0.069 + 0.4 \cong 0.47 \geqq 0.2$

c. Checking Computations of Radiators.—Since all temperatures are now determined from the foregoing analysis, the required heating surface of the radiation may be checked. In radiator 2, for example,

$$F_{R_2} = \frac{5{,}000}{1.4\left(\frac{t_f + t_r}{2} - 70\right)} = \frac{5{,}000}{1.4(158 - 70)} = \frac{5{,}000}{1.4 \times 88}.$$
$$= 40.6 \text{ sq. ft.}$$

This result exceeds the 39.5 sq. ft. as found from the preliminary computation.

e. *Loft Heating Systems*.[1]

Example 15.

Let the temperature of the water leaving the boiler be 185° F. and the temperature drop in the radiators 30° F. Assume the heat loss from the riser and the main return

[1] For systems in which the main return passes over doors, etc., see "Circulating Forces due to Heat Loss in Hot-water Heating Systems," from WIERZ, "Über die Kräfte durch Rohrabkühlung in Warmwasser heizungen," *Gesundh.-Ing.*, 1925.

to be negligible. The distribution main, downfeed riser, and riser connections are to be uninsulated. The insulated main return is run under the floor. Additional assumptions are: room temperature, 70° F.; center of boiler to be 16 in. higher than center of return main; other details shown on sketch of Fig. 173.

1. *Preliminary Computation of the Piping.* *a.* Determination of the least-favored Current Circuit and Its Pipe Sizes.

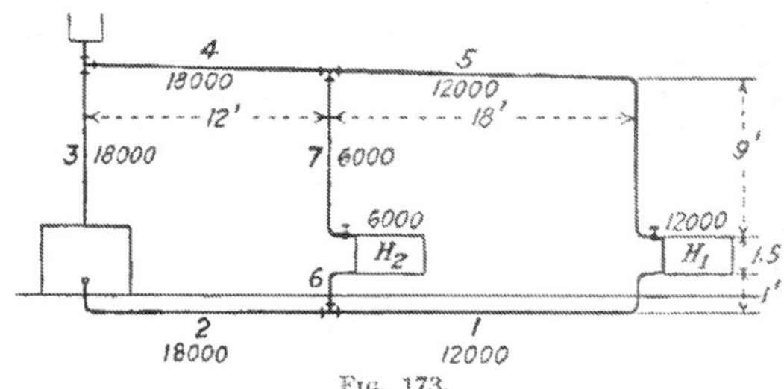

FIG. 173.

In accordance with Table XV the available circulating head for:

Radiator 1	Radiator 2
0.7 in. w.c.	0.27 in. w.c.

Deducting 15 per cent for the flow temperature of 185° F. (see Table XV).

Radiator 1	Radiator 2
0.595 in. w.c.	0.23 in. w.c.

Deducting 50 per cent for resistances of fittings, there remains

∼0.3 in. w.c.	∼0.12 in. w.c.

Frictional loss per foot of pipe,

$$R_1 = \frac{0.3}{79} = 0.0038 \text{ in. w.c.} \qquad R_2 = \frac{0.12}{43} = 0.0028 \text{ in. w.c.}$$

In the system under consideration there are two distinct current circuits, *i.e.,* that for radiator 1 and that for radiator 2. Each must be computed on the basis of its corresponding pressure head. From the foregoing it will be observed that the circuit of radiator 2 is the least favored therefore the sections 2, 3, and 4 must be proportioned for this pressure head.

b. Preliminary Computation of Circuit of Radiator 1.

Available pressure head less resistances of fittings	= 0.3	in. w.c.
Head used in pipe sections 2, 3, and 4 = 33 × 0.0028	= 0.092	in. w.c.
	0.208	
Remaining head for pipe sections 1 and 5	≅ 0.21	in. w.c.
Length of pipe sections 1 and 5	= 46 ft.	
Pressure loss per foot of pipe = $\frac{0.21}{46}$	= 0.0046	in. w.c.

With the aid of Chart 1 the diameter may now be approximated and tabulated in Summary F.

Summary F

Number of pipe section	B.t.u. required	Length, feet	Diameter d, inches	Computation												
				Original values					Corrected values							
				R per foot	v	lR	$\Sigma\zeta$	Z	d	R per foot	v	lR	$\Sigma\zeta$	Z	lR	Z
(a)	(b)	(c)	(d)	(e)	(f)	(g)	(h)	(i)	(j)	(k)	(l)	(m)	(n)	(o)	(p)	(q)

Current circuit radiator 2

Approximate pressure head 0.23 in. w.c. Pressure loss per foot = 0.0028 Actual pressure head 0.31 in. w.c.

(a)	(b)	(c)	(d)	(e)	(f)	(g)	(h)	(i)	(j)	(k)	(l)	(m)	(n)	(o)	(p)	(q)
2	18,000	12	1½	0.0023	0.24	0.028	3.0	0.03								
3	18,000	9	1¼	0.0023	0.24	0.021	3.0	0.03								
4	18,000	12	1½	0.0023	0.24	0.028	0.0	0.00								
7	6,000	9	1	0.0023	0.18	0.021	8.5	0.05								
6	6,000	1	1	0.0023	0.18	0.002	6.0	0.035								

$\Sigma(l \cdot R + Z)_2^5 = 0.100 + 0.145 = \sim 0.25$ in. w.c.

coinciding sufficiently with both the approximate and actual pressure head.

Current circuit radiator 1

Approximate pressure head 0.59 in. w.c. Pressure loss per foot = 0.0046 Actual pressure head 0.48 in. w.c.

(a)	(b)	(c)	(d)	(e)	(f)	(g)	(h)	(i)	(j)	(k)	(l)	(m)	(n)	(o)	(p)	(q)
1	12,000	19	1¼	0.0026	0.22	0.049	7.0	0.061								
5	12,000	27	1¼	0.0026	0.22	0.070	9.5	0.084								

$\Sigma(lR + Z) = 0.119 + 0.145 = 0.264$ in. w.c.

$\Sigma(lR + Z)_2^4 = 0.137$

$= 0.401 = \Sigma(lR + Z)_1^5$

2. *Checking Computation of the Piping.* *a.* Determination of All Temperatures Including the Heat Losses.

Number of pipe section	Q	l	d	$\frac{k}{V}$	$l \times \frac{k}{V}$	$1 - \eta$	t_f	t_{room}	$t_f - t_{room}$	ϑ	t_r
4	600	12	1½	1.1	13.2	1.0	185	70	115	2.5	182.5
5	400	27	1¼	0.96	25.9	1.0	182.5	70	112.5	7.3	175.2
7	200	9	1	0.86	7.7	1.0	182.5	70	112.5	4.3	178.2

$$\text{Average temperature in radiator 1} = \frac{175.2 + 145.2}{2} = 160.2°\text{ F.}$$

$$\text{Average temperature in radiator 2} = \frac{178.2 + 148.2}{2} = 163.2°\text{ F.}$$

COEFFICIENT OF RESISTANCE OF FITTINGS

Pipe section 1, d = 1¼ in	
Half radiator	ζ = 1.5
3 ells	= 4.5
Tee (straightway)	= 1.0
	$\Sigma\zeta$ = 7.0
Pipe section 2, d = 1½ in	
Part of boiler	ζ = 1.0
2 ells	= 2.0
	$\Sigma\zeta$ = 3.0
Pipe section 3, d = 1½ in	
Remainder of boiler	ζ = 1.5
Tee (branch)	= 1.5
	$\Sigma\zeta$ = 3.0
Pipe section 4	
	$\Sigma\zeta$ = 0
Pipe section 5, d = 1¼ in	
Tee (straightway)	ζ = 1.0
2 ells	= 3.0
Angle radiator valve	= 4.0
Half radiator	= 1.5
	$\Sigma\zeta$ = 9.5
Pipe section 6, d = 1 in	
Half radiator	ζ = 1.5
2 ells	= 3.0
Tee (branch)	= 1.5
	$\Sigma\zeta$ = 6.0
Pipe section 7, d = 1 in	
Tee (branch)	ζ = 1.5
Angle radiator valve	= 4.0
Half radiator	= 1.5
1 ell	= 1.5
	$\Sigma\zeta$ = 8.5

b Determination of the Pressure Head for Circuit of Radiator 2

$$\text{Average temperature of main 7} = \frac{182.5 + 178.2}{2} \cong 180.4°\text{ F}$$

Therefore

$$\begin{aligned} H &= 0.192\{9(s_{180.4} - s_{185}) + 1.5(s_{168.2} - s_{185})\} \\ &= 0.192\{(9 \times 0.1) + (1.5 \times 0.47)\} = 0.192(0.9 + 0.71)\} \\ &\cong 0.31 \text{ in w c} \end{aligned}$$

c Determination of the Head for the Circuit of Radiator 1.—Pipe section 5 has a temperature drop of 7.3° F and a length of 27 ft. For a 9-ft length the drop will be

$$\frac{7.3}{27} \times 9 \cong 2.4°\text{ F}$$

Terminal temperature of pipe section 5		= 175.2° F
Initial temperature for a length of 9 ft	= 175.2 + 2.4	= 177.6° F
Average temperature as a consequence		= 176.1° F

$$\begin{aligned} H &= 0.192\{9(s_{176.4} - s_{185}) + 1.5(s_{160.2} - s_{185})\} \\ &= 0.192\{(9 \times 0.19) + (1.5 \times 0.53)\} \\ &= 0.192(1.7 + 0.8) = 0.192 \times 2.5 \\ &\cong 0.48 \text{ in w c} \end{aligned}$$

d Greenhouse Heating

Example 16

Heat to be supplied by coil = 2 × 20,000	= 40,000 B t u per hour
Size of pipe in section 1	= 2 in
Length of pipe section 1 = 2 × 220 ft	= 440 ft
Size of pipe in section 2	= 1½ in
Length of pipe section 2	= 30 ft
Flow temperature	= 175° F
Return temperature	= 145° F

The problem is How much must the center of the boiler be below the center of the coil to insure satisfactory operation? (see Fig 174)

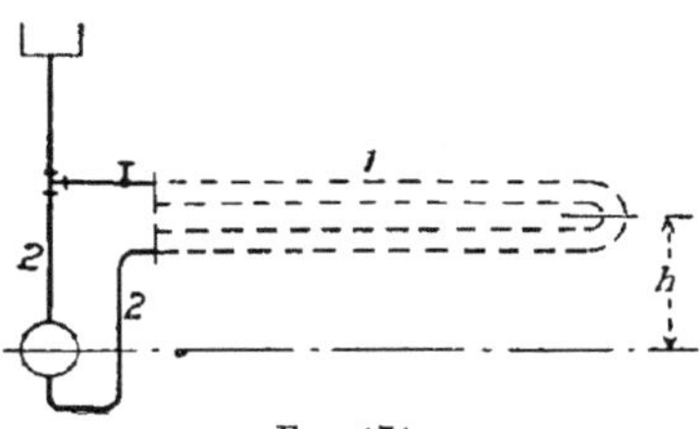

FIG 174

Solution: For approximation of the height

Frictional resistance per foot of 2-in pipe when transmitting 20,000 B t u per hour (Chart 1)	= 0 0007 in w c
$l \times R$	= 0 154 in w c
Frictional resistance per foot of 1½-in pipe when transmitting 40,000 B t u per hour	= 0 01 in w c
$l \times R$	= 0 3 in w c
ΣlR for entire system = 0 3 + 0 154	≅ 0 45 in w c

Since this is but 50 per cent of the total head, the remaining 50 per cent being used up in frictional resistance of fittings,

$$\Sigma(lR + Z) = 0\ 9 \text{ in w c}$$

The circulating pressure head per foot of height for temperature difference of 175° F in flow and 145° F in return is

$$H(\text{per foot}) = 0\ 192(S_{145} - S_{175}) = 0\ 192 \times 0\ 6$$
$$\cong 0\ 12 \text{ in w c}$$

The center of the boiler must therefore be lowered a distance of $\frac{0\ 9}{0\ 12} = 7\ 5$ ft

Check. Resistance coefficients for the fittings

Pipe section 1, d = 2 in

Entrance into distributing manifold	ζ = 1 0 (velocity change)
Bend (long radius)	= 1 0
Exit from manifold	= 1 0
	$\Sigma\zeta$ = 3 0

Pipe section 2, d = 1½ in

Boiler	ζ = 2 5
3 long-sweep ells	= 1 5
Valve	= 9 0
Tee (branch)	= 1 5
	$\Sigma\zeta$ = 14 5

SUMMARY G

Number	B.t.u.	l	d	R per foot	v	lR	$\Sigma\zeta$	Z
1	20,000	220′	2″	0.0007	0.15	0.15	3.0	0.012
2	40,000	30′	1½″	0.01	0.50	0.30	14.5	0.66

$\Sigma(lF + Z)_1{}^2 = 0.45 + \quad 0.672 =$
$1.122 \cong 1.12$ in. w.c.

From Summary G $\Sigma(lR + Z)_1{}^2 = 1.12$ in. w.c. Hence the vertical distance between the center of the boiler and the center of the coil should be

$$\frac{1.12}{0.12} \cong 9.3 \text{ ft.}$$

The latter figure should be used instead of the result 7.5 ft. as found in the approximation.

c. Forced-circulation Heating System.

Example 17.

Assume that the gravity circulating head is negligible. The pressure loss in each of the buildings *A*, *B*, and *C* is taken as 3 ft. w.c. The velocities and sizes of the

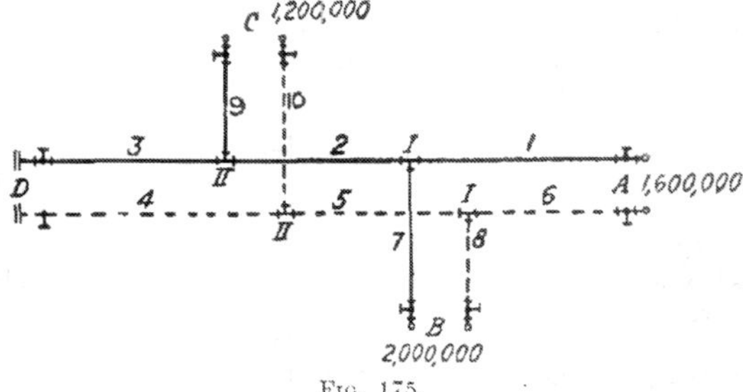

FIG. 175.

supply and distribution mains are to be chosen so that there remains a head of 40 ft. at the district heating power house at *D*. Assume further that the temperature drop is 190 − 150 = 45° F. Additional data are to be taken from Fig. 175 and Summary H. It is desired to find the pipe sizes.

1. *Preliminary Computations of the Pipe Sizes.*

Total head at heating station *D*....................	= 480	in. w.c.
Less loss of head in buildings.....................	= 36	in. w.c.
	444	in. w.c.
Resistance of fittings (= 10 per cent from Table XVI)	= 44.4	
Available head for pipe friction....................	399.6	in. w.c.

a. Circuit 1, 2, 3, 4, 5, and 6.—The total length of the least-favored circuit 1, 2, 3, 4, 5, and 6 is 2,400 ft. The pressure drop per foot is $\frac{399.6}{2,400} = 0.166$ in. w.c.

b Circuit 7, 8 and 9, 10 —The pressure at junction I is

$H = \pm(18 + 400 \times 0.166) = \pm(18 + 66.4)$	$= \pm 84.4$	in w c
Less pressure drop in *B*	$= \pm 18$	
Therefore head available for $\Sigma(lR + Z)$ of pipe section 7	$=$ 66 4	in w c
Length of pipe section 7	$=$ 120 ft	
Hence the pressure drop per foot of pipe in section $7 = \frac{66.4}{120}$	$\cong$ 0 55	in w c
Pressure at junction II is $H = (84.4 + 350 \times 0.166)$	$= \pm 142.5$	in w c.
Less pressure drop in *C*	$= \pm 18$	
Available head for friction	± 124.5	in w c.

The pressure drop per foot of pipe in section $9 = \frac{124.5}{160} = 0.778$ in w c.

The approximation of the pipe sizes may now be made with the aid of Chart 2

RESISTANCE OF FITTINGS

Pipe section 1, $d = 3$ in.		Pipe section 3, $d = 4\frac{1}{2}$ in	
Gate valve	$\zeta = 0.5$	Expansion bend	$\zeta = 4.0$
Expansion bend (4 ells)	$= 4.0$	Gate valve	$= 1.0$
			$\Sigma\zeta = 5.0$
Tee (straightway)	$= 1.0$	Pipe section 4 same as 3	
	$\Sigma\zeta = 5.5$	Pipe section 5 same as 2	
		Pipe section 6 same as 1	
Pipe section 2, $d = 4$ in		Pipe section 7, $d = 2\frac{1}{2}$ in	
Expansion bend	$\zeta = 4.0$	Tee (branch)	$\zeta = 1.5$
Tee (straightway)	$= 1.0$	Gate valve	$= 0.5$
	$\Sigma\zeta = 5.0$		$\Sigma\zeta = 2.0$
		Pipe section 8 same as 7	
		Pipe sections 9 and 10 same as 7	

2. *Checking Computation* *a* Circuit of Building *A*

From Summary H $\Sigma(lR + Z)_1{}^6$	$= 441.4$ in w c
Pressure drop in building	$= 36.0$
Total pressure head required	$= 477.4$ in w c.

This value is sufficiently near the pressure of 480 in w c available at *D*

b Circuit of Building *B* —From Summary II,

$\Sigma(lR + Z)_{7\,8}$	$= 146$ in w c.
$\Sigma(lR + Z)_{2\,3\,4\,5}$	$= 307.6$
Pressure drop in *B*	$= 36$
	489 6 in. w c

This value is in substantial agreement with the head available

c Circuit of Building *C* —From Summary G,

$\Sigma(lR + Z)_9{}^{10}$	$= 214$ in w c.
$\Sigma(lR + Z)_{3\,4}$	$= 180.6$
Pressure drop in *C*	$= 36$
	430 6 in w c

The available pressure of 480 in w c is in excess of that computed for building *C* given above If the designer thinks it advisable to alter the pipe sizes to insure closer agreement, pipe sections 9 or 10 may be made a trifle smaller.

SUMMARY H

Number of pipe section	B.t.u. 45° drop	B.t.u. 30° drop	Length l, feet	Diameter d, inches	Computation											Difference	
					Original values					Corrected values							
					R per foot	v	lR	$\Sigma\zeta$	Z	d	R per foot	v	lR	ζ	Z	lR	Z
(a)	(b)	(c)	(d)	(e)	(f)	(g)	(h)	(i)	(j)	(k)	(l)	(m)	(n)	(o)	(p)	(q)	(r)
					Circuit 1, 2, 3, 4, 5, 6												
					Pressure loss per foot = 0.166 in. w.c.												
1	1,600,000	1,066,600	400	3	0.14	3.3	56	5.5	10.9								
2	3,600,000	2,400,000	350	4	0.14	4.0	49	5.0	14.5								
3	4,800,000	3,200,000	450	4½	0.16	4.5	72	5.0	18.3								
4	4,800,000	8,200,000	450	4½	0.16	4.5	72	5.0	18.3								
5	3,000,000	2,400,000	350	4	0.14	4.0	49	5.0	14.5								
6	1,600,000	1,066,600	400	3	0.14	3.3	56	5.5	10.9								
					$\Sigma(lR + Z)_1^6$ =		354	+	87.1	= 441.4 in. w.c.							
					Circuit 7, 8												
					Pressure loss per foot = 0.55 in. w.c.												
7	2,000,000	1,333,300	120	2½	0.50	6.0	60	2.0	13.0	3	0.2	4	20				
8	2,000,000	1,333,300	120	2½	0.50	6.0	60	2.0	13.0								
					$\Sigma(lR + Z)_7^8$ =		120	+	26.0	= 146 in. w.c.							
					Circuit 9, 10												
					Pressure loss per foot = 0.778 in. w.c.												
9	1,200,000	800,000	160	2	0.6	5.5	96	2.0	11.0								
10	1,200,000	800,000	160	2	0.6	5.5	96	2.0	11.0								
					$\Sigma(lR + Z)_9^{10}$ =		192	+	22	= 214 = in. w.c.							

C. HIGH-PRESSURE STEAM HEATING

Determination of the Required Boiler Heating Surface

The design of high-pressure steam boilers is beyond the scope of this work For such data the reader is referred to the texts on that subject. Attention, however, should be drawn to the significance of Eqs (42) and (43) when considering the heating-up period and the steady state.

Determination of the Required Radiation

The computation of the necessary radiation is given by Eq 35 previously developed,

$$F_d = \frac{W}{K(t_d - t)} \qquad (35)$$

The notation has been given on page 182

Determination of the Required Pipe Sizes

1. Theory.

The following theory developed by Dr Wierz[1] simplifies the computations for determining pipe sizes in high- and low-pressure steam piping systems. It permits of a method similar to that employed in previous chapters

a. Circulating Pressure Head.—The circulating pressure head is the difference between the initial pressure and the stipulated terminal pressure

b. Frictional Resistance. (1) Without Reference to Heat Losses —The analysis is based on the general formula Eq 38 (p. 186),

$$R = \frac{p_2 - p_1}{l} = a\frac{v^n}{d^m}$$

Fritsche[2] pointed out that in the case of air piping and for constant values of n and m the value of a is proportional to the density of the air, *i e*, expressed mathematically

$$a = bs^{0\ 852} \qquad (57)$$

Numerous tests, published in *Bulletin* 21 of the Research Laboratory, Charlottenburg, show that this relationship is also valid in the case of steam It was furthermore determined that the value of m may be taken as 1 281 with sufficient accuracy for practical purposes

These experiments were conducted for a large number of pipe sizes and for pipes of various materials, *e g*, copper, brass, riveted steel, and cast iron. It was found that the value of m was independent of the nature

[1] "Simplified Analysis of Pipe Sizes for Steam Systems," from Brabbée-Wierz, "Vereinfachtes zeichnerisches und rechnerisches Verfahren zur Bestimmung der Durchmesser von Dampfleitungen," *Bull* 23 of the Research Laboratory, Charlottenburg

[2] "Resistance in Straight Cylindrical Pipe Lines," from Fritsche, "Untersuchungen uber den Stromungswiderstand in geraden, zylindrischen Rohrleitungen," Dresden, 1907

of the pipe surface. In other words the roughness of the pipe surface had practically no effect on the value of m. On the other hand, however, the exponent n of the velocity factor increases rapidly with increasing roughness of the pipe surface In *Bulletin* 23 of the Research Laboratory, Charlottenburg, it was shown that for steam piping $n = 1.853$

The general equation (Eq (38.), p. 186) therefore takes the form,

$$\frac{p_2 - p_1}{l} = 0.458\ 10^{-4}\ s^{0.852}\ \frac{v^{1.853}}{d^{1.281}} \tag{58}$$

where the pressure drop per unit of length is measured in pounds per square inch

The density of the steam, measured along the circuit within the pipe, is variable, so that Eq (58) is valid only when applied to a differential element of length; *e g*,

$$\frac{dp}{dl} = 0.458\ 10^{-4}.s^{0.852}.\ \frac{v^{1.853}}{d^{1.281}} \tag{59}$$

Due to the loss of heat from the piping, let it be assumed that the steam is saturated along the piping circuit. Application may then be made of the Mollier-Zeuner equation

$$s = \frac{p^r}{c} \tag{60}$$

or

$$s = 0.003p^{0.9375}. \tag{61}$$

Furthermore from page 186, Eq (40) is

$$Q = 3{,}600\frac{\pi d^2}{4}.\frac{1}{144}\ vs,$$

where Q is expressed in pounds per hour. From this results

$$v = \frac{Q}{25\ \frac{\pi d^2}{4}s}. \tag{62}$$

Substituting the values of s and v in Eq (59), it follows that[1]

$$\left.\begin{aligned}\frac{dp}{dl} &= \frac{kQ^{1.853}}{d^{4.987}\ p^{0.9375}}\\ k &= \frac{0.458\ 10^{-4}}{0.003(6.25\pi)^{1.853}}\end{aligned}\right\} \tag{63}$$

(2) Influence of the Heat Loss —Due to the loss of heat from the piping there is a decrease in the volume of steam along the length l of the piping circuit Assuming the volume shrinkage constant, then

$$\frac{dQ}{dl} = q,$$

or

$$Q_2 - Q_1 = ql, \tag{64}$$

[1] When two similar quantities have exponents of 1.853 and 0.852 on division (*i e*, subtraction of exponents), the slight difference 0.001 may be neglected so that the resultant exponent = 1.

in which Q_2 and Q_1 respectively represent the quantity of steam in pounds per hour flowing through the initial cross-sectional area (i e , when $l = 0$ and $p = p_2$) and through the terminal cross-sectional area of the pipe (i e , when $l = l$ and $p = p_1$)

q = weight of condensation per foot of pipe in pounds per hour due to the heat loss

Transposing Eqs (63) and (64) there results

$$\int_{p_1}^{p_2} p^{0\ 9375}\, dp = \frac{k}{qd^{4\ 987}}\int_{Q_1}^{Q_2} Q^{1\ 853}\, dQ,$$

which after integrating becomes

$$\frac{p_2^{1\ 9375} - p_1^{1\ 9375}}{1.9375} = \frac{k}{qd^{4\ 987}}\ \frac{Q_2^{2\ 853} - Q_1^{2\ 853}}{2\ 853}. \tag{65}$$

Let

$$\left.\begin{aligned} p_2^{1\ 9375} &= B_2 \\ p_1^{1\ 9375} &= B_1 \\ 1\ 9375k &= K \\ 2\ 853 &= n \end{aligned}\right\} \tag{66}$$

Substituting these in Eq (65), then

$$\frac{B_2 - B_1}{l} = \frac{K}{d^{4\ 987}}\ \frac{Q_2^n - Q_1^n}{qln} \tag{67}$$

(a) Insulated Steam Lines with Low Heat Loss —Equation (67) may be rewritten

$$\frac{B_2 - B_1}{l} = \frac{K}{d^{4\ 987}}\ \frac{Q_1^n}{qln}\left[\left(\frac{Q_2}{Q_1}\right)^n - 1\right]. \tag{68}$$

From Eq (64)

$$\frac{Q_2}{Q_1} = 1 + \frac{ql}{Q_1}$$

Hence

$$\left(\frac{Q_2}{Q_1}\right)^n - 1 = \left(1 + \frac{ql}{Q_1}\right)^n - 1.$$

Expanding the right-hand member by the binomial theorem, it becomes

$$\frac{n}{1}\frac{ql}{Q_1} + \frac{n(n-1)}{1\ 2}\left(\frac{ql}{Q_1}\right)^2 +$$

Due to the smallness of ql in comparison to Q_1 (low heat loss), the subsequent terms may be neglected in the above expansion Hence

$$\left(\frac{Q_2}{Q_1}\right)^n - 1 = \frac{nql}{Q_1}\left[1 + (n-1)\frac{ql}{2Q_1}\right] \tag{69}$$

The expression in the brackets of Eq (69) may be further simplified. If R represents the expression

$$R = \left(1 + \frac{ql}{2Q_1}\right)^{n-1},$$

then upon expansion into a series

$$R = 1 + \frac{n-1}{1}\frac{ql}{2Q_1} + \frac{(n-1)(n-2)}{1 \cdot 2}\frac{q^2l^2}{4Q_1^2} + \cdots$$

Since ql is small compared to $2Q_1$, terms beyond the second may be neglected. Consequently

$$1 + (n-1)\frac{ql}{2Q_1} = R = \left(1 + \frac{ql}{2Q_1}\right)^{n-1} \tag{70}$$

Substituting the latter in Eq. (69), there results

$$\left(\frac{Q_2}{Q_1}\right)^{n-1} - 1 = \frac{nql}{Q_1}\left(1 + \frac{ql}{2Q_1}\right)^{n-1} = \frac{nql}{Q_1^n}\left(Q_1 + \frac{ql}{2}\right)^{n-1} \tag{71}$$

Introducing these values in Eq. (68), it follows that

$$\frac{B_2 - B_1}{l} = \frac{K}{d^{4.987}}\left(Q_1 + \frac{ql}{2}\right)^{n-1},$$

or

$$\frac{B_2 - B_1}{l} = \frac{0.00012}{d^{4.987}}\left(Q_1 + \frac{ql}{2}\right)^{1.853}. \tag{72}$$

According to Zaruba,[1] Eq. (72) may also be used in proportioning pipe sizes for superheated steam. In this case the weight of the steam must be multiplied by the factor $\left(\frac{V_s}{V}\right)^{\frac{1}{1.853}}$ where

V_s = average specific volume of the superheated steam.

V = average specific volume of the saturated steam.

(b) Steam Coils.—For steam coils $Q_1 = 0$, and therefore from Eq. (64) $Q_2 = ql = N$, where N represents the heating effect of the coil. Hence

$$\frac{B_2 - B_1}{l} = \frac{K}{d^{4.987}}\frac{Q_2^n}{qln} = \frac{K}{d^{4.987}}\frac{qlQ_2^{n-1}}{qln}$$

or

$$\frac{B_2 - B_1}{l} = \frac{K}{d^{4.987}}\frac{N^{1.853}}{2.853}$$

$$= \frac{0.00012}{d^{4.987}}\left(\frac{N}{\sqrt[1.853]{2.853}}\right)^{1.853}$$

or finally

$$\frac{B_2 - B_1}{l} = \frac{0.00012}{d^{4.987}}\left(\frac{N}{1.76}\right)^{1.853} \tag{73}$$

(c) Resistance of Pipe Fittings.—The resistance in pounds per square inch introduced by pipe fittings is determined by Eq. (39) (p. 186), i.e.,

$$Z_1 = 0.007\Sigma\zeta\frac{v^2}{2g}s$$

[1] "Pipe Sizes for Superheated Steam," from ZARUBA, "Rohrweiten fur uberhitzten Dampf," *Gesundh.-Ing.*, 1917.

2. The Compilation and Use of Charts 3 and 4.

a General—In Eqs (72) and (73) let

$A = 0.00012$ $\qquad m = 4.987$

$Q = Q_1 + \frac{ql}{2}$ or its corresponding $\frac{N}{1.76}$ $\qquad n = 1.853$

With these abbreviations both equations take the form

$$\frac{B_2 - B_1}{l} = A\frac{Q^n}{d^m} \tag{74}$$

It will be observed that Eq (74) is similar in form to Eq (38) (p 186), used in the design of hot-water heating systems The difference consists of replacing the pressures p_2 and p_1 by the terms B_2 and B_1 As a consequence it is possible to compile a chart with the aid of the following expressions

$$\left.\begin{aligned} B_2 &= p_2^{1.9375} \\ B_1 &= p_1^{1.9375} \\ \frac{B_2 - B_1}{l} &= \frac{0.00012}{d^{4.987}}\left(Q_1 + \frac{ql}{2}\right)^{1.853} \quad \text{(for insulated steam pipes).} \\ \frac{B_2 - B_1}{l} &= \frac{0.00012}{d^{4.987}}\left(\frac{N}{1.76}\right)^{1.853} \quad \text{(for steam coils)} \\ Z &= 0.007\,\Sigma\zeta\frac{v^2}{2g}s. \\ v &= \frac{Q}{25\frac{\pi d^2}{4}s} \end{aligned}\right\} \tag{75}$$

In this connection the following points are to be noted

(1) In the case of hot-water heating systems it is possible to regard v as a constant whereas in the case of high-pressure steam heating systems v is a function of s The tables therefore are based upon a definite value of $s = 0.1$ If it is desired to find the velocity corresponding to the pressure p_n, the tabular values must be divided by $10s_n$

(2) The table should not be consulted for the value of Q_1 but the value of $Q_1 + \frac{ql}{2}$ used for insulated steam lines and the value of $\frac{N}{1.76}$ used for steam coil.

(3) In the case of hot-water heating systems the values of Z for $\Sigma\zeta$ varying from 1 to 15 were given For steam heating systems this is not possible since Z is a function of s Accordingly the values of Z for $\zeta = 1$ and for pressures $p = 14$ to 150 lb per square inch absolute are given which must moreover be multiplied by $\Sigma\zeta$

As a consequence the right-hand side of Chart 3 contains the following

(1) The quantity of steam transmitted in pounds per hour is given in the horizontal row 1.

(2) The average steam velocity in feet per second for $s = 0.1$ in the horizontal row 2

(3) The quantities $\frac{B_2 - B_1}{l}$ (i.e., per foot of pipe)

The left-hand side contains the following

(1) The average steam velocity in feet per second for values of $s = 0.1$ lb. per cubic foot

(2) The values of Z for pressures ranging from 14 to 150 lb per square inch when $\Sigma\zeta = 1$.

(3) The values of ζ for pipe-fitting resistances

On the same sheet will be found an auxiliary table, which shows the values of B for the corresponding values of p

b Preliminary Design for Cost Computations—As in previous cases the least-favored circuit is first considered The circulating pressure head is found from which the resistances of the pipe fittings are subtracted according to data given in Table XVI This determines the necessary initial and terminal pressures p_2 and p_1 available for overcoming the frictional resistance

From Chart 3 the corresponding values of B_2 and B_1 are taken, from which the value $\frac{B_2 - B_1}{l}$ is found by dividing by the length of the current circuit.[1] Proceeding horizontally from this value in the main table of Chart 3 to the steam quantity in the horizontal columns 1, the pipe size is given at the head of the column so determined

In steam lines where the heat loss is slight it may be neglected in estimating the steam flow. For steam coils the value of N must be divided by 1.76 and the quotient so found used in referring to the chart

c. Check Computation for Final Design—After determining the pipe sizes from the preceding, it is possible to determine the heat losses of the pipe circuit under consideration For piping effectively insulated it may be assumed that

$$q = \frac{30D}{\lambda}$$

where

q = condensation in pounds per hour per foot.
D = pipe diameter in inches
λ = latent heat of vaporization in B.t.u per pound

The quantity of steam Q_1 desired at the end of the pipe circuit is increased by an amount $\frac{ql}{2}$, and the sum so found is ascertained in columns 1 of Chart 3 Proceeding horizontally to the left, the value of $\frac{B_2 - B_1}{l}$

[1] It should be observed that it is permissible to distribute the pressure drop unevenly along the circuit if desired

is found and from this by multiplication with l, $B_2 - B_1$ is determined Since the value of B_1 corresponds to the known pressure p_1 at the end of the pipe circuit, B_2 is found by subtraction With the aid of the auxiliary table of Chart 3 the value of p_2 is obtained

To make allowance for the resistance through the fittings it is necessary to know the flow of steam in pounds per hour and also its pressure. Immediately below the quantity of steam Q, in column 2 of Chart 3, is given the steam velocity for $s = 0.1$ For the steam velocity so determined, together with the pressure existing in the fitting, the table of fitting resistance on the left will immediately disclose the value of Z in pounds per square inch for $\zeta = 1$. This in turn must be multiplied by $\Sigma\zeta$

The sum of the resistances through pipes and fittings should be equal or slightly less than the circulating pressure head In steam coils the value of $\frac{N}{1.76}$ is used instead of the effective quantity of steam N.

3. Return Lines.

Steam at 212° F occupies a volume approximately 1,600 times larger than that of the corresponding weight of water The return pipe sizes may not be proportioned solely on the basis of the corresponding volumes of the steam and its condensate, but must be designed to carry certain quantities of air Due to the involved uncertainties, return lines are proportioned according to empirical rules Table XVII, compiled by Rietschel, may be used in this connection

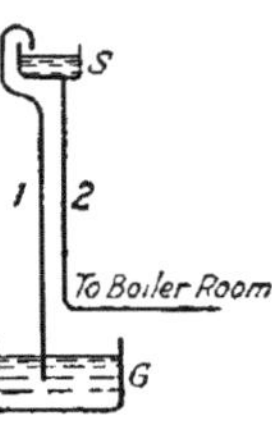

Fig 176

In the case of a district heating system, where the condensation flows to a receiver G (Fig 176) from which it is pumped to a second tank S whence it flows back to the boiler, the line 1 must be designed as pump discharge line and line 2 as a warm-water pipe in accordance with the formula

$$h = \Sigma(lR + Z)$$

In this formula h is the vertical distance in feet between the water levels in the two tanks

4. Numerical Examples.

a Computing the Required Boiler Heating Surface —In view of the scope of the subject the reader is referred to the texts on boiler practice for a discussion of the sizes and types of boilers

b Computing the Required Heating Surface

Example 18.

It is assumed that a dryer having a temperature of 90° F is to be supplied with 20,000 B t u per hour from a high-pressure steam coil The heating surface is to be of ribbed cast-iron pipe 3 in in diameter The inlet gage pressure is to be 30 lb per square inch, and the outlet gage pressure is 8 lb per square inch.

Solution: In accordance with Eq. (35) (p. 182), the heating surface is

$$F_d = \frac{W}{K(t_d - t)}.$$

From Table III

t_i = initial temperature of steam corresponding to 44.7 lb. per square inch absolute = 274° F.

t_r = final temperature of steam at outlet of pipe corresponding to 22.7 lb. per square inch absolute = 235° F.

t_d = average steam temperature $= \frac{t_i + t_r}{2} = 254.5°$ F.

From Table IX

K = 1.3 B.t.u. per square foot per degree Fahrenheit.

Substituting these values in the above equation,

$$F_d = \frac{20{,}000}{1.3(254.5 - 90)} = 93.5 \text{ sq. ft. of heating surface.}$$

Since 3-in. ribbed cast-iron pipe has a heating surface of about 5.5 sq. ft. per lineal foot of pipe, there will be required 93.5 ÷ 5.5 = 17 lin. ft. of pipe.

c. Piping Computations.

Example 19.

The mains of a district heating system as shown in Fig. 177 are to be designed. Assume that steam distribution headers for various mains are located at A, B', C' and D' while at E is located the distribution header at the plant. The steam pressure at E is 100 lb. per square inch absolute and at A, B', C', and D' it is 35 lb. per square inch absolute. The steam requirements and the length of mains may be taken from the following table. The piping is assumed to be good insulated.

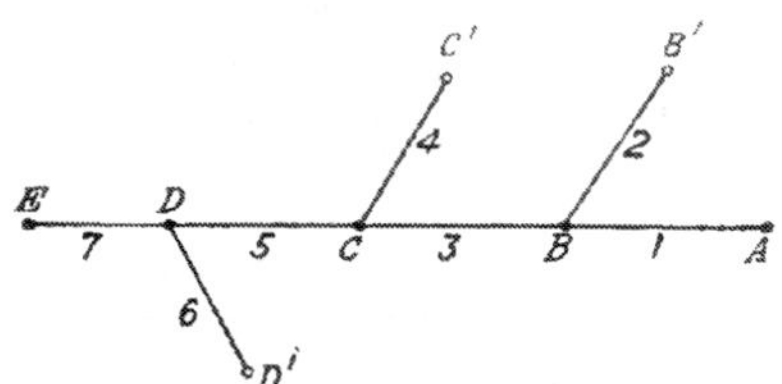

FIG. 177.

Number of the pipe section	Steam required, Q_1 in pounds per hour	Length of pipe in feet
1	3,150	660
2	4,420	290
3	7,570	290
4	3,370	300
5	10,940	510
6	4,640	360
7	15,580	690

1. *Piping Assumptions.*

Initial pressure = 100 lb. per square inch.
Terminal pressure = 35 lb. per square inch.

Difference = 65 lb. per square inch.

The average distance between buildings is assumed to be 300 ft. In accordance with Table XVI the magnitude of resistances of the fittings will be 10 per cent. Due to the layout of the system, however, there is a probability of using expansion bends in each of the pipe sections so that the total resistance of the fittings may reach 20 per cent. (The use of steam separators would further increase the magnitude of the resistance of the fittings.) The pressure available for overcoming pipe friction is therefore 65 lb. per square inch minus 20 per cent or 52 lb. per square inch. With the stipulated terminal pressure of 35 lb. per square inch, the initial pressure must be

$$35 + 52 = 87 \text{ lb. per square inch.}$$

The heat loss may be neglected for the present. The length of the least-favored circuit (pipe sections 1, 3, 5, and 7) is 2,150 ft.

The values of B_2 and B_1 corresponding to the pressures p_2 and p_1 (*i.e.*, 87 and 35 lb., respectively) may be taken from the auxiliary table of Chart 3 as follows:

$$B_2 = 5{,}720.$$
$$B_1 = 980.$$
$$\frac{B_2 - B_1}{l} = \frac{4{,}740}{2{,}150} = 2.2.$$

Based upon the value so found the results on consulting Chart 3 are as follows:

Number of pipe section	Weight of steam Q registered in pounds per hour	Assumed diameter d in inches
1	3,150	3
3	7,570	4
5	10,940	4½
7	15,580	5

For the pipe sections 1, 3, 5, and 7 the values of $B_2 - B_1$ are as follows:

(1) $660 \times 2.2 = 1{,}452.$
(3) $290 \times 2.2 = 638.$
(5) $510 \times 2.2 = 1{,}122.$
(7) $690 \times 2.2 = 1{,}518.$

The value of $\frac{B_2 - B_1}{l}$ for the other pipe sections is as follows:

(2) $\frac{1{,}452}{290} = 5.0.$

(4) $\frac{2{,}090}{300} = 6.96.$

(6) $\frac{3{,}212}{360} = 8.92.$

Hence the diameters for the respective sections will be

$d_2 = 3$ in.
$d_4 = 2\frac{1}{2}$ in.
$d_6 = 2\frac{1}{2}$ in.

2 *Checking Computation of the Piping*—As an example of the method to be employed in the checking computation for the various pipe sections, that for section 1 will be given in detail The checking computation for the entire piping layout is tabulated in Summary I, p 241

Let the first half of the pipe section 1 (from the valve to the expansion bend) be designated by 1*a*, and let the second half (from the expansion bend to the branch) be 1*b* (Fig 178)

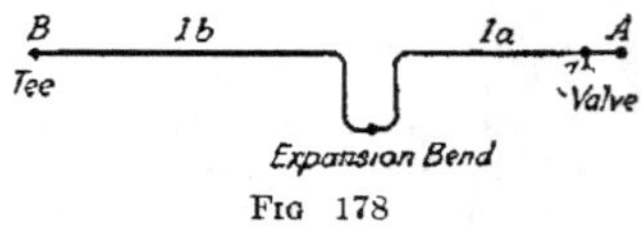

FIG 178

a Resistances of Fittings at Inlet to Pipe Section 1*a*

Valve	$\zeta = 7\ 0$
Diameter	$d = 3$ in
Pressure	$p = 35$ lb per square inch
Weight of steam	$Q = 3{,}150$ lb

From Chart 3 it will be found that the steam velocity is $v = 180$ ft per second and that the pressure drop caused by the valve is $Z = 7\ 0 \times 0\ 418 = 2\ 93$ lb per square inch

b Frictional Resistance of Pipe Section 1*a*—The condensation per lineal foot of pipe as indicated on page 235 in pounds per hour is

$$q = \frac{30D}{\lambda}.$$

Due to the small heat loss, the value of λ (latent heat of vaporization), corresponding to the average steam pressure of the entire piping layout, may be used for each individual pipe section

For $p_2 = 100$ lb per square inch	$\lambda = 888$ (see Table III).
For $p_1 = \ 35$ lb per square inch	$\lambda = 939$
Average	$\lambda = 914$ B t u per pound

The additional data required for the determination of the frictional resistance of pipe section 1*a* are as follows

$$\frac{ql}{2} = \frac{3\ 5 \times 30 \times 330}{914 \times 2} \cong 19 \text{ lb}$$ [1]

$$Q = 3{,}150$$

$$Q + \frac{ql}{2} = 3{,}169$$

$$p_1 = p + Z = 35 + 2\ 93 = 37\ 93 \text{ lb per square inch.}$$

From the pressure table in Chart 3 the following data will be found:

$$B_1 = 1{,}146 \text{ (corresponding to } p = 37\ 93)$$

$$B_2 - B_1 = \ 528^2$$

$$B_2 = 1{,}674$$

$$p_2 = \ 45\ 9 \text{ lb per square inch}$$

[1] External diameter taken (see Table I).

[2] According to Auxiliary Table 5 Weight of steam = 3,169 lb

Pipe diameter = 3 in

$$\frac{B_2 - B_1}{l} = 1\ 6$$

$$l = 330 \text{ ft}$$

$$B_2 - B_1 = 1\ 6 \times 330 = 528$$

c. Resistance of Fitting at Inlet of Pipe Section 1*b*.

Expansion bend	$\zeta = 4.0 = 4$ ells.
Diameter	$d = 3$ in.
Absolute pressure	$p = 45.9$ lb. per square inch.
Weight of steam	$Q = 3{,}150 + ql = 3{,}188.$
Z (determined as under *a*)	$Z = 1.31$ lb. per square inch.

d. Friction Resistance of Pipe Section 1*b*.

$$Q + \frac{ql}{2} = 3{,}188 + 19 = 3{,}207 \text{ lb.}$$

$$p_1 = p + Z = 45.9 + 1.31 = 47.21 \text{ lb. per square inch.}$$

$$B_1 = 1{,}765 \text{ (corresponding to } p_1 = 47.21).$$

$$B_2 - B_1 = \quad 528.$$

$$B_2 = 2{,}293.$$

$$p_2 = 54.16.$$

e. Resistance of Fittings at Outlet of Pipe Section 1*b*.

$$\text{Tee (through) } \zeta = 1.0.$$
$$d = 3 \text{ in.}$$
$$p = 54.16.$$
$$Q = 3{,}188 + ql = 3{,}226 \text{ lb.}$$
$$Z = 0.28 \text{ lb. per square inch.}$$

The pressure at *B* therefore is 54.16 + 0.28 = 54.44 lb. per square inch.

In a similar manner the checking computations are made for the remaining pipe sections. These are compiled in Summary I. The example shows that for the lowest outside temperature the heat loss from the piping may be neglected.

3. *Condensate Pipe Lines.*—In considering the steam mains of a district system, Table XVII may not be used. The condensation line is to be dimensioned in accordance with the rules given for hot-water lines. The circulating pressure may be caused by temperature difference or by the potential energy available when an elevated tank is employed in the line.

Example 20.

In the countercurrent apparatus shown in Fig. 179, 2,000,000 B.t.u. per hour is to be supplied by a number of ½-in. tubes each 13 ft. long and connected in parallel.

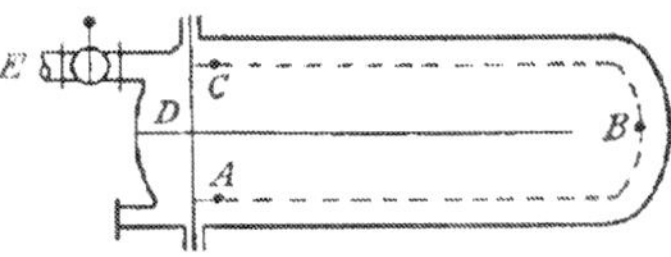

Fig. 179.—Counter current apparatus.

The pressure at *A* is to be 14.7 lb. per square inch absolute and the initial pressure at *E* must not exceed 32 lb. per square inch absolute. The average water temperature is to be 90° F. It is required to check the initial pressure. Assume one-half the steam is condensed from *C* to *B*. Assume also that the coefficient of heat transmission shall be 200 B.t.u. per square foot per hour per degree Fahrenheit and that the pressure in the coils at *C* is 20 lb. per square inch absolute. Therefore the average steam temperature in the coils is 220° F. corresponding to a latent heat of 965 B.t.u. per hour.

SUMMARY I

Number of pipe section		Length l, feet	Diameter d, inches	Weight of steam at start of pipe section Q_1, pounds per hour	Heat-loss for pipe length l $q \cdot l$, pounds per hour	Weight of steam at end of pipe section $Q_2 = Q_1 \div q \cdot l$, pounds per hour	Resistance of fitting at start of pipe section				Frictional resistance of pipe section						Resistance of fitting at end of pipe section				Pressure at point noted, pounds per square inch
							Type of resistance	ζ	Absolute pressure p_1 pounds per square inch	Pressure loss Z, pounds per square inch	Average steam weight $Q_1 + \frac{q \cdot l}{2}$, pounds per hour	Initial pressure for friction $p_1 = P + Z$, pounds per square inch	Auxiliary quantity B_1	$B_2 - B_1$	Auxiliary quantity B_2	End pressure for friction p_2, pounds per square inch	Type of resistance	ζ	Absolute pressure P, pounds per square inch	Pressure loss Z, pounds per square inch	
1	a	330	3	3,150	38	3,188	Valve	7.0	35	2.93	3,169	37.93	1,146	528	1,674	45.9					
	b	330	3	3,188	38	3,226	Expansion bend	4.0	45.9	1.31	3,207	47.21	1,765	528	2,293	54.16	Tee (run)	1.0	54.16	0.28	*B*: 54.4
2	a	145	3	4,420	17	4,437	Valve	7.0	35	6.10	4,429	41.1	1,337	435	1,772	47.8					
	b	145	3	4,437	17	4,454	Expansion bend	4.0	47.3	2.65	4,446	49.95	1,957	435	2,392	55.4	Y (branch)	1.0	55.4	0.57	*B*: 56.0
3	a	145	4	7,680	21	7,701	. . . .	. . .	. . . . . .	. . . . .	7,691	56	2,440	290	2,730	59.8					
	b	145	4	7,701	21	7,722	Expansion bend	4.0	59.3	2.13	7,712	61.43	2,919	290	3,209	64.6	Tee (run)	1.0	64.6	0.49	*C*: 65.1
4	a	150	2½	3,370	14	3,384	Valve	7.0	35	8.12	3,377	43.12	1,478	750	2,228	52.35					
	b	150	2½	3,384	14	3,398	Expansion bend	4.0	53.35	3.15	3,391	56.5	2,480	750	3,230	64.80	Tee (branch)	1.5	64.8	0.98	*C*: 65.8
5	a	255	4	11,120	38	11,158	. . . .	. . .	. . . . . .	. . . . .	11,139	65.8	3,330	1,020	4,350	73.6					*
	b	255	4½	11,158	42	11,200	Expansion bend	4.0	75.5	2.27	11,179	77.77	4,602	638	5,240	83.1	Tee (run)	1.0	83.1	0.5	*D*: 83.6
6	a	180	2½	4,640	17	4,657	Valve	7.0	35	14.5	4,649	49.5	1,925	1,440	3,365	66.2					
	b	180	2½	4,657	17	4,674	Expansion bend	4.0	66.2	4.56	4,666	70.76	3,826	1,440	5,266	83.3	Tee (branch)	1.5	83.3	1.4	*D*: 84.7
7	a	345	5	15,874	63	15,937	. . . .	. . .	. . . . . .	. . . . .	15,906	84.7	5,434	863	6,297	91.4					
	b	345	5	15,937	63	16,000	Expansion bend	4.0	91.4	2.3	15,969	93.7	6,618	863	7,481	99.9	Valve	7.0	99.9	3.7	*E*: 103.6†

* Pipe section 5*a* at first was assumed 4½ in. in diameter. This gave excessive pressure difference at *D* so that 4 in. is advisable.

† The slight difference compared to the stipulated value of *E* may be neglected.

Solution Heat emission of a single pipe is

$$W = FK(t_d - t) = \frac{13}{4\ 55}\ 200(220 - 90) \cong 74{,}300 \text{ B t.u. per hour}$$

$$\text{Number of tubes} = \frac{2{,}000{,}000}{74{,}300} \cong 27.$$

$$\text{Steam transmitted per tube} = N = \frac{74\ 300}{965} = 77 \text{ lb per hour}$$

1 *Resistance of a Single Tube* *a* Frictional Resistance —It is assumed that one-half of the steam, 38 5 lb per hour, has been condensed in passing from C to B and from B to A Consider first the section from A to B The heating effect N is that equivalent to 38 5 lb of steam According to Eq (73) this is to be divided by 1 76 before consulting Chart 3, *i e*,

$$\frac{N}{1\ 76} = \frac{38\ 5}{1\ 76} = 21.9.$$

The following data will be obtained from the chart for the value 21 9

$$\frac{B_2 - B_1}{l} = 1\ 3$$

$$B_2 - B_1 = 1\ 3 \times 6\ 5 = 8\ 45$$
$$B_1 = 183$$
$$B_2 \cong 192$$
$$p_2 = 15\ 1 \text{ lb per square inch absolute.}$$

b Resistance of Fitting at B —With the aid of Chart 3 let

$$\zeta = 2\ 0$$

At B, 38 5 lb are transmitted with a velocity of 80 ft per second The pressure at B = 15 1 lb per square inch and therefore for

$$\Sigma\zeta = 1,\ Z = 0\ 18$$
$$\Sigma\zeta = 2,\ Z = 0\ 36 \text{ lb. per square inch}$$

The pressure at B for the section C-B = 15.46 lb per square inch Q_1 at B = 38 5 and ql = 38 5 Therefore

$$Q_1 + \frac{ql}{2} = 57\ 8.$$

For this value

$$\frac{B_2 - B_1}{l} = 8\ 0$$

$$B_2 - B_1 = 8\ 0 \times 6\ 5 = 52$$
$$B_1 = 202$$
$$B_2 = 254$$
$$p_2 \cong 17\ 5 \text{ lb per square inch}$$

This to be contrasted with the assumed value of 20 lb per square inch

2 *Resistance at Inlet of the Tube*

$$\zeta = 1\ 0 \text{ (assumed)}$$
$$Q = 77 \text{ lb per hour.}$$
$$v = 160 \text{ ft per second}$$
$$p_2 \cong 17\ 5 \text{ lb per square inch}$$
$$Z \cong 0\ 63 \text{ lb per square inch}$$

The pressure at D is therefore

$$p \cong 17\ 5 + 0\ 63 = 18\ 13 \text{ lb per square inch}$$

3 *Resistance Due to Eddying in Chamber D*—Based on the velocity of steam flow

$$\zeta = 1\ 0 \text{ (assumed).}$$

This resistance may be considered in combination with the valve resistance merely by increasing the coefficient of resistance of the valve by 1

4 *Resistance Due to the Valve*—A 2-in globe valve of the usual type is assumed Due to considerations discussed under heading 3

$$\zeta = 7\ 0 + 1\ 0 = 8\ 0$$

$$Q = \frac{2{,}000{,}000}{965} = 2{,}070 \text{ lb per hour}$$

$$v \cong 300 \text{ ft per second}$$

The pressure at the valve outlet is 18 13 lb per square inch
The pressure at the valve inlet is 32 lb per square inch
The average pressure is 25 1 lb per square inch
From Chart 3

$$Z \text{ for } \Sigma\zeta = 1 \text{ is } 1\ 59$$
$$\Sigma\zeta = 8 \text{ is } 12\ 72$$

$$p = 18\ 13 + 12\ 72 = 30\ 9 \text{ lb per square inch}$$

Attention is directed to the considerable pressure drop through the valve

D LOW-PRESSURE STEAM HEATING SYSTEMS

COMPUTATION OF BOILER HEATING SURFACE

To determine the steam requirements of an installation Eqs (42) and (43) previously discussed apply Thus

During the heating-up period,

$$F_1 = \frac{1.1\{(W + Z)z + 0.12B(t_1 - \delta)\}}{K_1 z}.$$

In the steady state,

$$F_2 = \frac{1\ 1W}{K_2}.$$

The maximum values of K_1 and K_2 may be assumed as

$$K_1 = 4{,}000 \text{ B t u per square foot}$$
$$K_2 = 3{,}000 \text{ B t u per square foot}$$

As previously indicated (p 192) intermittent operation is assumed for outside temperatures of 32° F and above while continuous operation is required for temperatures below 32° F Hence the heating surface must be determined for both conditions, and the larger of the two values must be used in choosing the installation

COMPUTATION OF RADIATOR HEATING SURFACE

The required radiation to be installed is found from Eq (35) previously given, *i e*,

$$F_d = \frac{W}{K(t_d - t)}.$$

The values of K may be taken from Table IX.

PIPE SIZES

1. Theory.

a Circulating Pressure Head.—The circulating pressure head for low-pressure heating systems ranges from 0.1 to 3.0 lb. per square inch. It is usual to assume the maximum pressure and design the system accordingly.

b Frictional Resistances.—The following theory[1] was developed by Dr. Wierz. For low-pressure steam heating systems the following assumptions may be made.

Average heat of vaporization $\lambda = 967$ B.t.u. per pound.
Average density of steam $s = 0.041$ lb. per cubic foot.

Consequently Eq. (63) (p. 231) becomes

$$\frac{dp}{dl} = \frac{0.458\ 10^{-4}}{0.003(6.25\pi)^{1.853}}\ \frac{Q^{1.853}}{d^{4.987}p^{0.9375}}.$$

Since

$$s = 0.003p^{0.9375},$$

therefore

$$\frac{dp}{dl} = \frac{0.458\ 10^{-4}}{(6.25\pi)^{1.853}}\ \frac{Q^{1.853}}{d^{4.987}s}$$

$$= \frac{0.458\ 10^{-4}}{0.041(6.25\pi)^{1.853}}\ \frac{Q^{1.853}}{d^{4.987}} \qquad (76)$$

Substituting from Eq. (64) the value $\frac{1}{dl} = \frac{q}{dQ}$ and transposing

$$dp = \frac{0.458\ 10^{-4}}{0.041(6.25\pi)^{1.853}}\ \frac{Q^{1.853}dQ}{qd^{4.987}} \qquad (77)$$

Integrating between the limits $l = 0$ and $l = l$, dividing by l and substituting the value $n = 2.853$, there results

$$\frac{p_2 - p_1}{l} = \frac{0.449\ 10^{-5}}{d^{4.987}}\ \frac{Q_2^n - Q_1^n}{nql}$$

$$= \frac{0.449\ 10^{-5}}{d^{4.987}}\ \frac{Q_1^n}{nql}\left[\left(\frac{Q_2}{Q_1}\right)^n - 1\right] \qquad (78)$$

From Eq. (71) it is known that

$$\left(\frac{Q_2}{Q_1}\right)^n - 1 = \frac{nql}{Q_1^n}\left(Q_1 + \frac{ql}{2}\right)^{n-1}$$

Substituting this in Eq. (78) it becomes

$$\frac{p_2 - p_1}{l} = 0.449\ 10^{-5}\ \frac{\left(Q_1 + \frac{ql}{2}\right)^{1.853}}{d^{4.987}}. \qquad (79)$$

[1] "Simplified Analysis of Pipe Sizes for Steam Installations," from BRABBÉE-WIERZ, "Vereinfachtes zeichnerisches oder rechnerisches Verfahren zur Bestimmung der Durchmesser von Dampfleitungen," *Bull.* 23, Research Laboratory, Charlottenburg, 1915.

Moreover since

$$Q = \frac{W}{967} \qquad \text{and} \qquad q = \frac{w}{967},$$

where

W = heat requirements in B t u per hour,
w = heat loss in the pipe section in B t u per lineal foot,

then Eq (79) becomes

$$\frac{p_2 - p_1}{l} = \frac{0\ 449\ 10^{-5}}{967^{1\ 853}} \cdot \frac{\left(W + \frac{wl}{2}\right)^{1\ 853}}{d^{4\ 987}}.$$

$$= 0\ 132\ 10^{-10} \frac{\left(W + \frac{wl}{2}\right)^{1\ 853}}{d^{4\ 987}} \tag{80}$$

(1) Low-pressure Steam Lines —For insulated low-pressure steam piping it may be assumed that

$$W + \frac{wl}{2} = 1\ 05W, \tag{81}$$

where

W = heat requirement at end of pipe section in B t u. per hour.

It should be remembered that an additional allowance for the heat loss is really a function of the pipe diameter and the existing pressure drop. A consideration of these elements, however, would complicate the computations without practical justification A number of installations have shown that with an insulation efficiency of 75 per cent an additional allowance of 5 per cent to the heat requirement for the heat loss is ample and does not increase the cost of the piping appreciably Should it be necessary under unfavorable conditions to increase the factor, the actual heat loss is easily determined in the checking computation It is seldom that a change in the pipe diameter will be required as a result of the checking computation

From Eqs (81) and (80) it follows that

$$\frac{p_2 - p_1}{l} = R = 0\ 132\ 10^{-10} \frac{(1\ 05W)^{1\ 853}}{d^{4\ 987}}. \tag{82}$$

For bare pipes the value of W should be increased an additional 10 per cent, making a total increase of 15 per cent

(2) Low-pressure Steam Coils —For low-pressure steam coils, in Eq (78) let $Q_1 = 0$; *i e*, $Q_2 = ql$ As a consequence

$$\frac{p_2 - p_1}{l} = \frac{0\ 449\ 10^{-5}}{d^{4\ 987}} \cdot \frac{(ql)^{2\ 853}}{nql}$$

$$= \frac{0\ 449\ 10^{-5}}{d^{4\ 987}} \cdot \frac{(ql)^{1\ 853}}{2\ 853} \tag{83}$$

or

$$\frac{p_2 - p_1}{l} = \frac{0\ 449\ 10^{-5}}{d^{4\ 987}} \left[\frac{ql}{\sqrt[1\ 853]{2\ 853}}\right]^{1\ 853}.$$

If M represents the effective heat transfer in B t u , then

$$\frac{p_2 - p_1}{l} = \frac{0\ 132\ 10^{-10}}{d^{4\ 987}}\left(\frac{M}{1\ 76}\right)^{1\ 853}$$

measured in pounds per square inch per foot

In Chart 5 the values of $1.05W$ have been introduced instead of the heat requirement W To use the table, the value 1 76 must be increased to 1 85, in which case the equation becomes

$$\frac{p_2 - p_1}{l} = \frac{0\ 132 \times 10^{-10}}{d^{4\ 987}}\left(\frac{M}{1\ 85}\right)^{1\ 853} \tag{84}$$

in pounds per square inch per foot of length of pipe

(3) Resistances of Fittings —The general equation Eq. (39) takes the form

$$Z = 0\ 007 \sum \zeta \frac{v^2}{2g} s = 0\ 007 \times 0\ 041 \sum \zeta \frac{v^2}{2g}$$

Since the density s may be assumed constant, all difficulties encountered in high-pressure systems due to the variable density are eliminated and the resistance of the fitting may be taken directly from Chart 5, as was done in the case of hot-water heating systems

2. Chart 5 and Its Application.

a General —From previous discussions the following group of equations may be written

For insulated steam piping

$$R = \frac{p_2 - p_1}{l} = 0.132 \times 10^{-10} \frac{(1\ 05W)^{1\ 853}}{d^{4\ 987}}$$

For steam coils

$$R = \frac{p_2 - p_1}{l} = \frac{0\ 132 \times 10^{-10}}{d^{4\ 987}}\left(\frac{M}{1\ 85}\right)^{1\ 853} \tag{85}$$

$$Z = 0\ 041 \times 0\ 007 \sum \zeta \frac{v^2}{2g}$$

On the right-hand side Chart 5 contains the following

(1) The values of W in horizontal row I When consulting for the quantity of heat W in this table, the values are in reality 5 per cent larger In this way the usual cases of heat loss from insulated steam piping is allowed for sufficiently to make additional computation unnecessary

(2) The steam velocities in feet per second in horizontal row II

(3) The frictional resistance R in pounds per square inch per foot of pipe

(4) The pipe diameters from ½ to 12 in.

On the left-hand side·

(1) The steam velocities in feet per second

(2) The resistances of fittings Z for $\Sigma\zeta 1$ to 15

(3) The resistance coefficients for the usual fittings

b Preliminary Design of Piping (1) When Piping Is Sufficiently Insulated —The analysis is begun with consideration of the least-favored circuit According to the most recent investigations[1] it is desirable to allow a pressure of 0 3 lb per square inch at the inlet of the valve of all radiators except in vapor heating systems In this case, however, the fitting resistances are one-third of the remaining pressure head[2] instead of one-half of it Dividing the frictional resistance of the pipe circuit by its length in feet, the pressure drop R per foot of pipe is obtained Ascertaining this value in Chart 5 and proceeding horizontally to the quantity of heat to be transmitted, the required diameter of pipe is found at the head of the column For long pipes in which the condensation flows against the steam (*e g*, risers passing through several floors), it is desirable to keep the pressure drop within certain limits This limit in practice is assumed between 0 2 and 0 5 lb per square inch per 100 ft of pipe In America the "Journal of the Society of Heating and Ventilating Engineers, Sept 1927, p. 568" specifies accepted velocities, which are not to be exceeded

(2) Bare Pipes —The method for the design of bare piping is similar to that for insulated pipes except that the quantity of heat is increased 10 per cent and then ascertained in the horizontal row I.

(3) Steam Coils —In the design of steam coils the procedure is the same as for bare pipes except that the heat effect of the coil must be divided by 1 85 and this quotient then sought in the Chart 5

c Checking Computation of the Piping (1) Insulated Piping — With the pipe diameters known from the previous discussion, the quantity of heat carried is found in the respective column of the chart, and at the extreme left the pressure drop in pounds per square inch per foot of pipe is given. Directly underneath the quantities of heat are the corresponding steam velocities, and these when transferred to the fitting resistance table will give the values of Z in connection with the coefficients of resistance ζ

(2) Bare Pipes —In the case of bare pipes the heat loss of the piping is determined with the aid of Table I, and one-half of this value is added to the quantity of effective heat required This value may then be found in Chart 5 Strictly speaking this procedure is not wholly correct since all the quantities in Chart 5 have been already increased by 5 per cent Since the uninsulated pipe is usually short, however, the

[1] "Pressure Relations in Low-pressure Heating Systems," from FRENKEL, "Über Druckverhaltnisse in Niederdruckdampfheizungen," *Bull* 31, Research Laboratory, Charlottenburg, 1921

[2] For district heating mains the values stipulated for hot-water heating systems may also be used in steam systems

suggested procedure will not increase the size and the cost appreciably. When it is desired to be more accurate, the tabular values should be corrected for the 5 per cent increase, and to this result one-half of the heat loss of the bare piping should be added. The chart may then be used as before.

(3) Steam Coils.—The computation for steam coils is carried out as indicated for insulated pipes with the exception that the effective heat quantities must be divided by 1.85 and this quotient then used in consulting Chart 5.

(4) Return Lines.—The condensation lines are proportioned as shown on page 312 and in accordance with values given in Table XVII.

Example 21.

Assume boiler gage pressure at 0.7 lb. per square inch and gage pressure at radiator valves 0.3 lb. per square inch. All piping is adequately insulated. The layout of the installation is shown in Fig. 180. Additional data may be taken from Summary J. The problem is to find the pipe sizes.

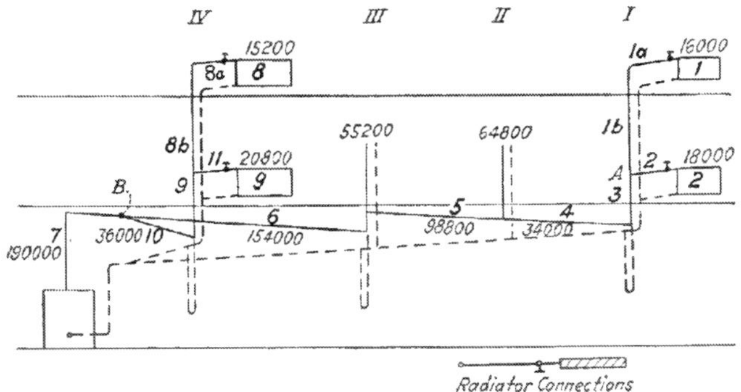

FIG. 180.

Solution: 1. *Piping Assumptions.* *a.* The Least-favored Current Circuit 1*a*, 1*b*, 3, 4, 5, 6, and 7.

Gage pressure at boiler	= 0.7	lb. per square inch.
Gage pressure at radiator valve	= 0.3	lb. per square inch.
Difference	= 0.4	lb. per square inch.
Allowance of one-third for resistances of fittings	= 0.13	lb. per square inch.
Available for pipe friction	= 0.27	lb. per square inch.

This allowance is for 1*a*, 1*b*, 3, 4, 5, 6, and 7 having a total length of 92 ft. Therefore the pressure drop $R = \frac{0.27}{92} = 0.00293$ lb. per square inch per foot of pipe. The value of d entered in Summary J is found according to the drop R with the aid of Chart 5.

SUMMARY J

Number of pipe section	Heat transmitted in B.t.u.	Length, feet	Assumed, inches	Computation data: Original values: R per foot, pounds per square inch	v, feet per second	lR, pounds per square inch	Σζ	Z, pounds per square inch	Corrected values: d, inches	R per foot, pounds per square inch	v, feet per second	lR, pounds per square inch	Σζ	Z, pounds per square inch	lR (g − n)	Z (i − p)
(a)	(b)	(c)	(d)	(e)	(f)	(g)	(h)	(i)	(k)	(l)	(m)	(n)	(o)	(p)	(q)	(r)
Radiator 1																
1a	16,000	7	¾	0.0040	40	0.0280	2.0	0.0143								
1b	16,000	12	1	0.0009	20	0.0108	1.0	0.00178								
3	34,000	5	1	0.0036	45	0.0180	1.5	0.0136								
4	34,000	16	1	0.0036	45	0.0576	1.0	0.00908								
5	98,800	16	1½	0.0036	60	0.0576	1.5	0.0241								
6	154,000	26	2	0.0020	50	0.0520	2.5	0.0279								
7	190,000	10	2	0.003	60	0.0300	2.0	0.0321								
		92														
$\Sigma(lR + Z)_1^7 = 0.254 + 0.123 = 0.377$ lb. per square inch, the remainder to be throttled down.																
Radiator 2																
2	18,000	12	¾	0.0050	45	0.06	1.5	0.0136								
$\Sigma(lR + Z)_2 + \Sigma(lR + Z)_3^7 = 0.0736 + 0.322 = 0.396$ lb. per square inch, the remainder to be throttled down.																
Radiator 8																
8a	15,200	6	¾	0.0036	36	0.0216	2	0.0116								
8b	15,200	12	¾	0.0036	36	0.0432	1.0	0.00578								
9	36,000	5	1	0.0040	45	0.0200	1.5	0.0136								
10	36,000	26	1	0.0040	45	0.1040	1.5	0.0136								
		49														
$\Sigma(lR + Z)_8^{10} = 0.189 + 0.0446 = 0.234$ lb per square inch.																
$\Sigma(lR + Z)_7 = 0.062$																
0.296 the remainder to be throttled down.																
Radiator 9																
11	20,800	6	¾	0.006	45	0.036	1.5	0.0136								
$\Sigma(lR + Z)_{11} = 0.0196$ lb. per square inch.																
$\Sigma(lR + Z)_7 + \Sigma(lR + Z)_8^{10} = 0.062 + 0.151 = 0.213$ lb. per square inch.																
$\Sigma(lR + Z)_{11} + \Sigma(lR + Z)_3^7 + \Sigma(lR + Z)_8^{10} = 0.263$ lb. per square inch, the remainder to be throttled down.																

b Radiator 2

At A the initial pressure is 0 055 lb per square inch to overcome the frictional resistance of pipe section 2 The frictional drop per foot of pipe is therefore $R = \frac{0\ 055}{12} = 0\ 0046$ lb per square inch The required diameter is found from Chart 5

c Radiator 8

The pressure drop due to friction in pipe sections 1*a*, 1*b*, 3, 4, 5, and 6 is 82 × 0 00293 = 0 240 lb per square inch. The total length of the circuit 8*a*, 8*b*, 9, and 10 is 49 ft The pressure drop per foot is therefore $\frac{0\ 24}{49} = 0\ 0049$ lb. per square inch With this pressure drop consult Chart 5 and enter the diameters for the respective pipe sections in Summary J Care must be taken, that the frictional resistance of these pipe sections in which the condensation flows in countercurrent to the steam does not exceed the values given on page 247

Exceptions are only permissible for short, vertical pipes effectively drained, *e g*, pipe section *q*, in which the original pressure drop of 0 0049 may be maintained In this case and for 36,000 B t u Chart 5 shows the required pipe diameter to be 1 in In pipe section 10 condensation flows in steam direction so that the pressure drop of 0 0049 can be upheld, and Chart 5 realizes a diameter of 1 in

From the foregoing analysis supplemented by data in Chart 5 the summary J is compiled

2 *Checking Computations of the Pipe Sizes* —The check results are presented in Summary J The influence of unavoidable variations in the boiler pressure, as well as the influence on the pressure caused by shutting off one or more radiators may be counteracted to some extent by using ⅛ to ¼-in pipe lines D, as shown in Fig 181, which will act as a throttle They are installed with a drip allowing for drainage so as to prevent damage by frost These pipe lines must be sufficiently elastic to allow for the temperature variations in the system (212 to 32° F) Excellent results have been obtained with them in the installation at the Research Laboratory, Charlottenburg.

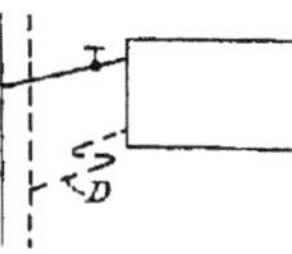

FIG 181

Example 22.

A dryer is to be supplied with steam from a low-pressure manifold The gage pressure at the manifold is 0 7 lb per square inch, and the gage pressure at the dryer is to be 0 5 lb per square inch The heat requirement of the dryer is 400 000 B t u. per hour The pipe is 160 ft long and $\Sigma\zeta = 8$ The piping is adequately insulated.

Preliminary Computation

Pressure drop available	= 0 2 lb per square inch
20 per cent allowance for resistances of fittings (Table XVI)	= 0 04
Pressure drop in pipe	0 16

Pressure drop per foot of pipe $= \frac{0\ 16}{160} = 0\ 001$ lb per square inch

From Chart 5 $d = 3\frac{1}{2}$ in

Check Computation.

R per foot = 0 0007 lb per square inch
$l \times R$ = 0 112 lb per square inch
Z = 0 057 lb per square inch

$lR + Z$ = 0 169 lb per square inch

Therefore the pipe size need not be altered

Example 23.

A low-pressure steam coil is to supply 40,000 B t u per hour to a room the temperature of which is 70° F The available pressure is 0 07 lb per square inch

Preliminary Computation: Choose a pipe diameter of 1½ in

$$\text{Surface } F \text{ per foot of pipe} = \frac{1}{2\ 01} \cong 0\ 5 \text{ sq ft (Table I).}$$

Coefficient of heat transmission K = 2 3 B t u per square foot per degree Fahrenheit per hour Table IX Therefore the length l of the pipe coil is

$$l = \frac{40\ 000}{0\ 5 \times 2\ 3(t_d - t)} = \frac{40{,}000}{0\ 5 \times 2\ 3(212 - 70)}$$

$$= 245 \text{ ft}$$

The coil may be installed in 12 lengths of 20 ft each, and the remaining surface may be considered supplied by the manifolds (headers)

Checking Computations.

$$d = 1\tfrac{1}{2} \text{ in}$$

$$l = 240 \text{ ft}$$

$$\Sigma\zeta(\text{for 1 valve and 11 headers}) = 7 + 11 \times 1.$$

$$= 18$$

$$\frac{M}{1\ 85} = \frac{40{,}000}{1\ 85} = 21{,}600 \text{ B t u}$$

$$R \text{ per foot} = 0\ 0002 \text{ lb per square inch}$$

$$l \times R = 0\ 048 \text{ lb per square inch}$$

$$Z = 0\ 0116$$

$$(lR + Z) \cong 0\ 06 \text{ lb per square inch against } 0\ 07 \text{ available}$$

E. VACUUM HEATING SYSTEMS

The theory developed for high-pressure steam mains on page 230 is also valid for pressures below atmosphere Piping for vacuum heating systems may be determined with the aid of Chart 4 which corresponds with Chart 3 for high-pressure steam. It differs, however, in that the values of the resistance of fittings and the values of B have been extended to be valid for pressures lower than 14 lb per square inch.

F. COMBINATION HEATING SYSTEMS

Steam Hot-water Heating Systems

The steam circuit of combined steam and water systems may be designed in accordance with principles discussed on page 243 while the water circuit may be proportioned as outlined on page 194

G. WARM-AIR HEATING SYSTEMS

In the design of warm-air systems the following items must be determined.

(1) The quantity of heat necessary for effective heating with due regard to the heat lost in the duct system

(2) The heat required for evaporation

(3) The size of the heater

(4) The volume of air to be handled for the proportioning of the ducts.

(5) The dimensions of the duct system

ITEMS 1 AND 2

The computations of warm-air systems are substantially the same as for fan systems to which reference should be made Attention might be drawn, however, to specific items

ITEM 3

Steam and water radiators are frequently used in warm-air systems as well as in ventilation systems. The following heat emission may be assumed for warm-air furnaces

Gravity systems, 600–800 B t u per square foot of smooth heating surface
Fan systems, 2,000 B t u per square foot of smooth heating surface.

ITEM 4

In the case of ventilating systems the volume of air to be handled is known and the proportions of the duct system are based thereon This is taken up in greater detail on page 265 For gravity warm-air heating systems, two conditions must be considered

1. When the Number of Air Changes is Indefinite.

The following elements are known or are assumed·

(1) The heat W in B t u per hour required to maintain the temperature of the room in extreme weather

(2) The temperature t' of the entering air (to be assumed not to exceed 150° F)

(3) The room temperature t in degrees Fahrenheit

The relation exists

$$W = \frac{L}{1 + \alpha t} \times 0\ 08635 \times 0\ 24\ (t' - t) \qquad (86)$$

in which

L = volume of air in cubic feet at $t°$ F which is to be heated from $t°$ F to $t'°$ F

$\frac{L}{1 + \alpha t}$ = reduction of the volume L occupied at $t°$ F to the volume occupied at 0° F (α = 0 00218 = coefficient of expansion of the air)

0 08635 = density of the air at 0° F. and at 14.7 lb. per square inch

$0\ 08635 \frac{L}{1 + \alpha t}$ = weight of the air L in pounds

0 24 = specific heat of air in B t u per pound

$0\ 08635 \times 0\ 24 \frac{L}{1 + \alpha t} = 0\ 0207 \frac{L}{1 + \alpha t}$ = heat necessary to raise the temperature of L cu ft of air 1° F

Consequently the total heat for the interval $(t' - t)$ is

$$W = 0.0207 \frac{L}{1 + \alpha t} (t' - t) \tag{87}$$

If the heat requirement of a room be W B t u and this must be carried by means of the air supply, the necessary volume of air at $t°$ F will be

$$L = \frac{W(1 + \alpha t)}{0.0207(t' - t)} \tag{88}$$

The value of W is that required for the lowest outside temperature though not the most unfavorable for the design of the duct system The proportioning of the duct system is based on the highest outside temperature for which the system must operate by gravity alone and be capable of maintaining the necessary room temperature $t°$ F The values in Eq (88) require a change in notation for this purpose as follows

The quantity of heat changes from W to W_1

The volume of air changes from L to L_1 (both at $t°$ F)

The temperature of entering air changes from t' to t_1' To proportion the ducts the following values of L_1 have been found from experience to give satisfactory results

$$L_1 = \begin{cases} 0.7L \text{ for an outside temperature of } 32° \text{ F} \\ 0.6L \text{ for an outside temperature of } 40° \text{ F} \\ 0.5L \text{ for an outside temperature of } 50° \text{ F} \end{cases} \tag{89}$$

Since L_1 is defined, it is possible to compute the temperature of the air in the duct with the aid of Eq (88) so that the relation becomes

$$t_1' = t + \frac{W_1(1 + \alpha t)}{0.0207L_1} \tag{90}$$

2 When the Number of Air Changes Is Specified.

In Eq (88), W is the heat loss determined for the lowest outside temperature and is therefore a constant quantity Again, t' may not be assumed higher than 150° F In view of this, any specified number of air changes L_v can represent only a minimum air change If L_v is greater than the volume of air L computed from Eq (88) and $t' = 150°$ F the design of the ducts must be based on the larger volume

On the other hand if L_v is less than L, L_1 as determined by Eq (89) is to be compared with L_v, and the larger of the two should be used in the computations. The entering air temperature t_1' therefore becomes

$$\left.\begin{aligned} t_1' &= t + \frac{W_1(1 + \alpha t)}{0.0207L_1} \text{ (when } L_1 \text{ is used)} \\ t_1' &= t + \frac{W_1(1 + \alpha t)}{0.0207L_v} \text{ (when } L_v \text{ is used)} \end{aligned}\right\} \tag{90}$$

From Eq (88) it is possible to make important deductions Consider a school having several classrooms Assume that all classes are fully occupied and that the air change L may be specified as L_v Since the exposures are different for the different rooms and since also the prevailing winds vary, it is apparent that the heat requirements for the different rooms depend upon the atmospheric conditions prevailing at the time From the operating conditions the initial temperature leaving the furnace is fixed and the same for all ducts Hence regulation to individual rooms is accomplished by regulating the temperature in its particular duct by mixing with cooler air in accordance with the heating requirements It will be seen that efficient operation from a central point is difficult This objection led to the disuse of this method of heating school houses throughout Europe

H. DISTRICT HEATING

District heating systems, depending upon their use of steam or hot water, are designed in accordance with principles previously discussed under those headings

PART IV
VENTILATION SYSTEMS

SECTION I

DETERMINATION OF THE REQUIRED AIR CHANGES

In the discussion of heating systems it was believed advisable to separate the theoretical treatment from its application to actual examples For the purposes of this section it is deemed desirable to combine the two

As outlined on page 129, it seems best for present purposes to assume the number of air changes in accordance with the results obtained in past experience In certain cases, however, it may be desirable to determine the necessary air supply for ventilation based upon the following standards

Heat standard in which the necessary volume of air is based on the highest permissible temperature

Humidity standard in which the necessary volume of air is determined from the allowable humidity

Carbon dioxide standard for which the required air supply limits the maximum carbon dioxide content

Pressure standard wherein the volume of air determines the maximum pressure to be created

A. HEAT STANDARD

In determining the quantity of heat to be dissipated, the following relation holds

$$W = W_1 + W_2 + W_3 \pm W_4 \qquad (91)$$

in which

W_1 = bodily heat emission (see p 123)

W_2 = heat emitted by illumination (see p 123)

W_3 = heat emitted by machinery, boilers, etc

W_4 = heat loss from the room (see p 174) The positive sign is used for summer weather and the negative sign for winter weather conditions

To remove the quantity of heat W by means of an air change L which when expressed in terms of the room temperature according to Eq (88) becomes

$$L = \frac{W(1 + \alpha t)}{0\ 0207(t - t')}, \qquad (92)$$

in which t' = temperature of entering air $t' < t$

For warm-air heating systems, Eqs (91) and (92) take the following forms.

$$\left.\begin{aligned} W &= W_4 - W_1 - W_2 - W_3 \\ L &= \frac{W(1 + \alpha t)}{0.0207(t' - t)} \text{ when } t' > t \end{aligned}\right\} \tag{93}$$

If the volume of air is desired for a temperature other than the room temperature (e g , in terms of the outside temperature), then

$$L_0 = \frac{L(1 + \alpha t_0)}{1 + \alpha t} \tag{94}$$

Example 24

Ventilation is to be provided for a hall the dimensions of which are Length 80 ft , width 46 ft , height 25 ft , making a content of 92,000 cu ft The capacity is to be 350 persons The inside temperature shall not exceed 70° F when the outside temperature is 50° F and air enters the room at 60° F Ventilation is to take place from floor to ceiling The heat loss from the hall is assumed to be 20,000 B t u per hour Electric illumination is to be provided by metallic filament lamps It is also assumed that $W_2 = W_3 = 0$ The problem is to find the volume of air required

It is assumed that the room is occupied to capacity so that in accordance with discusson on page 123

$$W = 200 \text{ B t u per hour per person.}$$

The total heat emitted by occupants is therefore

$$200 \times 350 = 70{,}000 \text{ B t u per hour}$$

From Eq (91)

$$\begin{aligned} W &= W_1 + W_2 + W_3 - W_4 \\ &= 70{,}000 - 20{,}000 = 50{,}000 \text{ B t u per hour} \end{aligned}$$

This then is the heat to be dissipated by ventilation

The required air volume is therefore (Eq 92)

$$L = \frac{50{,}000(1 + 70\alpha)}{0.0207(70 - 60)} \cong 278{,}400 \text{ cu ft at } 70° \text{ F}$$

Table XIX may be used in computations of this kind as follows For a temperature difference of 10° F and a room temperature of 70° F , it is found that 5,520 cu ft of air are necessary to remove or to supply 1,000 B t u Since 50,000 B t u are to be removed, the volume of air will be 5,520 × 50 = 276,000 cu ft per hour at 70° F , which checks well with the above figure

If the air change L is to be expressed in terms of the cubical contents J of the room, and if

$$\begin{aligned} \text{Air change} &= L = 276{,}000 \text{ cu ft per hour at } 70° \text{ F} \\ \text{Cubical contents} &= J = 92{,}000 \text{ cu ft} \end{aligned}$$

L will equal $3J$, signifying that the room air is changed three times during 1 hour

B HUMIDITY STANDARD

Let it be assumed that

L = volume of air in cubic feet at room temperature which passes through the room per hour

J = volume of room in cubic feet

A = weight of moisture emitted per person in pounds per hour

n = number of persons

m = water content of room in pounds per cubic foot at any instant

m_1 = water content of room in pounds per cubic foot before occupancy

m_2 = water content of room in pounds per cubic foot after occupancy.

a = water content of outside air at a temperature t_0 lb per cubic foot.

z = duration of occupancy in hours.

The water vapor introduced in the time dz is

$$\left(La\frac{1 + \alpha t_0}{1 + \alpha t} + nA\right)dz.$$

The water vapor dissipated during the same interval is

$$L\, m\, dz$$

In the steady state therefore

$$Jdm = \left(La\frac{1 + \alpha t_0}{1 + \alpha t} + nA\right)dz - Lmdz,$$

or

$$\int\frac{dz}{J} = \int\frac{-dm}{Lm - La\dfrac{1 + \alpha t_0}{1 + \alpha t} - nA}$$

It is to be noted that when

$$z = 0, \quad m = m_1$$
$$z = z, \quad m = m_2,$$

consequently

$$\frac{z}{J} = \frac{1}{L}\log_e\frac{Lm_1 - La\dfrac{1 + \alpha t_0}{1 + \alpha t} - nA}{Lm_2 - La\dfrac{1 + \alpha t_0}{1 + \alpha t} - nA}$$

If the natural logarithm is developed into a series and only the first term is used, there results

$$L = \frac{\dfrac{J}{z}(m_1 - m_2) + nA}{\dfrac{m_1 + m_2}{2} - a\dfrac{1 + \alpha t_0}{1 + \alpha t}}$$

If, when the room is first occupied, $m = a\dfrac{1 + \alpha t_0}{1 + \alpha t}$, which may usually be assumed, then

$$L = \frac{2nA}{m_2 - a\dfrac{1 + \alpha t_0}{1 + \alpha t}} - \frac{2J}{z} \qquad (95)$$

Table XX is based on Eq. (95) for use under usual conditions and simplifies the computation.

Example 25.

Use the data in example 24 and in addition assume that the hall is to be occupied from 6 to 8 hr Table XX includes the material to solve for the new conditions

Volume of hall	=	92,000 cu ft
Occupancy	=	350 persons
Volume of room air per person	$\cong$	263 cu ft

From Table XX

Air change L_1 per person = 930 cu ft per person per hour

Total air change L = 930 × 350 = 325,500 cu ft of 70° F per hour

C CARBON DIOXIDE STANDARD

Assume the following notation:

L = volume of air in cubic feet reduced to the volume at room temperature which passes through the room.

J = volume of the room in cubic feet

p_1, p_2 = the carbon dioxide contents in cubic feet per cubic foot of room air, having a temperature t, at the beginning of the ventilation and after a lapse of z hr.

a = carbon dioxide content in cubic feet per cubic foot of outside air of a temperature of t_0° F.

n = number of sources carbon emitting dioxide into the room

k = amount of carbon dioxide emitted per hour in cubic feet by one of the sources

z = duration of occupancy in hours

In a manner similar to that used in the previous section,

$$\frac{z}{J} = \frac{1}{L}\log_e \frac{Lp_1 - La\dfrac{1+\alpha t_0}{1+\alpha t} - nk}{Lp_2 - La\dfrac{1+\alpha t_0}{1+\alpha t} - nk} \tag{96}$$

When the above equation is expanded into a series and all but the first term is eliminated, then

$$L = \frac{\dfrac{J}{z}(p_1 - p_2) + nk}{\dfrac{p_1 + p_2}{2} - a\dfrac{1+\alpha t_0}{1+\alpha t}}. \tag{97}$$

In most cases it may be assumed that

$$p_1 = a\frac{1+\alpha t_0}{1+\alpha t},$$

in which case Eq (97) may be further simplified to

$$L = \frac{2nk}{p_2 - a\dfrac{1+\alpha t_0}{1+\alpha t}} - \frac{2J}{z} \tag{98}$$

The steady state is quickly attained, and in this case it may be assumed in Eq (97) that $p_1 = p_2$, whence

$$L = \frac{nk}{p_2 - \alpha \dfrac{1 + \alpha t_0}{1 + \alpha t}} \tag{99}$$

and for $t_0 = t$

$$L = \frac{nk}{p_2 - a} \tag{100}$$

Equation (96) may be used to determine the air change of a room due to gravity At the beginning carbon dioxide (CO_2) is developed and uniformly distributed throughout the room Later when the CO_2 producing sources are removed, $nk = 0$ in Eq 96 and it is formed thus:

$$L = \frac{J}{z} \log \frac{p_1 - a \dfrac{1 + \alpha t_0}{1 + \alpha t}}{p_2 - \alpha \dfrac{1 + \alpha t_0}{1 + \alpha t}} \tag{101}$$

The values of p_1 and p_2 at the beginning and at the end of the time z are determined, from which the magnitude of L may then be computed

Air changes based upon the carbon dioxide standard are always smaller than those computed under headings A and B In consequence they are not considered further

D PRESSURE STANDARD

Based on investigations of Kiell,[1] Dietz[2] assumes a quantity of air as standard which in the case of the lowest outside temperature is still capable of maintaining a pressure slightly above atmospheric By this means objectionable cold drafts are avoided

[1] "Pressure Ventilation of Public Lecture Rooms in Nurnberg, ' from KRELL, SR "Uberdruckluftung ohne Ventilatorbetrieb des Sitzungssaales der stadtischen Kollegien in Nurnberg," *Gesundh -Ing* , 1906

[2] "Text on Heating and Ventilation,"from DIETZ, ' Lehrbuch der Luftungs und Heizungstechnik," R Oldenbourg, Munich-Berlin, 1920

SECTION II

FORMULAS FOR THE DESIGN OF DETAILS OF VENTILATING SYSTEMS

A. FILTERS

Baffle Filters

Baffle-type filters offer very little resistance to the flow of air, and this resistance may usually be neglected in computations.

Through Filters

In tests conducted at the Research Laboratory, Charlottenburg, it was found that the resistance of filters may be determined by means of the following formulas.

For cloth filters after a longer period in service,

$$Z = 0.17v^{1.46} \tag{102}$$

For stone filters 8 in. thick having stones from 3 to 4 in. in diameter, when in a dry state,

$$Z = 0.09v^{1.88} \tag{103}$$

When wetted by a water spray,

$$Z = 0.12v^{1.96} \tag{104}$$

In the equations

Z = resistance of filter in inches of water column

v = velocity of air determined from the total filtering area in feet per second.

If the volume of air per second is known, the filter surface in square feet may be computed from the formula

$$f = \frac{L}{v}. \tag{105}$$

Example 26

140,000 cu. ft. of air per hour is to be passed through a water spray and stone filter (Fig. 182). The stones of the filter are to be about 3 to 4 in. in size and the filter surface sufficiently extended that the resistance shall not exceed $Z = 0.18$ in. w.c.

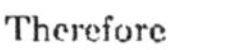

Fig. 182

From Eq. 104

$$Z = 0.12v^{1.96}$$

Therefore

$$v = \sqrt[1.96]{\frac{Z}{0.12}} = \sqrt[1.96]{\frac{0.18}{0.12}} = 1.23 \text{ ft. per second}$$

Now with Eq 105

$$f = \frac{140,000}{3,600 \quad 1\ 23} \cong 31\ 6 \text{ sq. ft.}$$

$$= \text{filter area required}$$

B. APPARATUS FOR HUMIDIFYING, WASHING, AND DRYING AIR

The resistances for these types of apparatus are given by the manufacturer

C HEATING SURFACE

DETERMINATION OF THE HEAT REQUIREMENTS

The heat required is the sum of the heat needed to warm the air and to evaporate the water

1. Heat Required for Warming Air.

The heat W_1 in B t u per hour required to warm the air is computed from Eq (87) (p 253):

$$W_1 = \frac{0\ 0207L_0}{1 + \alpha t_0}(t' - t_0)$$

where

L_0 = volume of air in cubic feet per hour at t_0° F.
t_0 = temperature of air entering the heater.
t' = temperature of air leaving the heater
α = coefficient of expansion = 0 00218

According to the design of the system, W_1 will be supplied by preheaters or final heaters which will be discussed under the subheading Computing the Required Heating Surface (p 264)

2. Heat Required for Humidification.

To obtain the heat required for humidification it is first necessary to determine the amount of water which must be evaporated. If therefore

L = volume of air in cubic feet per hour at room temperature t° F
p = degree of humidity desired for the room air in per cent
g = vapor content in pounds per cubic foot of room air at t° F (saturation assumed)
p_0 = relative humidity of outside air in per cent
g_0 = vapor content at saturation in pounds per cubic foot of outside air at t_0° F

Then the total weight in pounds of moisture in the inside air is

$$\frac{Lpg}{100}$$

while the same weight of outside air contains

$$L\frac{1 + \alpha t_0}{1 + \alpha t}\frac{g_0 p_0}{100}$$

in pounds of vapor The weight A in pounds of moisture that must be added to attain the desired relative humidity in the room will be

$$A = \frac{L}{100}\left(pg - \frac{1 + \alpha t_0}{1 + \alpha t} p_0 g_0\right) \qquad (106)$$

The heat W_2 in B t u. required to evaporate A lb of water is

$$W_2 = Aq' \qquad (107)$$

A may be taken from Table XXI q' represents the total heat of the vapor, *i e*, heat of the liquid plus the latent heat of vaporization The heat of the liquid in this case represents the heat required to raise the water from entering water temperature to the wet-bulb temperature of the air passing through the sprays. Table III gives the values of the heat of vaporization

Example 27

The ventilation requirements of a building are 3,500,000 cu ft. of air per hour at a temperature of 70° F The atmospheric conditions are

Outside, 0° F, p = 80 per cent
Inside, 70° F., p = 50 per cent

What quantity of heat is required for heating and humidifying?

Solution. The volume of outside air required is

$$L_0 = \frac{3,500,000}{1 + 70\alpha} \cong 3,040,000 \text{ cu ft per hour at } 0° \text{ F.}$$

From Eq (87)

$$W_1 = 0\,0207 \times 3,040,000(70 - 0)$$
$$\cong 4,400,000 \text{ B t u per hour}$$

From Eq (107)

$$W_2 = Aq'$$

From Table XXI A = 0 527 lb per 1,000 cu ft of air and for 3,500,000 cu ft, it is 0 527 × 3,500 ≅ 1,850 lb

Assume that the temperature of the water enters from the service connection at 50° F This must be heated to about 59° F corresponding to the wet-bulb temperature and then evaporated The water vapor is then superheated to 70° F If the specific heat of superheated steam is taken at 0 5, then the heat required is

$$q' = 1,850(9 + 1,058) + 0\,5 \times 1,850 \times 11$$
$$= 1,850 \times 1,067 + 0\,5 \times 1,850 \times 11$$
$$\cong 1,984,000 \text{ B t u per hour}$$

The total heat required is

$$W_1 = 4,400,000$$
$$+ W_2 = 1,984,000$$
$$= 6,384,000 \text{ B t u per hour.}$$

Computing the Required Heating Surface

1. Cast-iron Vento Sections.

Figure 100 shows a Vento unit which is extensively used. The amount of heating surface required for any set of conditions is obtained in the manufacturer's data, and is found in Tables X *a*, *b*, *c*. At the end of the book An example will show how the data have to be used.

Example 28.

It is required that 48,000 cu ft of air per minute be heated from 60 to 140° F The steam pressure in the heater is to be 5 lb per sq in , the corresponding temperature of which is about 227° F

From Table X *b* it will be found that the requirements may be met by the regular section 5-in spacing and a heater five stacks deep Moreover, it will be seen that the condensation is 1 23 lb per square foot per hour and that the air velocity (measured at 70° F) through the heater is 1,200 ft per minute The required free area will be

$$\frac{48,000}{1,200} = 40 \text{ sq ft}$$

Table X*a* shows that for 5-in centers and 40 sq ft , a double deck of heater will be required each having 22 sections The total heating surface will be

$$22 \times 5 \times 2 \times 16 = 3,520 \text{ sq ft}$$

The friction loss for this heater, according to Table X*c*,

$$= 0\ 376 \text{ in w c}$$

2. Unit Heaters.

A typical type of unit heater is shown in Fig 101 As in the previous case the designer must refer to the manufacturer for the necessary data which are given in a form similar to Table XI

Example 29

Suppose 10,000 cu ft of air per minute are to be heated from 60 to 120° F Let the steam pressure at the heater be 5 lb per square inch From Table XI a heater size 9 is to be chosen which will supply the heating requirement of 670,000 B t u per hour.

The necessary power for air movement through the heater will be furnished by an ⅙-hp motor which is an integral part of the unit

In a similar way the engineering data from other manufacturers are used for their special product

D 4 DUCT SYSTEMS

Loss of Heat from the Ducts

In the discussion to follow let

G = quantity of air to be transmitted in pounds per hour
F = exposed heat-emitting surface of the duct in square feet
t_2 = initial temperature of the air in the duct in degrees Fahrenheit
t_1 = terminal temperature of the air in the duct in degrees Fahrenheit
t_z = temperature of the air surrounding the duct in degrees Fahrenheit
c = average specific heat of air = 0 24 B t u per pound
K = coefficient of heat transmission in B t u per square foot per degree Fahrenheit per hour (see Table XII)

Since the heat lost by the air must equal that transmitted through the duct enclosure,

$$Gc(t_2 - t_1) = FK\left(\frac{t_1 + t_2}{2} - t_z\right), \tag{108}$$

from which it follows that

$$t_2 = \frac{(2Gc + FK)t_1 - 2FKt_z}{2Gc - FK} \tag{109}$$

Example 30.

Air at average velocities of 6, 20, and 30 ft per second is to be transmitted through a sheet-iron pipe The pipe is to be 5 in in diameter, 30 ft long, and made of 20-gage metal The temperature t_1 of the air leaving the duct is to be 100° F and the surrounding air temperature is assumed at 60° F It is required to find the initial temperature when uninsulated and when insulated at the usual efficiency of 75 per cent

Assume that in the uninsulated pipe the air velocity of $v = 6$ ft per second

The surface F of the uninsulated pipe is

$$F = \frac{(5 + 2 \times 0\ 038) \times 3\ 14 \times 30}{12} = \frac{5\ 076 \times 3\ 14 \times 30}{12}$$

$$\cong 39\ 9 \text{ sq ft}$$

Density of air s_1 at 100° F	= 0 0709 lb per cubic foot
Temperature t_2 assumed	= 134° F
Density of air s_2 at 134° F	= 0 0668 lb per cubic foot
Average density s_m	= 0 0689 lb per cubic foot

The weight G of the air to be transmitted is

$$G = 3{,}600 \times \frac{\pi d^2}{4} v s_m \ (d \text{ in feet})$$

$$= 3{,}600 \times \frac{\pi \times 5^2}{4} \times \frac{1}{144} \times 6 \times 0\ 0689$$

$$\cong 203 \text{ lb per hour.}$$

$$\frac{t_1 + t_2}{2} - t_1 = \frac{100 + 134}{2} - 60 = 57° \text{ F}$$

From Table XII, $K = 0\ 71$ B t u per hour per square foot per degree Fahrenheit Substituting these values in Eq (109),

$$t_2 = \frac{(2 \times 203 \times 0\ 24 + 39\ 9 \times 0\ 71)100 - 2 \times 39\ 9 \times 0\ 71 \times 60}{2 \times 203 \times 0\ 24 - 39\ 9 \times 0\ 71}$$

$$\cong 133° \text{ F}$$

This result is close to the value 134 previously assumed It will be observed that since the initial and terminal air temperatures in the pipe are 133 and 100° F, respectively, there is a drop of 33° in 30 ft of pipe at a velocity of 6 ft per second In a longer pipe there would be a greater temperature drop, and the arithmetical mean temperatures of Eq (108) will not be sufficiently accurate For such cases Eq (110) to follow should be used, i e,

$$Gc(t_2 - t_1) = \frac{FK[(t_2 - t_z) - (t_1 - t_z)]}{\log_e \dfrac{t_2 - t_z}{t_1 - t_z}} \tag{110}$$

If in the example chosen the length were 70 ft, t_2 would have exceeded 200° F It is apparent therefore that with low air velocities ducts suffer considerable heat loss Insulation of ducts must as a consequence be given due attention

If this example is calculated for velocities of 6, 20 and 30 ft per sec for both insulated and uninsulated ducts using the assumptions given, the following results are found

$v_m = 6$ ft per second	Insulated duct $t_2 \cong 106°$ F
$v_m = 20$ ft per second	Uninsulated duct $t_2 \cong 111°$ F Insulated duct $t_2 \cong 103°$ F
$v_m = 30$ ft per second	Uninsulated duct $t_2 \cong 108°$ F Insulated duct $t_2 \cong 102°$ F.

When a small heat loss is desired, ducts should be insulated, and high air velocities should be chosen

Duct Sizes

1. General Theory.

The group of equations discussed on page 187 apply also in the design of the duct system.

$$\left.\begin{aligned} H &= 0.192h(s'' - s') \text{ in w.c} \\ R &= \frac{p_2 - p_1}{l} = a\frac{v^n}{d^m} \text{ in w c} \\ Z &= 0\ 192\Sigma\zeta\frac{v^2}{2g}s \text{ in w c} \\ Q &= 25\frac{\pi d^2}{4}vs \text{ lb. per hour} \end{aligned}\right\} \quad (41)$$

2. The Effective Pressure Head.

In the following, three distinct cases require discussion: (1) when the average duct temperature is known, (2) when the average duct temperature is to be determined, and (3) when the gravity pressure head may be neglected in comparison with the pressure head of a fan Cases 1 and 2 apply to gravity ventilation systems, and case 3 to a fan system.

a. When the Average Duct Temperature Is Known —A known average duct temperature is obtained in the case of a gravity system in which the discharged air is not heated This is also true when the discharged air is heated in a way to secure a known and reasonably uniform average temperature Let therefore,

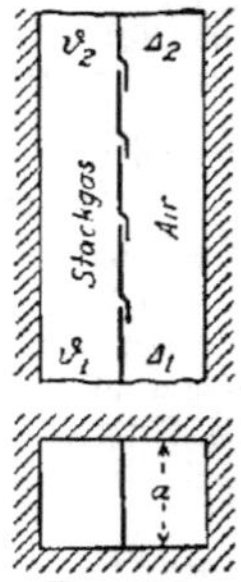

t' = outside temperature in degrees Fahrenheit

s' = density of air at the temperature t'

t'' = average temperature in duct in degrees Fahrenheit

s'' = density of air at the temperature t''.

Then

$$H = 0\ 192h(s'' - s')$$

in inches of water column pressure in which h is the vertical height of the duct in feet

Fig 183

b When the Average Duct Temperature Is to Be Determined —Suppose it is desired to determine the duct temperature for a case illustrated in Fig 183 where the chimney gases are used to heat the discharged air. The heat emitted by the flue gas in B.t.u. is

$$W = Gpc(\vartheta_1 - \vartheta_2) \quad (111)$$

in which

G = weight of flue gas in pounds per pound of fuel used.

p = weight of fuel burnt per hour

c = specific heat of the flue gas

ϑ_1 = initial temperature of flue gas in degrees Fahrenheit.

ϑ_2 = terminal temperature of flue gas in degrees Fahrenheit.

The heat absorbed by the air (Eq (87), p 253) is

$$W = 0.0207\frac{L}{1 + \alpha t}(\Delta_2 - \Delta_1), \tag{112}$$

wherein

L = volume of air in cubic feet per hour at $t°$ F flowing through the ducts

Δ_2 = final temperature of the air in degrees Fahrenheit

Δ_1 = initial temperature of the air in degrees Fahrenheit.

Transposing Eq (33) (p 182), the quantity of heat W transmitted by the flue gas to the air through the surface F in square feet is[1]

$$W = FK\left(\frac{\vartheta_1 + \vartheta_2}{2} - \frac{\Delta_1 + \Delta_2'}{2}\right) \tag{113}$$

In the above equation Δ_2' is used instead of Δ_2 It is assumed that only $1/n$th part of the air is heated by the heat flowing through the partition, and this air is raised to a temperature Δ_2' instead of Δ_2 Consequently

$$W = 0.0207\frac{L(\Delta_2' - \Delta_1)}{n(1 + \alpha t)} \tag{114}$$

Since the gain of heat by the air must equal the loss of heat from the flue gas, Eq (114) must equal Eq (112) so that $\Delta_2' = n(\Delta_2 - \Delta_1) + \Delta_1$, and Eq (113) becomes

$$W = FK\left(\frac{\vartheta_1 + \vartheta_2}{2} - \Delta_1 - \frac{n(\Delta_2 - \Delta_1)}{2}\right) \tag{115}$$

As a general rule it may be assumed that $n = 2$ so that

$$W = FK\left(\frac{\vartheta_1 + \vartheta_2}{2} - \Delta_2\right) \tag{116}$$

In the steady state the value of W determined by Eqs (111), (112), and (116) must be identical Hence the average temperature of the air in the duct may be determined as well as the average density in pounds per cubic foot, and finally the pressure head H from the relation

$$H = 0.192h(s' - s)$$

In the latter equation s' = density of the outside air and h = the vertical height of the duct

Example 31

Determine the pressure head of vitiated air in a duct which is warmed by the heat of an adjacent chimney Assume that the heat-emitting side of the chimney be as shown in Fig 183 The kitchen range connected to the chimney burns $p = 14$ lb anthracite coal per hour

[1] This equation may be used with sufficient accuracy instead of the logarithmic form in Eq (31).

Furthermore,

G = 22 lb of air per pound of coal (i e , 100 per cent excess air).
Q = 270 cu ft flue gas at 32° F per pound of fuel
C = specific heat of 0 25 B t u per pound per degree Fahrenheit
ϑ_1 = 680° F
Δ_1 = 70° F
Height of chimney = 50 ft
Chimney width (dimension a in Fig 183) = 8 in
Cross-sectional area of chimney 8 by 8 in = 0 44 sq ft
Volume of air flowing through duct = 18,000 cu ft per hour at 70° F
Outside temperature = 32° F

Solution From Eq (111)

$$W = 22 \times 14 \times 0\,25(680 - \vartheta_2) = 52{,}360 - 77\vartheta_2.$$

From Eq (112)

$$W = \frac{0\,0207L(\Delta_2 - \Delta_1)}{1 + \alpha t}$$
$$= \frac{0\,0207 \times 18{,}000(\Delta_2 - 70)}{1 + 0\,00218 \times 70} = 323\Delta_2 - 22{,}610$$

To use Eq (116) it is necessary to determine the value of K from Table XII For this purpose it is necessary to know the temperature difference,

$$\frac{\vartheta_1 + \vartheta_2}{2} - \Delta_2$$

Assume that ϑ_2 = 530° F. and Δ_2 = 100° F Therefore

$$\frac{\vartheta_1 + \vartheta_2}{2} - \Delta_2 > 500° \text{ F.}$$

The velocity of the flue gas is

$$v = \frac{14 \times 270(1 + \alpha \times 530^1)}{3{,}600 \times 0.44 \times (1 + \alpha \times 32)} \cong 4\,8 \text{ ft per second.}$$

From Table XII therefore

K = 0 67 B t u per square foot per hour per degree Fahrenheit

Equation (116) may then be written

$$W = 50 \times 0\,67 \times 0\,67\left(\frac{680 + \vartheta_2}{2} - \Delta_2\right) = 7{,}631 + 11\,2\vartheta_2 - 22.4\Delta_2$$

Three equations have resulted, having three unknown quantities

$$W = 52{,}360 - 77\vartheta_2 \qquad (a)$$
$$W = 323\Delta_2 - 22{,}610. \qquad (b)$$
$$W = 7{,}631 + 11\,2\vartheta_2 - 22\,4\Delta_2 \qquad (c)$$

Combining Eqs (a) and (b),

$$52{,}360 - 77\vartheta_2 = 323\Delta_2 - 22{,}610, \text{ or } \quad 4\,2\Delta_2 + \vartheta_2 = 974$$

Combining Eqs (a) and (c),

$$52{,}360 - 77\vartheta_2 = 7{,}631 + 11\,2\vartheta_2 - 22\,4\Delta_2, \text{ or } -0\,25\Delta_2 + \vartheta_2 = 507$$
$$4\,45\Delta_2 = 467$$
$$\Delta_2 \cong 105° \text{ F.}$$

[1] Basing on ϑ_{min}.

Inserting the value of Δ_2 found in the equation

$$4.2\Delta_2 + \vartheta_2 = 974,$$

there then results

$$4.2 \times 105 + \vartheta_2 = 974.$$
$$\vartheta_2 \cong 533$$

This result is sufficiently close to the original assumptions of 530°F. Had ϑ_2 been so low as to affect the chimney draft seriously, it would have been necessary to decrease the heat-emitting surface of the chimney.

The following results were also obtained.

$$\frac{\vartheta_1 + \vartheta_2}{2} \cong 607° \text{ F.} \quad \frac{\Delta_1 + \Delta_2}{2} \cong 88° \text{ F}$$

The pressure head available is

$$H = 0.192 \times 50(s_{32} - s_{88}) = 0.192 \times 50(0.0807 - 0.0724).$$
$$\cong 0.08 \text{ in. w.c.}$$

It must be determined whether this pressure head is sufficient to discharge the 18,000 cu. ft. per hour. This computation is given on page 276.

c. *When the Gravity Pressure Head May Be Neglected in Comparison with the Pressure Head of the Fan.*—If the fan discharges directly into

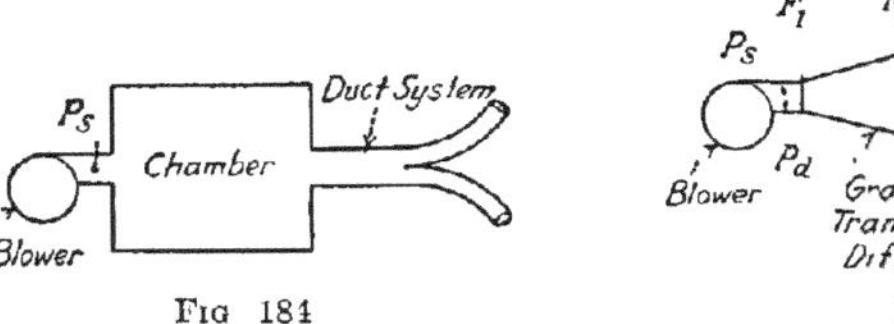

FIG. 184

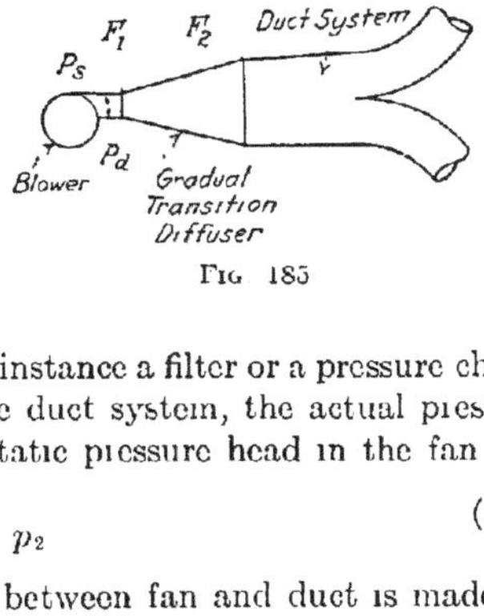

FIG. 185

a comparatively large chamber, as for instance a filter or a pressure chamber (see Fig. 184), before entering the duct system, the actual pressure head may be assumed equal to the static pressure head in the fan discharge. Therefore

$$H = p_2 \tag{117}$$

On the other hand if the connection between fan and duct is made by gradually increasing the area by means of a diffuser shown in Fig. 185, then it may be assumed that a part of the dynamic pressure pd is available to overcome the resistance of the duct. The following equation may then be written

$$H = p_s + \eta(p_{d_{F_1}} - p_{d_{F_2}}) = p_s + 0.192s\eta\left(\frac{v_1^2}{2g} - \frac{v_2^2}{2g}\right) \tag{118}$$

in which

p_s = static pressure in inches of water column for cross-section F_1

$p_{d_{F_1}} = 0.192s\frac{v_1^2}{2g}$ = dynamic pressure in inches of water column for the cross-section F_1.

$p_{d_{F_2}} = 0.192s\frac{v_2^2}{2g}$ = dynamic pressure in inches of water column for the cross-section F_2

v_1 = average velocity of air in feet per second for cross-section F_1.

v_2 = average velocity of air in feet per second for cross-section F_2

η = efficiency of the diffuser

3. Frictional Resistance of Sheet-steel Ducts.

a Round and Rectangular Ducts.—The resistance R in the group of Eq. (41) is subject to further development. The factor a in the formula for frictional resistance allows for the influence of the density and viscosity of the fluid. In the case of air the effect of the viscosity may be neglected. Extensive studies in the Research Laboratory, Charlottenburg,[1] resulted in setting the equation

$$R = \frac{p_2 - p_1}{l} = bs^{0.852}\frac{v^n}{d^m} \qquad (119)$$

In this equation the influence of the density of the air was considered with due reference to the tests of Fritsche.[2]

Tests at the above-mentioned laboratory showed the following points:

(1) m is independent of the pipe diameter at least within ½ to 40 in. inside diameter.

(2) m is independent of the roughness of the pipe and includes very smooth brass and copper pipes as well as the usual riveted sheet-metal ducts.

(3) m is independent of the form of the duct, e.g., there is no appreciable difference in the values of m for either round or rectangular cross-sections.

(4) m = 1,281.

(5) n is dependent upon the smoothness of the pipe to a considerable extent. In view of this the tests were made on pipes commonly used in ventilating practice, and it was found that n = 1.924. As a consequence Eq. (119) becomes

$$R = \frac{p_2 - p_1}{l} = bs^{0.852}\frac{v^{1.924}}{d^{1.281}} \qquad (120)$$

[1] "Simplified Analysis of Duct Sizes in Ventilating Work," from BRABBÉE-BRADTKE, "Vereinfachtes zeichnerisches und rechnerisches Verfahren zur Bestimmung der Rohrleitungen von Luftungs- und Luftheizanlagen," R. Oldenbourg, Munich-Berlin, 1915. See also references there given.

[2] See footnote, p. 230.

(6) For practical purposes medium moist air of 70° F can be assumed, in which case

$$bs^{0.852} = 0.000128 \tag{121}$$

and

$$R = 0.000128 \frac{v^{1.927}}{d^{1.281}}. \tag{122}$$

in which

$$s = 0.0749 \text{ pound per cubic feet}$$

(7) For temperatures much in excess of 70° F as in the case of warm-air heating systems, the values of R from Eq (122) must be multipled by $\left(\frac{s}{0.0749}\right)^{0.852}$. For convenience in computation this expression (for different values of s) is given in Table XXII

(8) Equation (122) is valid not only for round but also for rectangular cross-sections when the factor d_e is introduced. This factor is the equivalent diameter for the rectangular section and is defined by the equation

$$d_e = \frac{2ab}{a + b} \tag{123}$$

in which a and b are the lengths of the sides of the rectangle in inches

b Brick Ducts of Round and Rectangular Section —Tests showed that the resistance of brick ducts is about double that for metal ducts under similar circumstances The values so chosen are sufficiently accurate for practical purposes

4. Resistances of Fittings

Reference is again made to the group of Eq. (41).,

$$Z = 0.192 \Sigma \zeta \frac{v^2}{2g} s$$

where Z is the pressure drop in inches of water column caused by the fittings In the particular case of air when $s = 0.0749$ pound per cubic foot, the equation becomes

$$Z = 0.192 \times 0.0749 \Sigma \zeta \frac{v^2}{2g} = 0.0144 \Sigma \zeta \frac{v^2}{2g} \tag{124}$$

For temperatures much in excess of 70° F, as for instance in warm-air heating systems, the values in Eq (124) must be multiplied by $\frac{s}{0.0749}$

Chart 6 contains values of ζ for the fittings most frequently used in ventilating work. Attention is directed to the values of the resistances of branch fittings

5 Application of Charts 6 and 7.

a General —The important equations in ventilating work are:

$R = 0.000128 \frac{v^{1.924}}{d^{1.281}}$ = resistance in inches of water column pressure per foot of round pipe of a diameter d in inches

$R = 0.000128\frac{v^{1.924}}{d_g^{1.281}}$ = resistance in inches of water column pressure per foot of rectangular pipe with an equivalent diameter of d_g in inches.

$Z = 0.0144\Sigma\zeta\frac{v^2}{2g}$ = resistances of fittings in inches of water column pressure for round and rectangular ducts.

$d_g = \frac{2ab}{a + b}$ = equivalent diameter for rectangular ducts in inches.

$L = \frac{\pi d^2}{4}\frac{v}{144}$ = quantity of air transmitted in cubic feet per second by circular pipe with air at 70° F.

From the group of equations given above Chart 6 was compiled. The right-hand portion of chart contains:

(1) Quantity of air L in cubic feet per minute transmitted by circular pipes (horizontal row I).

(2) Average velocity in feet per minute in circular pipes (horizontal row II).

(3) Actual and equivalent diameters from 2 to 100 in.

(4) Frictional resistance in inches of water column pressure per 100 ft. of pipe of diameter d'' or the equivalent round section d_g'' for rectangular pipe.

The left-hand side of the table includes:

(1) Average velocities of air in feet per minute for circular pipes.

(2) Resistances of fittings in inches of water column pressure for values of $\Sigma\zeta$ of from 1 to 9.

(3) Values of ζ for the resistances of fitting most frequently used in ventilating work.

Chart 7 contains the equivalent diameters of round section for rectangular sections of sides a and b ranging from 2 to 100 in. and also the corresponding duct areas.

b. Preliminary Design of Duct System. (1) General.—The design of the ducts of a ventilating system considers first the least-favored circuit. The circulating pressure head H is determined as outlined on page 267. From this the frictional resistances of the fittings are deducted in accordance with the values given in Table A following:

TABLE A

Inside dimensions of ducts in inches	Ratio of the resistances of fittings to the total resistance of the circuit in per cent	
	Sheet-metal ducts	Brick ducts
2–6	40	30
4–12	60	50
8–24	80	70
16–45	90	80
over 40	95	85

Special attention is directed to the influence of the resistance of fittings particularly for the larger duct sizes It will be remembered that fittings are called ells, tees, changes of directions, changes in diameter, shutters, etc In some cases it might be better to consider the friction of the pipe as a comparatively small percentage of the friction of the fittings, especially in the large sizes In ventilating practice the opposite, though incorrect, is assumed frequently, so that failures in operation are readily traceable to insufficient consideration of the fittings Likewise puzzling flow phenomena may be due to the magnitude of the resistances of the fittings

(2) Circular Ducts —After deducting the circulating pressure head H, required to overcome the frictional resistance of the fittings, the balance is available for overcoming the friction of the pipes. This rest is divided by the length of the duct and multiplied by 100 for the frictional resistance per 100 ft of pipe The value of R so found is looked up in Chart 6 Proceeding horizontally to the right, the value of L will be found Directly above is indicated the required pipe diameter For large circular ducts reference should be made to the next heading

(3) Rectangular Ducts —For rectangular ducts the procedure is somewhat different and as follows.

Determine the circulating pressure head and also from Table A the magnitude of the resistance of the fittings Z For a specific case let it be assumed that $Z = 0\ 152$ in w c For the duct section under consideration determine the sum of the coefficients of resistance, say $\Sigma\zeta = 5$, and from Chart 6 find the corresponding value of v, which for the particular example chosen is 700 ft per minute Since the volume of air is given or assumed and since the velocity in each section of the duct is now known, the cross-sectional area is found by dividing the volume by the velocity With the aid of Chart 7 the lengths of the sides of the duct are readily determined

It will be observed that it is possible to choose a and b in many ways Those values are selected which best fit the installation

It is not essential to assume equal velocities in all the duct sections Indeed it may be advisable to increase the velocity in the larger ducts In any case the value of Z corresponding to $\Sigma\zeta$ must be equal or slightly less than the value allotted to the resistance of the fittings as given in Table A (p 273)

For large circular ducts in which the resistance of the fittings is more important than the pipe friction it is desirable to depart from the procedure of the previous heading and treat as outlined in this section

c Checking Computations for the Ducts (1) General —If ventilating systems are designed for fan operation instead of gravity circulation, the problem is somewhat different. In this case the duct sizes are frequently dictated by architectural features so that the necessary circulating head must be computed to insure delivery of the amount of air needed In this connection the result is a simplification in the treatment of the problem and will be considered in what follows

(2) Circular Ducts —For a given size of duct in Chart 6 the volume of air to be transmitted and its velocity is indicated At the extreme left in the same row the frictional resistance per 100 ft of pipe will be found Multiplying the latter quantity by $\frac{l}{100}$ gives the frictional resistance of the duct From the table of resistances of fittings the value of Z for $\Sigma\zeta$ is obtained corresponding to the velocity previously determined. The pressure head must be

$$H \geqq \Sigma(lR + Z)$$

(3) Rectangular Ducts —With the sides of the rectangular duct known or assumed, Chart 7 will assist in determining the cross-sectional area of the duct which in connection with the volume transmitted permits calculation of the velocity From the assumed layout of the building the values of $\Sigma\zeta$ may be computed and with the aid of Chart 6 the resistances Z in inches of water column for each duct section determined The next step is to ascertain the equivalent diameter from Chart 7 It is then necessary to figure the frictional resistance for this diameter corresponding to the velocity previously ascertained. Finally the pressure head must satisfy the following relation

$$H \geqq \Sigma(lR + Z)$$

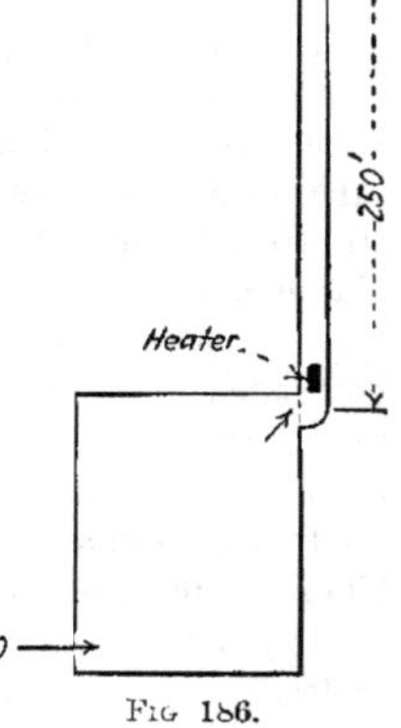

Fig 186.

Example 32.

The kitchen of a restaurant (see Fig 186) located in a 25-story building is to be ventilated by gravity circulation Instead of a sheet-metal duct a tile-lined flue of equal friction is used The average temperature within the duct is to be 110° F while that prevailing outside is to be 70° at maximum capacity for gravity circulating conditions. At temperatures above 70° outside a blower will be used The dimensions of the kitchen are 30 by 20 by 10 and the system is to be designed for six air changes per hour The problem is to find the required dimensions of the duct

$$\begin{aligned} \text{Volume of kitchen} &= 30 \times 20 \times 10 = 6{,}000 \text{ cu ft} \\ \text{Air volume in duct} &= 6 \times 6{,}000 \\ &= 36{,}000 \text{ cu ft per hour} \\ &= 600 \text{ cu ft per minute} \end{aligned}$$

Circulating head Eq (37) (p 185),

$$\begin{aligned} H &= 0\ 192h(s'_{70} - s''_{110}) \\ &= 0\ 192 \times 250(0\ 0749 - 0\ 0696) \\ &= 0.25 \text{ in w c.} \end{aligned}$$

Assuming that the size of the duct will range from 4 to 12 in, then from Table A (p 273) the magnitude of the resistance of the fittings may be assumed to be 60 per cent of the total head The available head for duct friction will therefore be 0 40 × 0 25 = 0 1 in w c.

Length of duct	= 250 ft
Frictional resistance per 100 ft	$= \frac{0\ 1}{2\ 5} = 0\ 04$ in w c

From Chart 6 it may be seen that for $R/100$ ft = 0 04 in w c and a volume of 600 cu ft per minute, the duct must be 14 in in diameter Choose 12-in pipe, however, since it is a convenient standard size

Checking Computation. From Chart 6 it is found that for a 12-in duct at 600 cu ft per minute there is a pressure drop of 0 08 in w c per 100 ft of duct at a velocity of 800 ft per minute The total duct friction is therefore 2 5 × 0 08 = 0 2 in w c

Resistances of fittings

Exit-loss to atmosphere	$\zeta = 1\ 0$
Heater restriction	$= 0\ 0$
Inlet grille total area = 1 5 duct area and free area 50 per cent of total	$= 0\ 6$
	$\Sigma\zeta = 1\ 6$

Therefore, for 800 ft per minute and $\Sigma\zeta = 1\ 6$, $Z = 0{\cdot}06$ in w c The total head is 0 2 + 0 06 = 0 26 in w c against a total of 0 25 in w c found above This is sufficiently accurate for the practical problem

Example 33

Consider the conditions discussed in Example 31 (p 268) It is desired to find the duct area to meet the ventilation requirements of 18,000 cu ft per hour

Solution.

Pressure head as determined in Example 31	= 0 08 in w c
Average duct temperature (Example 31)	= 88° F
Duct area 8 by 14 in	≅ 0 78 sq ft

Duct resistance

Equivalent diameter $\frac{2ab}{a + b}$	$d_e \cong 10$ in

Resistances

Loss at exit of duct	$\zeta = 1\ 0$
Short-radius ell (rectangular)	$= 2\ 0$
Grille	$= 2\ 0$
	$\Sigma\zeta = 5\ 0$

$$\text{Volume of air to be transmitted} = \frac{18{,}000}{60} = 300 \text{ cu ft per minute}$$

$$\text{Velocity of air in duct} = \frac{300}{0\ 78} \cong 400 \text{ ft per minute}$$

For an equivalent diameter of 10 in and a velocity of 400 ft per minute, R/ft = 0 00025, so that lR = 50 × 0 00025 = 0 0125 in w c Since it has been decided to use a brick duct,

$$lR = 2 \times 0\ 0125 = 0\ 025 \text{ in w c}$$

The fitting resistances for $\Sigma\zeta = 5\ 0$, and for a velocity of v = 400 ft per minute, Z = 0 0496 in w c Finally

$$\Sigma(lR + Z) = 0\ 025 + 0\ 0496 \cong 0\ 075 \text{ in w c}$$

This should be equal or less than the available pressure head As the latter is 0.08 in. w.e., the solution is correct

In this computation it was assumed that the opening through which the air enters the room offers a negligible resistance to the flow Should the resistance be appreciable, it should be included in the value of $\Sigma(lR + Z)$. In this case it would mean an increase in the duct area

Example 34

The ventilating duct system shown in Fig 187 is to transmit the volume of air in cubic feet as indicated at the outlet of the ducts and at a temperature of 70° F The blower is connected to discharge into duct section 8 by means of a diffuser of 90 per cent efficiency The blower is to produce a static pressure of 0.64 in. w.e. The blast area of the fan outlet is $F_1 = 0.27$ sq ft and the cross-sectional area at the outlet of the diffuser assumed to be $F_2 = 0.5$ sq ft

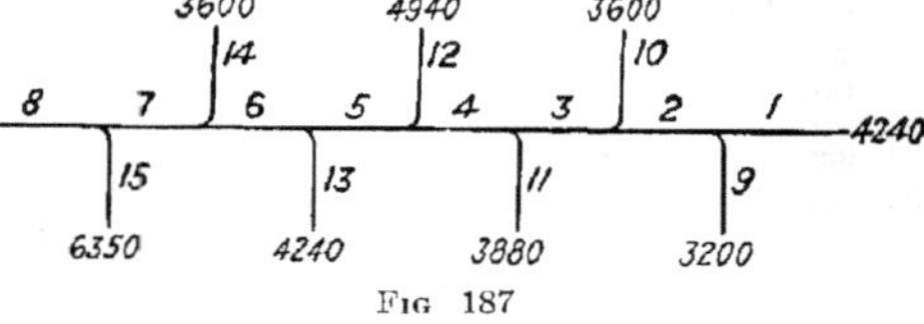

Fig 187

The problem is to proportion the ducts They are to be (1) round and (2) rectangular In the latter case it shall be assumed that one dimension of the ducts must not exceed 6 in. The velocity is to be chosen so that (a) it is equal in all ducts and (b) that it increases towards duct section 8

Solution *First Case*—Beginning with the least-favored circuit the pressure head from Eq (118) is

$$H = p_s + \eta(p_{dF_1} - p_{dF_2})$$

$$p_s = 0.64 \text{ in w e}$$

$$\eta = 0.90$$

$$v_1 = \frac{568}{0.27} \cong 2{,}104 \text{ ft per minute.}$$

$$v_2 = \frac{568}{0.5} \cong 1{,}136 \text{ ft per minute.}$$

$$p_{dF_1} = \frac{2{,}104^2}{64.4} \times \frac{0.192 \times 0.0749}{3{,}600}$$

$$p_{dF_2} = \frac{1{,}136^2}{64.4} \times \frac{0.192 \times 0.0749}{3{,}600}$$

Therefore

$$p_{dF_1} \cong 0.27 \text{ in w e}$$

$$p_{dF_1} \cong 0.08 \text{ in w e}$$

$H = 0.64 + 0.9(0.27 - 0.08)$	$\cong 0.81$ in w e
Deducting 60 per cent for resistances of fittings[1]	$\cong 0.49$ in w e
Remaining head available duct friction	$= 0.32$ in w e.
Total length of duct	$= 141$ ft
Pressure drop[2] per foot $= \frac{0.32}{141}$	$= 0.00227$ in w e
Pressure drop per 100 ft	$= 0.227$ in w e

[1] As in the previous example duct sizes range from 4 to 6 in and from Table A the resistances of fittings is about 60 per cent

[2] It should be observed that a different pressure distribution might have been chosen arbitrarily in place of the uniform distribution in the present example

For reasons of economy in construction it is best to maintain uniform duct dimensions as far as possible. With this object in view Chart 6 gives the diameters entered in Summary K.

Checking Computation: The checking computation is carried out and the results entered in columns g, h, and i of Summary K. It should be noted that the duct sizes remain unchanged. The duct at section 8 is 10 in. in diameter or of a cross-sectional area of 0.55 sq. ft., while the area of the diffuser was assumed to be $F_2 = 0.5$ sq. ft. This will cause a slight increase in the pressure head but will not require revision in the dimensions of the ducts.

SUMMARY K

Number of duct section	Volume, cubic feet per minute	l, feet	v, feet per minute	$\Sigma\zeta$	d, inches	R per foot, inches water column	lR, inches water column	Z, inches water column
(a)	(b)	(c)	(d)	(e)	(f)	(g)	(h)	(i)
1	71	13	400	1.5	6	0.0005	0.0065	0.0149
2	124	12	600	1.0	6	0.0012	0.0144	0.0223
3	184	17	900	1.0	6	0.0025	0.0425	0.0502
4	249	22	700	1.0	8	0.0010	0.0220	0.0304
5	331	17	1,000	1.0	8	0.0020	0.0340	0.0619
6	402	14	1,200	1.0	8	0.0030	0.0420	0.0892
7	462	26	1,400	1.0	8	0.0035	0.0910	0.1210
8	568	20	1,000	1.0	10	0.0017	0.0340	0.0619
		141					0.288 +	0.452 ≅ 0.74 in. w.c.

Computation of a Typical Duct: Let duct 12 (length 20 ft.) be used as an example. The pressure at the inlet end of the duct section 12 is $\sum_1^4 (lR + Z)$, the value of which may be taken from Summary K and is about 0.203 in. w.c.

Assuming diameters to be within the range of 2 to 6 in., Table A shows the magnitude of resistance for fittings to be 40 per cent of the total. Therefore

Available remaining head for friction . . . . = 0.12 in. w.c.

Pressure head per foot $= \frac{0.12}{20}$ . . . . = 0.006 in. w.c.

Given			Assumed		Checking computation				
Number	l	L	R per foot	d, inches	R per foot	lR	v	$\Sigma\zeta$	Z
12	20	82	0.006	4	0.004	0.08	900	3	0.151

$$\Sigma(lR + Z)_{12} = 0.08 + 0.151 \cong 0.23 \text{ in. w.c.}$$

It will be noted that $\Sigma(lR + Z)_{12}$ for duct 12 is slightly higher than $\Sigma(lR + Z)_1^4$. In view of the fact, however, that $\Sigma(lR + Z)_5^8 + \Sigma(lR + Z)_{12} \cong 0.77$ in. w.c. against 0.81 in. w.c. pressure head created by the blower, this diameter may be used safely.

Solution: *Second Case.*—The computations for the rectangular ducts one dimension of which does not exceed 6 in. are entered in Summaries L and M. In Summary M the velocities in all ducts are the same, whereas in Summary L the velocities are chosen so that the smaller duct areas are enlarged somewhat while the larger ones are decreased. For practical purposes this procedure is advantageous at times.

SUMMARY L

Number of duct section	l, feet	$\Sigma\zeta$	v, feet per minute	Z, inches water column	L, cubic feet per minute	F, square feet	$a \times b$,* inches	$d_e = \frac{2ab}{a+b}$, inches	R per foot, inches water column	LR, inches water column
1	13	1.5	900	......	71	0.079	6 × 2	3	0.006	0.078
2	12	1.0	900	......	124	0.138	6 × 3¼	4	0.004	0.048
3	17	1.0	900	......	184	0.205	6 × 5	5	0.003	0.051
4	22	1.0	900	......	249	0.277	6 × 6¾	6	0.0025	0.055
5	17	1.0	900	......	331	0.368	6 × 8¾	7	0.0020	0.034
6	14	1.0	900	......	402	0.447	6 × 10¾	8	0.0017	0.0238
7	26	1.0	900	......	462	0.513	6 × 12¼	8	0.0017	0.0442
8	20	1.0	900	......	568	0.632	6 × 15¼	9	0.0014	0.028

$\Sigma\zeta = 8.5$ $\Sigma Z = 0.427$ $\Sigma lR = 0.362$ in. w.c.

$\Sigma(lR + Z)_1{}^8 \cong 0.79$ against 0.81 in. w.c.

* The areas represented by the chosen values of a and b vary slightly from the areas determined under F. In other words, it would now be necessary to redetermine the values of v and Z however the difference is negligible.

SUMMARY M

Number of duct section	l, feet	$\Sigma\zeta$	v, feet per minute	Z, inches water column	L, cubic feet per minute	F, square feet	$a \times b$, inches	$d_e = \frac{2ab}{a+b}$, inches	R per foot, inches water column	lR, inches water column
1	13	1.5	700	0.0456	71	0.101	6 × 2½	4	0.0025	0.0325
2	12	1.0	700	0.0301	124	0.177	6 × 4¼	5	0.0020	0.0240
3	17	1.0	800	0.0396	184	0.23	6 × 5½	6	0.0020	0.0340
4	22	1.0	900	0.0502	249	0.277	6 × 6¾	6	0.0025	0.0550
5	17	1.0	1,000	0.0619	331	0.331	6 × 8	7	0.0025	0.0425
6	14	1.0	1,000	0.0619	402	0.402	6 × 9¾	7	0.0025	0.0350
7	26	1.0	1,000	0.0619	462	0.462	6 × 11	8	0.0020	0.0520
8	20	1.0	1,200	0.0892	568	0.473	6 × 11¼	8	0.0025	0.0500

$\Sigma Z = 0.441$ in. w.c. $\Sigma lR = 0.325$

$\Sigma(lR + Z)_1{}^8 = 0.77$ in. w.c. against 0.81 in. w.c.

The duct sections remain unchanged.

E. FANS

As previously mentioned on page 3, it is not usually within the province of the heating engineer to design the fans. It is necessary however for him to determine the volume of air and the delivery pressure. Upon these requirements a fan[1] of the desired type and size is selected from the manufacturer's data.

[1] "Fan Ventilation in Buildings," from BRABBÉE, "Drucklüftung in Gebäuden," *Zeit. des Ver. deut Ing.*, p. 31 *et seq.*, 1908.

In example 34 for instance, the fan would be specified as follows·
(1) The fan blows directly into a pressure chamber·

Air volume . . . . . . $\cong$ 34,000 cu ft per hour at 70° F.
Static pressure = $\Sigma(lR + Z)$. = 0.74 in w c.

(2) The fan blows directly into a diffuser Velocity of air leaving blower is $v_1 \cong 2{,}104$ ft per minute Velocity of air entering main duct is $v_2 \cong 1{,}136$ ft per minute Efficiency of diffuser is 90 per cent

$$\text{Available dynamic pressure therefore} = 0\ 9\left(\frac{v_1^2 s}{2g} - \frac{v_2^2}{2g}s\right)\ 0\ 192$$

$$= 0\ 9\left(\frac{2{,}104^2\ \ 0.0749 \cdot 0\ 192}{64\ 4\ \ 3{,}600} - \frac{1{,}136^2\ \ 0\ 0749\ \ 0\ 192}{64\ 4\ \ 3{,}600}\right)$$

$$= 0\ 17 \text{ in w c}$$

Static pressure = 0 74 − 0 17 = 0 57 in w c
Air volume = 34,000 cu ft per hour at 70° F

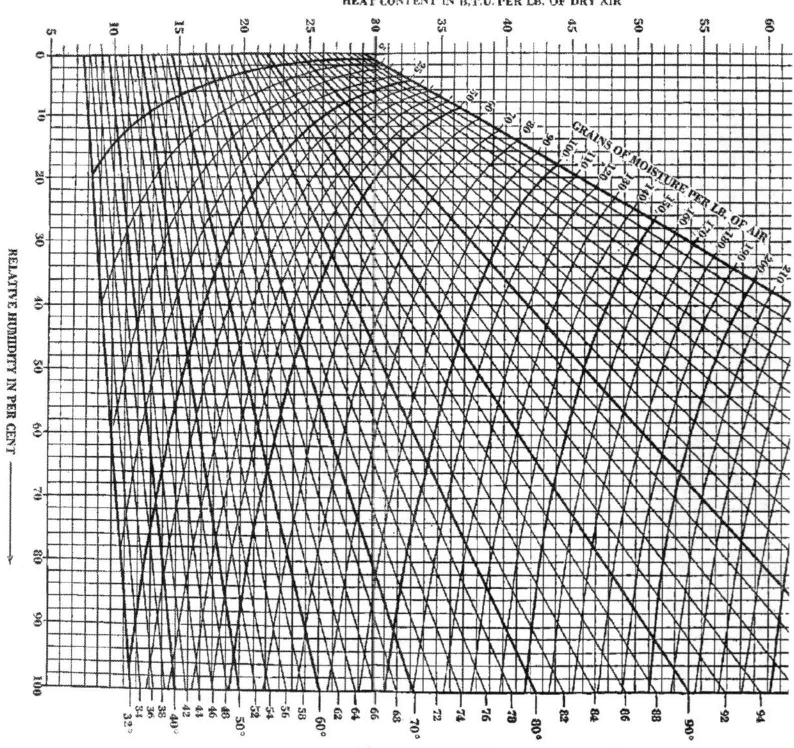
HEAT CONTENT IN B.T.U. PER LB. OF DRY AIR
5 10 15 20 25 30 35 40 45 50 55 60
GRAINS OF MOISTURE PER LB. OF AIR
RELATIVE HUMIDITY IN PER CENT
0 10 20 30 40 50 60 70 80 90 100
32° 34 36 38 40° 42 44 46 48 50° 52 54 56 58 60° 62 64 66 68 70° 72 74 76 78 80° 82 84 86 88 90° 92 94
TEMPERATURES IN °F

SECTION III

COOLING AND DRYING AIR

The design of cooling plants due to the scope of this text will be limited to the determination of the heat to be abstracted and to the extent of the cooling surface required The heat to be abstracted consists of the heat to be absorbed for cooling the air and, when the temperatures drop below the dew point, the latent heat of the condensed water The sum of these quantities of heat may be taken from the curves in Fig. 188 [1]

Example 35.

Assume the following

Temperature of outside air . . .	. 90° F
Humidity of outside air	80 per cent
Temperature of room	72° F
Humidity of air in room ,	55 per cent

The temperature and humidity chart is used as follows The temperature 90° F is found as the temperature ordinate to the right of the chart Proceed then along the line of equal temperature until it bisects the vertical line representing 80 per cent humidity Continue horizontally to the left from this point of intersection and find the heat content of the outside air ≅ 47 5 B t u per pound If, instead of proceeding horizontally, the curved line is followed (lines of equal water content) to its intersection with the moisture scale, it will be found that

Water content of outside air ≅ 175 gr per pound of air

The same method is employed for finding the properties of the air within the room.

Heat content of air in the room	= 27 B t u per pound
Moisture content of air in the room	= 65 gr per pound

For 1 lb of air, therefore (47 5 − 27 0) = 20 5 B t u and (175 − 65) = 110 gr of moisture must be removed

Assuming that the air required per hour is 140,000 cu ft. at 70° F , therefore

Heat to be removed	= 140,000 × 0 0749 × 20 5 ≅	215,000 B t u per hour
Moisture to be removed	= 140,000 × 0 0749 × 110	≅ 1,154,000 gr
		≅ 165 lb per hour

With the aid of the psychrometric chart another type of problem may be solved. For instance suppose it is desired to know to what temperature the outer air must be cooled so that when brought to saturation and reheated to a required temperature it will have a predetermined relative humidity For instance, to what temperature must the outer air in this

[1] "The Humidity of Air" from MARR, "Die Feuchtigkeit der Luft," *Gesundh -Ing* , 1915

example be cooled so that it will have a relative humidity of 55 per cent at 72° F ?

To obtain the desired answer, follow the 72° temperature line to its intersection with the 55 per cent humidity line Proceed along the line of equal moisture content to the right and find at 100 per cent (saturation) a temperature of about 55° F. In other words saturated air at about 55° F. when heated to 72° F without addition of moisture will attain a relative humidity of 55 per cent.

Suppose it is desired to find the dew point of the outer air in this example Follow the 90° temperature line to its intersection with 80 per cent relative humidity Proceed along the line of constant moisture content to the saturation line (100 per cent relative humidity) and there find the dew point to be about 83° F

COOLING SURFACE

For large volumes of air it will be observed that large quantities of water are necessary

If fin-surfaced pipe is used, the required cooling surface from Eq (32) is

$$F_g = \frac{W}{K[(t' - t_2) - (t'' - t_1)]} \log_e \frac{t' - t_2}{t'' - t_1}$$

The value of K depends largely upon the air velocity, and is also influenced by the nature and the form of the cooling surface

For low air velocities,

$K = 2\ 5$ B t u per square foot per degree Fahrenheit per hour if cooling surface is dry.

$K = 3\ 0$ B t u per square foot per degree Fahrenheit per hour if cooling surface is wet

For high velocities of the air, the value of K increases rapidly For instance at 23 ft per second, $K = 10\ 0$ B t u per square foot per degree Fahrenheit per hour In many cases it is much more satisfactory to avoid metallic cooling surfaces and use mixtures of cooling liquid (water) and air An installation of this type is described on page 144

SECTION IV

CHIMNEY COMPUTATION

APPROXIMATION

For most purposes in heating practice the chimney design may be approximated.[1] For this purpose the following formulas are proposed:

According to Redtenbacher,[2]

$$f = 0.0096\frac{Gp}{\sqrt{h}}. \tag{125}$$

According to Reiche[3] (valid for chimneys over 35 ft. high),

$$f = \frac{q}{4} \text{ for anthracite}$$

$$f = \frac{q}{6} \text{ for soft coal}$$

For an assumed fuel consumption of 100 lb. per hour:

q = 7.5 sq. ft. for anthracite
q = 10–12 sq. ft. for soft coal
q = 6.5–9 sq. ft. for coke.
$h \geqq 25d$ but always $h \geqq 35$ ft.

The following notation applies:

G = weight of combustion gases in pounds per pound of fuel
p = weight of fuel burned per hour
h = height of chimney in feet
d = inside diameter or the length of the side (square chimney) at the top cross-section in feet
f = inside cross-sectional area at top of chimney in square feet
q = total grate area in square feet.

Example 36

Determine the cross-sectional area of a chimney for a coke-fired boiler installation. Assume the following:

p = 33 lb. per hour
h = 70 ft.
G = 21.5 lb. of air per pound of fuel (100 per cent excess air)

[1] "Investigations on Chimney Draft," from DEINLEIN, "Untersuchungen über den Schornsteinzug," *Bay. Rev. Vereins*, pp. 11, 24, 41, 1912.

[2] "Machine Design," from REDTENBACHER, "Der Maschinenbau," Mannheim, 1863.

[3] "Steam Boilers," from VON REICHE, "Dampfkessel," Leipzig, 1876.

According to Redtenbacher,

$$f = \frac{0.0096 \times 21.5 \times 33}{\sqrt{70}} \cong 0.82 \text{ sq. ft.}$$

According to Reiche,

$$q = \frac{33 \times 9}{100} \cong 2.97.$$

$$f = \frac{q}{4} = \frac{2.97}{4} \cong 0.74 \text{ sq. ft}$$

An accurate chimney computation may be made by considering the chimney as a hot-air duct for a gravity system (the usual type of chimney) or as a forced-air system (induced draft)

TABLES

TABLE I.—DIAMETER, WEIGHT, EXTERNAL SURFACE AND VALUES OF $\frac{k}{l'}$

Pipe diameter				Nominal weight, lb. per ft.		Length of pipe in ft. per sq. ft. external surface l'	Values of $\frac{k}{l'}$ for uninsulated pipe when the difference between the average water temperature and the average temperature of the surrounding air =					
Nominal size, inches	External, inches	Internal,* inches	Internal standard, inches	Standard	Extra strong		Under 70°	Over 70° to 90°	Over 90° to 110°	Over 110° to 130°	Over 130° to 150°	Over 150°
½	0.840	0.546	0.622	0.850	1.087	4.547	0.48	0.51	0.53	0.55	0.57	0.57
¾	1.050	0.742	0.824	1.130	1.473	3.637	0.60	0.63	0.66	0.69	0.72	0.72
1	1.315	0.957	1.049	1.678	2.171	2.904	0.76	0.80	0.83	0.86	0.90	0.90
1¼	1.660	1.278	1.380	2.272	2.996	2.301	0.78	0.83	0.87	0.96	1.0	1.0
1½	1.900	1.500	1.610	2.717	3.631	2.010	0.90	0.95	1.0	1.1	1.1	1.2
2	2.375	1.939	2.067	3.652	5.022	1.608	1.1	1.2	1.3	1.4	1.4	1.5
2½	2.875	2.323	2.469	5.793	7.661	1.328	1.3	1.4	1.5	1.7	1.7	1.7
3	3.500	2.900	3.068	7.575	10.252	1.091	1.6	1.7	1.8	2.0	2.0	2.0
3½	4.000	3.364	3.548	9.109	12.505	0.955	1.8	2.0	2.1	2.3	2.3	2.3
4	4.500	3.826	4.026	10.790	14.983	0.849	1.9	2.1	2.2	2.2	2.2	2.2
4½	5.000	4.290	4.506	12.538	17.611	0.764	2.1	2.4	2.5	2.5	2.5	2.5
5	5.563	4.813	5.047	14.617	20.778	0.687	2.3	2.6	2.8	2.8	2.8	2.8
6	6.625	5.761	6.065	18.974	28.573	0.577	2.8	3.0	3.0	3.0	3.0	3.0
7	7.625	6.625	7.023	23.544	38.048	0.501	3.2	3.4	3.4	3.4	3.4	3.4
8	8.625	7.625	7.981	28.544	43.388	0.443	3.6	3.8	3.8	3.8	3.8	3.8
9	9.625	8.625	8.941	33.907	48.728	0.397	4.0	4.3	4.3	4.3	4.3	4.3
10	10.75	9.750	10.020	40.483	54.735	0.355	4.5	4.8	4.8	4.8	4.8	4.8
11	11.75	10.750	11.000	45.557	60.075	0.325	4.9	5.2	5.2	5.2	5.2	5.2
12	12.75	11.750	12.000	49.562	65.415	0.299	5.4	5.7	5.7	5.7	5.7	5.7

* For extra heavy pipe.

TABLE II.—LATENT HEAT OF EVAPORATION OF WATER FOR TEMPERATURE TO 212° F.

Temperature	Latent heat of evaporation	Temperature	Latent heat of evaporation	Temperature	Latent heat of evaporation
32	1,073.4	100	1,035.6	170	995.8
40	1,068.9	110	1,030.0	180	989.9
50	1,063.3	120	1,024.4	190	983.9
60	1,057.8	130	1,018.8	200	977.8
70	1,052.3	140	1,013.1	210	971.6
80	1,046.7	150	1,007.4	212	970.4
90	1,041.2	160	1,001.6		

Table III.—Pressure, Temperature, Etc. of Water Vapor

Pressure in lb. per sq. in. ab.	Temperature	Latent heat of evaporation	Total heat	Density s in lb. per cu. ft.	Volume of steam, cu. ft. per lb. $\frac{1}{s}$
1	101.83	1,034.6	1,104.4	0.00300	333.0
2	126.15	1,021.0	1,115.0	0.00576	173.5
3	141.52	1,012.3	1,121.6	0.00845	118.5
4	153.01	1,005.7	1,126.5	0.01107	90.5
5	162.28	1,000.3	1,130.5	0.01364	73.33
6	170.06	995.8	1,133.7	0.01616	61.89
7	176.85	991.8	1,136.5	0.01867	53.56
8	182.86	988.2	1,139.0	0.02115	47.27
9	188.27	985.0	1,141.1	0.02361	42.36
10	193.22	982.0	1,143.1	0.02606	38.38
11	197.75	979.2	1,144.9	0.02849	35.10
12	201.96	976.6	1,146.5	0.03090	32.36
13	205.87	974.2	1,148.0	0.03330	30.03
14	209.55	971.9	1,149.4	0.03569	28.02
14.7	212.00	970.4	1,150.4	0.03732	26.79
15	213.0	969.7	1,150.7	0.03806	26.27
16	216.3	967.6	1,152.0	0.04042	24.79
18	222.4	963.7	1,154.2	0.04512	22.16
20	228.0	960.0	1,156.2	0.04980	20.08
22	233.1	956.7	1,158.0	0.05445	18.37
24	237.8	953.5	1,159.6	0.05907	16.93
26	242.2	950.6	1,161.2	0.0636	15.72
28	246.4	947.8	1,162.6	0.0682	14.67
30	250.3	945.1	1,163.9	0.0728	13.74
35	259.3	938.9	1,166.8	0.0841	11.89
40	267.3	933.3	1,169.4	0.0953	10.49
45	274.5	928.2	1,171.6	0.1065	9.39
50	281.0	923.5	1,173.6	0.1175	8.51
55	287.1	919.0	1,175.4	0.1285	7.78
60	292.7	914.9	1,177.0	0.1394	7.17
65	298.0	911.0	1,178.5	0.1503	6.65
70	302.9	907.2	1,179.8	0.1612	6.20
75	307.6	903.7	1,181.1	0.1721	5.81
80	312.0	900.3	1,182.3	0.1829	5.47
85	316.3	897.1	1,183.4	0.1937	5.16
90	320.3	893.9	1,184.4	0.2044	4.89
95	324.1	890.9	1,185.4	0.2151	4.65
100	327.8	888.0	1,186.3	0.2258	4.429
105	331.4	885.2	1,187.2	0.2365	4.230
110	334.8	882.5	1,188.0	0.2472	4.047
115	338.1	879.8	1,188.8	0.2577	3.880
120	341.3	877.2	1,189.6	0.2683	3.726
125	344.4	874.7	1,190.3	0.2791	3.583
130	347.4	872.3	1,191.0	0.2897	3.452
135	350.3	869.9	1,191.6	0.3002	3.331
140	353.1	867.6	1,192.2	0.3107	3.219
145	355.8	865.4	1,192.8	0.3213	3.112
150	358.5	863.2	1,193.4	0.3320	3.012

TABLE IVa.—COEFFICIENT OF HEAT CONDUCTIVITY

Material	Remarks	Density S in lb. per cu. ft.	Temperature t in °F.	Moisture in %	Heat Conductivity c, B.t.u. per sq. ft. per ft. per hr. per °F.
Brick	....	101	123	0.00	0.275
		101	...	0.08	0.288
		101	...	0.90	0.402
		101	110	1.81	0.550
Cardboard	....	....	under 32°	.....	0.107
Cement	Portland	....	193	.....	0.171
Concretes:					
Pumice stone concrete	....	50	32	10.3	0.161
Slag concrete	....	50	32	10.3	0.161
Concrete	....	100	32	.....	0.483
Concrete	....	144	32	10.2	0.698
Claybrick	mixed with straw	94	...	dry	0.235
Claybrick	unbaked	111	...	7.4	0.402
Claybrick	unbaked	....	...	10.0	0.536
Claywall	stamped	119	...	5.7	0.348
Cemented wood	dry	45	32	0.4	0.074
Cemented wood	wet	52	32	12.4	0.094
Felt	....	....	below 32	.....	0.021
Felt	dark grey wool felt	9.4	104	.....	0.036
Feathers	intermixed with air	....	48	.....	0.013
Air	....	0.0807	32	.....	0.013
Glass	....	....	...	.....	0.538
Gypsum	artificially dried for 3 wks.	78	77	.....	0.248
Gravel	....	116	68	.....	0.214
Lime sandstone	....	103	75	15.3	0.536
Linoleum	....	74	32	.....	0.101
Linoleum	....	74	68	.....	0.107
Marble	white	....	86	.....	1.44
Marble	black	....	86	.....	1.66
Natural sandstone	grey	141	50	.....	0.892
Plaster	....	113	68	1.4	0.389
Plaster	....	117	68	2.0	0.309
Sand	....	95	68	0	0.187
Sand	....	102	68	11.3	0.653
Slate	....	....	below 32	.....	0.194
Water	....	62	39	.....	0.312
Wool	hairwool	5.6	34	.....	0.021
Wool	hairwool	5.6	142	.....	0.029
Woods:					
Oak ∥ to grain	....	51	122	.....	0.249
Oak ⊥ to grain	....	52	59	.....	0.120
Pine ∥ to grain	....	34	68	.....	0.201
Pine ⊥ to grain	....	34	59	.....	0.087
Fir ∥ to grain	....	28	131	.....	0.147
Fir ⊥ to grain	....	....	140	.....	0.062
Teak ⊥ to grain	....	40	59	.....	0.101
Teak ∥ to grain	....	38	59	.....	0.220
Shavings	pine	....	....	.....	0.030
Sawdust	....	13	122	.....	0.037

TABLE IV*b*.—COEFFICIENTS OF HEAT CONDUCTIVITY OF INSULATING MATERIALS

Material	Remarks	Spec. weight in lbs. per cu. ft.	Temperature in °F.	Heat conductivity *C*, B.t.u. per sq. ft. per ft. per hr. per °F.
Asbestos		36	32	0.087
Asbestos		36	120	0.103
Asbestos		36	212	0.112
Asbestos		36	300	0.117
Asbestos		36	400	0.121
Asbestos		36	500	0.123
Asbestos		36	600	0.125
Asbestos		36	700	0.127
Bark		21.4	.....	0.038
Blast furnace slag		22.5	120	0.064
Cork (granulated)		5.3	70	0.028
Cork (compressed)	previously subjected to heat	2.9	70	0.022
Cork sheet		6.2	.....	0.023
Cork sheet		12.5	.....	0.028
Cork sheet		18.7	.....	0.034
Cork sheet		24.9	.....	0.039
Cork sheet		31.2	.....	0.046
Cotton		5.1	32	0.032
Cotton		5.1	120	0.036
Cotton		5.1	212	0.040
Fire clay brick		....	400	0.340
Fire clay brick		....	1,100	0.440
Fire clay brick		....	1,800	0.550
Hair wool		5.6	18	0.021
Insulation comp	loose	25.3	32	0.040
Insulation comp		25.3	120	0.047
Insulation comp		25.3	212	0.051
Insulation comp		25.3	300	0.053
Insulation comp		25.3	400	0.054
Insulation comp	mixed with water and then dried	43.1	300	0.067
Insulation comp	mixed with water and then dried	43.1	450	0.080
Infusorial earth	loose	21.9	32	0.035
Infusorial earth		21.9	120	0.040
Infusorial earth		21.9	212	0.044
Infusorial earth		21.9	300	0.047
Infusorial earth		21.9	400	0.050
Infusorial earth		21.9	500	0.051
Infusorial earth		21.9	550	0.052
Infusorial earth		21.9	650	0.053
Infusorial earth	mixed with water and then dried	36.2	300	0.056
Infusorial earth	mixed with water and then dried	36.2	650	0.082
Linoleum		73.8	32	0.101
Magnesia stone		....	1,100	0.860
Magnesia stone		....	1,800	0.970
Peat	slabs	12.0	32	0.032
Peat		23.2	32	0.042
Peat		45.5	32	0.064
Peat		51.8	32	0.950
Compressed horse hair		10.7	70	0.030
Pulverized cork		10.1	32	0.021
		10.1	120	0.028
		10.1	212	0.032
		10.1	300	0.035
		10.1	400	0.037
Reeds		4.7	.....	0.027
Silicate stones		....	400	0.380
Silicate stones		....	1,100	0.590
Silicate stones		....	1,800	0.800
Sheep wool		8.5	32	0.022
Sheep wool		8.5	120	0.028
Sheep wool		8.5	212	0.034
Silk		6.3	32	0.026
Silk		6.3	120	0.030
Silk		6.3	212	0.034

TABLE V.—RADIATION CONSTANT C OF BUILDING MATERIALS

Material	C	Material	C
Black body	0.162	Zinc, dull	0.036
Cotton	0.128	Tin	0.0077
Glass	0.155	Oil paint	0.130
Wood	0.126	Paper	0.134
Brick	0.156	Plaster	0.156
Metals:		Lampblack	0.156
Wrought iron, dull oxidized	0.156	Sand	0.127
Wrought iron, highly polished	0.047	Shavings	0.124
		Silk	0.130
Cast iron, rough oxidized	0.154	Water	0.112
Copper, slightly polished	0.028	Wool	0.130
Brass, dull	0.036		
Silver	0.0046		

TABLE VI.—HEAT TRANSMISSION COEFFICIENTS
(American Gas Association)

Frame Walls

Clapboards, Sheathing, Studding, Cork, Plaster

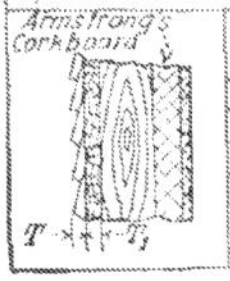

Armstrong's Corkboard

T	T_1	K ½" Cork	K 2" Cork
7/16"	3/4"	.12	.10

Frame Walls

Clapboards, Celotex, Studding, Celotex and Plaster

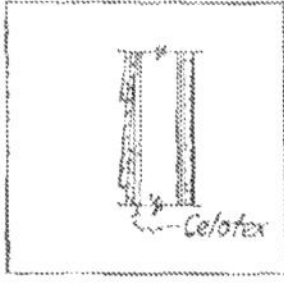

T	T_1	K
		0.25

Board Walls

Plain Tongue and Grooved Board

T	T_1	K
1"		.50

Frame Walls

Clapboard, Sheathing, Sawdust, L. & P.

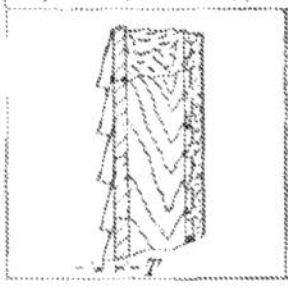

T	T_1	K
7/16"	3/4"	.15

Frame Walls

Clapboards, Paper, Sheathing, Back Plaster L. & P.

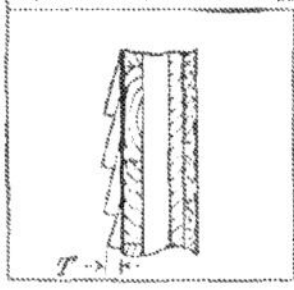

T	T_1	K
7/16"	3/4"	.22

Board Walls

Corrugated Iron, T. & G. Boards

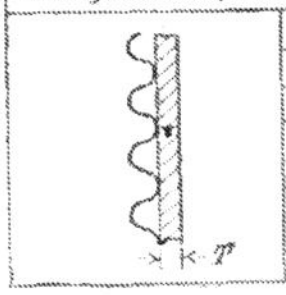

T	T_1	K
1"		.48

Stucco Walls

Frame and Stucco Construction

T	T_1	K
		.45

Frame Walls

Clapboards, Paper, Sheathing, Studding, L. & P.

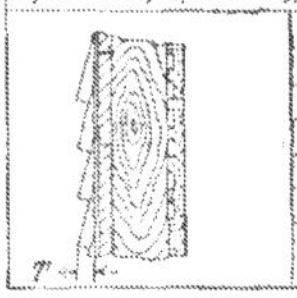

T	T_1	K
7/16"	3/4"	.24

Board Walls

Corrugated Iron, Paper, T. & G. Boards

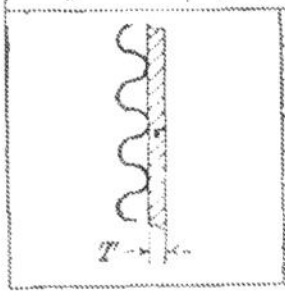

T	T_1	K
1"		.42

Frame Walls

Clapboards, Studding, Lath and Plaster

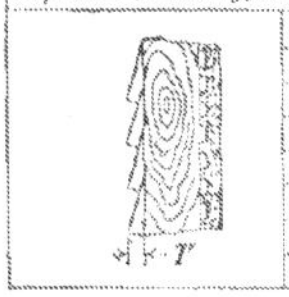

T	T_1	K
7/16"		.44

Frame Walls

Clapboards, Paper, Sheathing, Studding 3⅝" of #17 INSULEX Insulation, L. & P.

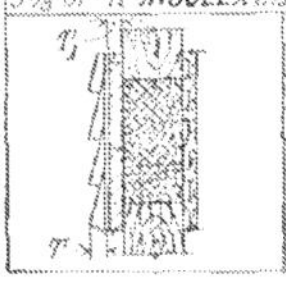

T	T_1	K
7/16"	3/4"	.074

Steel Walls

Unlined Corrugated Iron

Area is Projected Area Not Corrugated Area

T	T_1	K
		1.5

Frame Walls

Clapboards, Paper, Studding, Lath and Plaster

T	T_1	K
7/16"		.31

Frame Walls

Clapboards, Sheathing, Studding, L. & P.

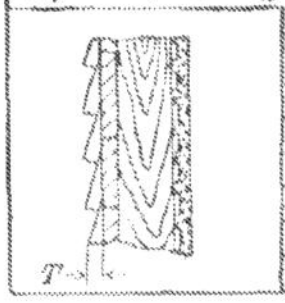

T	T_1	K
7/16"	3/4"	.28

Steel Walls

Sheet Metal, Siding Unlined

T	T_1	K
		1.2

TABLE VI.—HEAT TRANSMISSION COEFFICIENTS.—(*Continued*)

Brick Walls

Plaster One Side Only

T	T_1	K
4"		.50
8"		.38
12"		.29
16"		.24
18"		.22
20"		.19
24"		.17

Brick Walls

Brick, Furring, Celotex and Plaster

Celotex

T	T_1	K
9"		0.15
13"		0.13
18"		0.11
24"		0.096

Sandstone or Granite Walls

Plain Sandstone, Granite or Marble

T	T_1	K
8"		.60
12"		.47
16"		.41
18"		.38
20"		.36
24"		.33
28"		.29

Brick Walls

Brick, Mortar, Cork, Plaster

Armstrong's Corkboard

T	K ½" Cork	K 2" Cork
4"	.14	.12
8"	.13	.11
12"	.12	.10
16"	.11	.09
20"	.10	.08
24"	.09	.08

Brick Walls

Brick, Hollow Tile and Plaster One Side

T	T_1	K
4"	4"	.30
8"	4"	.26
12"	4"	.22
16"	4"	.20
18"	4"	.19
20"	4"	.18
24"	4"	.16

Sandstone Walls

Sandstone, Fur, Lath and Plaster

T	T_1	K
8"		.50
12"		.40
16"		.35
18"		.33
20"		.31
24"		.28
28"		.25

Brick Walls

Furring Lath and Plaster

T	T_1	K
4"		.33
8"		.27
12"		.23
16"		.21
18"		.20
20"		.19
24"		.18

Brick and Sandstone Walls

Plain with Sandstone Face

T	T_1	K
4"	4"	.42
8"	4"	.35
12"	4"	.29
16"	4"	.24
20"	4"	.19
24"	4"	.17
28"	4"	.15

Limestone Walls

Plain Limestone

T	T_1	K
8"		.62
12"		.50
16"		.43
18"		.41
20"		.39
24"		.35
28"		.31

Brick Walls

Brick, Studding, Cork, Plaster

Armstrong's Corkboard

T	K ½" Cork	K 2" Cork
4"	.13	.10
8"	.12	.10
12"	.11	.09
16"	.10	.09
20"	.10	.08
24"	.10	.08

Brick and Sandstone Walls

Plastered

T	T_1	K
4"	8"	.35
8"	8"	.29
12"	8"	.24
16"	8"	.20
20"	8"	.18
24"	8"	.16
28"	8"	.14

Limestone Walls

Limestone, Fur, Lath and Plaster

T	T_1	K
12"		.42
16"		.36
18"		.34
20"		.32
24"		.30
28"		.28
30"		.26

Brick Walls

2" Furring Strips Filled with #12 INSULEX Insulation, Lath and Plaster

T		K
4"		.12
8"		.11
12"		.10
16"		.09
18"		.096
20"		.094
24"		.091

Brick and Sandstone Walls

Furred and Plastered

T	T_1	K
4"	12"	.30
8"	12"	.25
12"	12"	.21
16"	12"	.19
20"	12"	.17
24"	12"	.15
28"	12"	.13

Concrete Walls

Concrete

T	T_1	K
8"		.60
12"		.46
16"		.40
20"		.36
24"		.32
28"		.28
30		.27

TABLE VI.—HEAT TRANSMISSION COEFFICIENTS.—(*Continued*)

Concrete Walls

Concrete, Fur, Lath and Plaster

T	T_1	K
8"		.50
12"		.40
16"		.34
18"		.32
20"		.31
24"		.28
28"		.25

Partition Walls

Stud Partition, L.& P. Both Sides

T	T_1	K
		.34

Intermediate Floors

Single Floor - No Plaster

T	T_1	K
3/4"		.30

Concrete Walls

Concrete, Furring, Celotex and Plaster

T	T_1	K
4"		.24
6"		.19
9"		.18
12"		.17
15"		.15
20"		.14

Basement Floors

Dirt

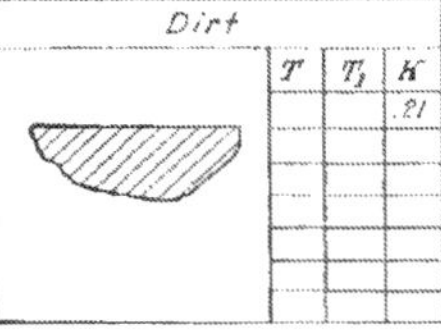

T	T_1	K
		.21

Intermediate Floors

Double Floor - No Plaster

T	T_1	K
1½"		.22

Concrete Walls

2" Furring Strips, filled with #12 INSULEX Insulation, Lath and Plaster

T	T_1	K
8"		.14
12"		.13
16"		.12
18"		.12
20"		.12
24"		.11
28"		.11

Basement Floors

Cement or Concrete on Dirt

T	T_1	K
		.31

Intermediate Floors

Single Floor, Lath and Plaster below Joist

T	T_1	K
3/4"		.21

Stucco Walls

Tile - Stucco Outside - Plaster Inside

T	T_1	K
4"		.40

Basement Floors

Wood Floor and Sleeper on Dirt

T	T_1	K
		.13

Intermediate Floors

Double Floor, Lath and Plaster below Joist

T	T_1	K
1½"		.15

Stucco Walls

Tile - Stucco Outside - Plaster Inside

T	T_1	K
8"		.30
12"		.25

Basement Floors

Wood on Sleepers, and Cinders on Dirt

T	T_1	K
		.12

Intermediate Floors

Wood Flooring, Filling, Air Space L. & P.

T	T_1	K
		.06

TABLE VI.—HEAT TRANSMISSION COEFFICIENTS.—(*Continued*)

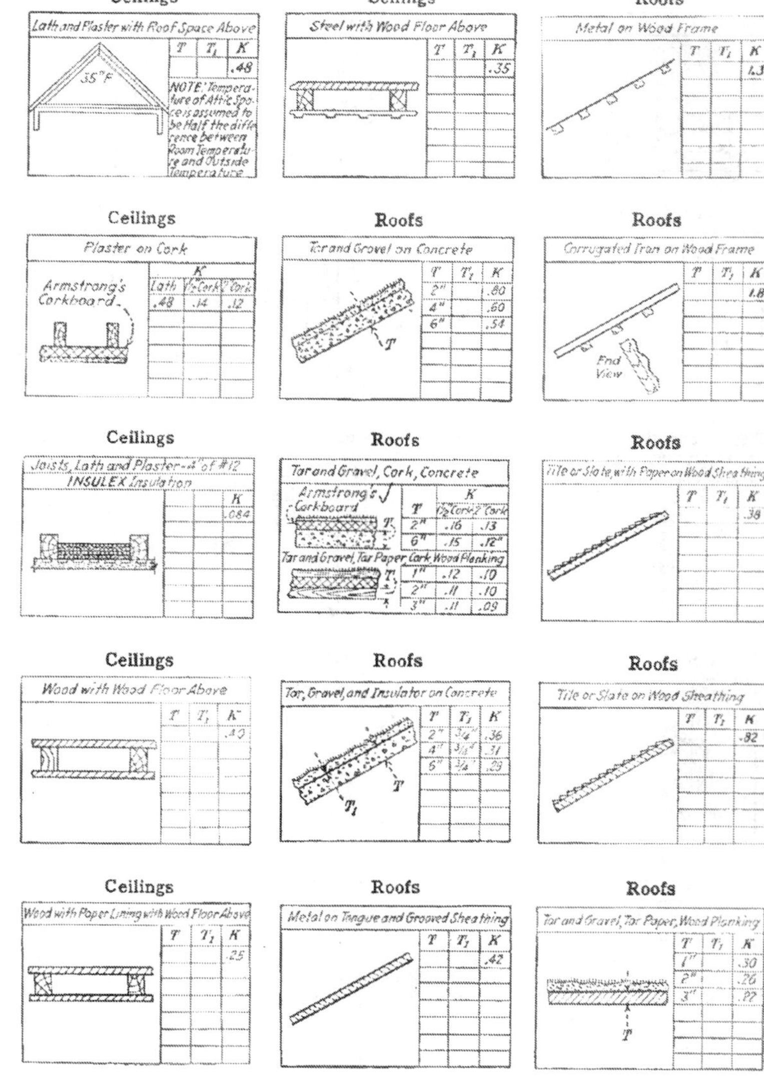

Ceilings — Lath and Plaster with Roof Space Above

T	T_1	K
		.48

35° F

NOTE: Temperature of Attic Space is assumed to be Half the difference between Room Temperature and Outside Temperature

Ceilings — Steel with Wood Floor Above

T	T_1	K
		.35

Roofs — Metal on Wood Frame

T	T_1	K
		1.3

Ceilings — Plaster on Cork

Armstrong's Corkboard

K		
Lath	1½" Cork	2" Cork
.48	.14	.12

Roofs — Tar and Gravel on Concrete

T	T_1	K
2"		.80
4"		.60
6"		.54

Roofs — Corrugated Iron on Wood Frame

End View

T	T_1	K
		1.8

Ceilings — Joists, Lath and Plaster—4" of #12 INSULEX Insulation

		K
		.084

Roofs — Tar and Gravel, Cork, Concrete

Armstrong's Corkboard

T	K	
	1½" Cork	2" Cork
2"	.16	.13
6"	.15	.12"

Tar and Gravel, Tar Paper, Cork, Wood Planking

T	K	
1"	.12	.10
2"	.11	.10
3"	.11	.09

Roofs — Tile or Slate, with Paper on Wood Sheathing

T	T_1	K
		.38

Ceilings — Wood with Wood Floor Above

T	T_1	K
		.40

Roofs — Tar, Gravel, and Insulator on Concrete

T	T_1	K
2"	¾"	.36
4"	¾"	.31
6"	¾"	.29

Roofs — Tile or Slate on Wood Sheathing

T	T_1	K
		.82

Ceilings — Wood with Paper Lining with Wood Floor Above

T	T_1	K
		.25

Roofs — Metal on Tongue and Grooved Sheathing

T	T_1	K
		.42

Roofs — Tar and Gravel, Tar Paper, Wood Planking

T	T_1	K
1"		.30
2"		.26
3"		.22

TABLE VI.—HEAT TRANSMISSION COEFFICIENTS.—(Continued)

Roofs

Wood Plank with Tar and Gravel Roofing on #24 INSULEX Insulation

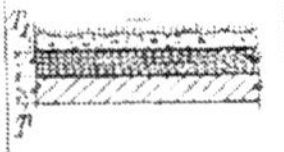

T	T_1	K
1"	2"	.17
2"	2"	.15
3"	2"	.14
1"	3"	.14
2"	3"	.13
3"	3"	.12

Roofs

Wood Shingles, Furring, Celotex, Rafters

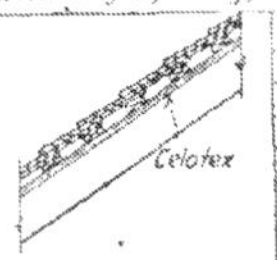

T	T_1	K
		.021

Windows

Double Window

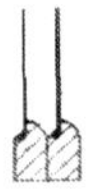

T	T_1	K
		.60

Roofs

Tar and Gravel Roofing, Celotex, Wood Planking

T	T_1	K
7/8"		0.27
1 3/4"		0.22
2 5/8"		0.18

Tar and Gravel Roofing, Celotex On Concrete

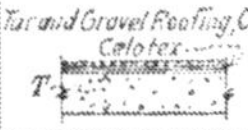

T	T_1	K
2"		0.33
4"		0.30
6"		0.28

Roofs

Shingles, Sheathing, Studding, Lath and Plaster

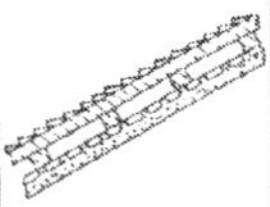

T	T_1	K
		.30

Skylight

Single Light

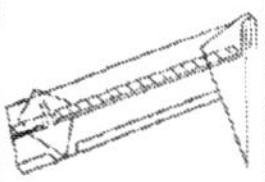

T	T_1	K
		1.3

Roofs

Tar and Gravel, Concrete, Hollow Tile

T	T_1	K
6"		.38
8"		.36

Roofs

Shingles, Sheathing, Rafters with 4" of #12 INSULEX Insulation Lath and Plaster

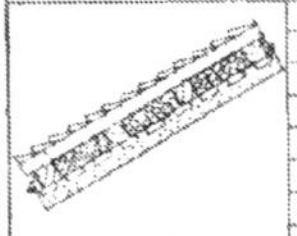

		K
		.075

Skylight

Single Monitor

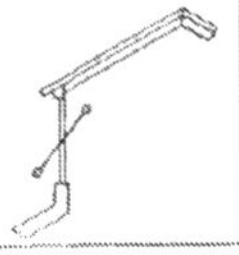

T	T_1	K
		1.3

Roofs

Tar and Gravel on Concrete and Suspended L & P

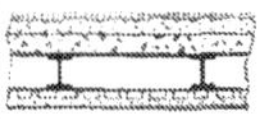

T	T_1	K
		.30

Roofs

Shingles, Sheathing Rafters, Cork, Plaster

T	K	
	1 1/2" Cork	2" Cork
3/4"	.12	.10

Doors

Single Door

T	T_1	K
All Glass		1.1
Upper Half Glass		.75
All Wood		.45

Roofs

Shingles, Sheathing and Studding

T	T_1	K
		.44

Windows

Single Window

T	T_1	K
		1.1

Doors

Inner Vestibule Door

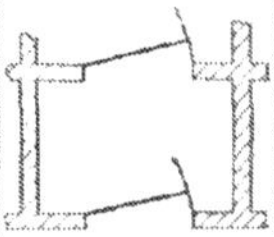

T	T_1	K
All Glass		.60
Upper Half Glass		.45
All Wood		.30

TABLE VII.—ALLOWANCES FOR SPECIAL CONDITIONS

To the heat loss as determined by computations previously given the following additions should be made

a Allowance for outside surfaces based on direction of exposure

For north, northeast, northwest, and east add 15%
For west, southeast, southwest add 10%

This also applies to vertical and oblique roofs having heated attics.

b Allowance for corner rooms etc

Corner rooms and others, in which the surface of the intersecting outside walls equals or is greater than one-half the surface of the remaining walls add 10%
Rooms with parallel outside walls add 10%
Corner rooms and others in which the surface of the intersecting outside walls equals or is greater than one-third the surface of the remaining walls add 5%
Corner rooms in which the one corner wall is half as long as the other add 5%

c. Allowance for wind

Outside walls exposed to cold wind, also all exposed walls of a heated attic add 10%
A wall is considered exposed to the wind if, within a radius of 120 ft there are no shielding obstructions of equal height For exceptionally high wind exposure, as *e g* wall surfaces at right angles to the direction of a street as for houses in the suburbs of a city add 15%

d Sudden temperature changes

Allowance to cover any sudden drop in temperature add . 10%

NOTE The allowances under headings *a*, *b*, *c* and *d*, are valid only for outside surfaces, and must be totaled for each case if necessary

e Allowance for high rooms

Rooms exceeding 12 ft in height receive an additional 2½% to the uncorrected heat loss for each additional 3 ft of height . 2½%
and not more than 20%

This correction is not required for hallways If the increase in temperature for the increase in height has been considered in the heat loss computations, no addition need be made

f Allowance for the heating-up period and for intermittent operation of heating surface

1 For continuous operation of the heating system (in consideration of banking at night) add 10%
2 For continuous operation during the day with low rate of driving at night (*i e* night attention) add 5%
3 For intermittent operation 13 to 15 hr. a day including heating-up period of not less than 3 hr add 15%
4 For intermittent operation 9 to 12 hrs a day including heating-up period as above add . 20%
5 For operation after comparatively long intervals of vacancy as for instance in entertainment halls add 30%

The allowances for the heating-up period and for intermittent operation are to be added to the sum of heat losses for other safety allowances.

g Exceptional allowances

1 For rooms with exceptionally heavy walls (*i e* thicker than 3 ft) an additional safety factor of 15 B t u for every square foot of wall area must be added to provide for heat absorption of walls

2 Churches and similar halls, rooms, etc which are used for only short periods and do not attain a steady state are to be designed according to method described elsewhere in the text

TABLE VIII.—HEAT TRANSMISSION COEFFICIENTS K FOR WATER HEATED SURFACES

Nature of the heating surface	Quantity of heat K emitted per sq. ft. per hour and °F. temp. difference if the average temp. difference between water and air equals					
	Below 70°	Over 70° to 90°	Over 90° to 110°	Over 110° to 130°	Over 130° to 150°	Over 150°
a. Steel heating surfaces						
1. Single vertical or horizontal pipe.						
For 1″ pipes and under	2.2	2.3	2.4	2.5	2.6	2.6
For pipes over 1″ to approx. 2″ nominal size	1.8	1.9	2.0	2.2	2.3	2.4
For pipes over 2″ to approx. 3½″ nominal size	1.7	1.9	2.0	2.2	2.2	2.2
For pipes over 3½″ to approx. 5″ nominal size	1.6	1.8	1.9	1.9	1.9	1.9
For pipes over 5″ nominal size	1.6	1.7	1.7	1.7	1.7	1.7
2. For pipes installed over one another in form of a pipe coil having a total height of up to 3 ft. The pipe centers are at least 1 pipe diameter apart.						
For pipes up to 1″ nominal size	1.8	2.0	2.2	2.3	2.3	2.4
For pipes over 1″ nominal size	1.4	1.6	1.7	1.8	1.8	1.8
3. Same as under 2 only allowing height to exceed 3′.						
For pipes up to 1″ nominal size	1.6	1.7	1.8	1.9	1.9	1.9
For pipes over 1″ nominal size	1.3	1.4	1.5	1.6	1.6	1.6
4. Header pipe coils consisting of a number of horizontal or vertical pipes.						
1 Row	1.5	1.6	1.7	1.8	1.9	1.9
2 Rows	1.1	1.2	1.3	1.4	1.4	1.4
4 Rows	1.0	1.1	1.2	1.2	1.2	1.2
5. Plate heater.						
Up to 3 ft. in height	1.5	1.6	1.7	1.8	1.9	1.9
Over 3 ft. in height	1.3	1.4	1.5	1.6	1.7	1.7
b. Cast iron heating surfaces						
Radiators: Minimum distance between sections not less than one inch.						
One and 2 col. radiators, 6 or more sections.						
Total height 26″ including legs	1.3	1.4	1.4	1.4	1.5	1.5
Total height 32″	1.2	1.3	1.3	1.4	1.4	1.4
Total height 45″	1.2	1.2	1.3	1.3	1.4	1.4
For radiators of 3–6 sections. These values are to be increased 5%.						

TABLE VIII.—HEAT TRANSMISSION COEFFICIENTS K FOR WATER HEATED SURFACES.—(*Continued*)

Nature of heating surface	Quantity of heat K emitted per sq. ft. per hour and °F. temp. difference if the average temp. difference between water and air equals					
	Below 70°	Over 70° to 90°	Over 90° to 110°	Over 110° to 130°	Over 130° to 150°	Over 150°
Radiators 3 col. 6 or more sections.						
Total height 26″	1.2	1.2	1.3	1.3	1.3	1.4
Total height 32″	1.1	1.2	1.2	1.3	1.3	1.3
Total height 45″	1.1	1.1	1.2	1.2	1.2	1.3
For radiators of 3–6 sections these values are to be increased 5%.						
Ribbed surface rad. (see Vol. I) 24″ high approx. with vertical fins spaced not less than 1¾″.						
Height of fin 0″	1.5	1.7	1.7	1.8	1.8	1.8
Height of fin ¾″	1.1	1.2	1.3	1.3	1.3	1.3
Height of fin 1½″	1.0	1.1	1.1	1.2	1.2	1.2
Height of fin 2″	0.92	1.0	1.1	1.1	1.1	1.1
Height of fin 2½″	0.92	0.92	1.0	1.0	1.0	1.0
Finned radiator with oblique fins spaced not less than ½″	0.82	0.92	1.0	1.0	1.0	1.1
Finned pipe with fin spacing of at least 1½″	0.82	0.92	1.0	1.0	1.1	1.1
Finned pipe radiator for which individual pipe lengths are installed in series immediately above one another. Spacing of fins at least ¾″.						
1 Pipe	0.72	0.92	0.92	1.0	1.0	1.0
3 Pipe } (The fins of the pipes partially	0.61	0.72	0.82	0.82	0.82	0.82
6 Pipe } interlock).	0.51	0.61	0.61	0.72	0.72	0.72
Finned pipe radiator as above but having pipe and fins of oval shape. Spacing of fins at least ½″.						
1 Pipe	1.0	1.1	1.2	1.3	1.3	1.3
3 Pipe (The fins do not interlock)	0.82	0.92	0.92	1.0	1.0	1.0
6 Pipe	0.61	0.72	0.82	0.82	0.82	0.82

TABLE IX.—HEAT TRANSMISSION COEFFICIENTS K FOR STEAM HEATED SURFACES

Type of heating surface	Quantity of heat, K, emitted per sq. ft. per hour and °F. temp. difference between the av. steam temperature and the air
a. Steel heating surfaces	
Single horizontal pipe.	
For pipes up to 1″ nominal diameter	2.7
For pipes from 1″ to 4″ nominal diameter	2.5
For pipes over 4″ nominal diameter	2.4
Single vertical pipe.	
(α) For low pressure steam heating.	
For pipes up to 1″ nominal diam	2.8
For pipes from 1″ to 4″ nominal diam	2.6
For pipes over 4″ nominal diam	2.5
(β) For high pressure steam heating.	
For pipes up to 1″ nominal diam	2.9
For pipes from 1″ to 4″ nominal diam	2.7
For pipes over 4″ nominal diam	2.6
For pipes installed over one another in form of a pipe coil having a total height up to 3 ft. The pipes centers are at least 1 pipe diameter apart.	
Pipes up to 1″ nominal size	2.6
Pipes over 1″ nominal size	2.3
Same as above only for heights exceeding 3 ft.	
Pipes up to 1″ nominal size	2.3
Pipes over 1″ nominal size	1.9
Header pipe coils consisting of a number of horizontal or vertical pipes.	
1 Row	2.4
2 Rows	1.8
4 Rows	1.6
Plate heater.	
Up to 3 ft. in height	2.5
Over 3 ft. in height	2.3
b. Cast iron heating surfaces	
Radiators: minimum distance between sections not less than one inch.	
(α) Low pressure steam heating.	
One and 2 col. radiators 6 or more sections.	
Total height 26″	1.7
Total height 32″	1.6
Total height 45″	1.6
For radiators of from 3–6 sections these values to be increased 5%.	
Radiators: 3 col. 6 or more sections.	
Total height 26″	1.5
Total height 32″	1.4

TABLE IX.—HEAT TRANSMISSION COEFFICIENTS K FOR STEAM HEATED SURFACES.—(*Continued*)

Type of heating surface	Quantity of heat, *K*, emitted per sq. ft. per hour and °F. temp. difference between the av. steam temperature and the air
Total height 45″	1.4
For radiator of from 3–6 sections these values to be increased 5%.	
(β) For high pressure steam heating.	
One and 2 col. radiators 6 or more sections.	
Total height 26″	1.8
Total height 32″	1.7
Total height 45″	1.6
For radiators of from 3–6 sections these values to be increased 5%.	
Radiators: 3 col. 6 or more sections.	
Total height 26″	1.6
Total height 32″	1.5
Total height 45″	1.4
For radiators of from 3–6 sections these values to be increased 5%.	
Ribbed surface radiator. 24″ approx. with vertical fins spaced not less than 1¾″.	
Height of fin 0″	2.3
Height of fin ¾″	1.6
Height of fin 1½″	1.5
Height of fin 2″	1.4
Height of fin 2½″	1.3
Finned radiators with oblique fins spaced not less than ½″	1.2
Finned pipe with fin spacing of at least 1½″	1.3
Finned pipe radiator for which individual pipe lengths are installed in series immediately above one another. Spacing of fins at least ¾″.	
1 Pipe	1.2
3 Pipes } (The fins partially interlock)	0.92
6 Pipes } (The fins partially interlock)	0.82
Finned pipe radiator as above but having pipe and fins of oval shape. Spacing of fins at least ½″.	
1 Pipe	1.4
3 Pipes } (The fins do not interlock)	1.1
6 Pipes } (The fins do not interlock)	0.92

TABLE Xa.—VENTO HOT-BLAST HEATERS
Regular section—ratings and free areas
60″ section (steam or water)—16 square feet. Height $60^{11}/_{16}$″. Width $9\frac{1}{8}$″

Number of sections in stack	Square feet of heating surface	Equivalent in lineal feet 1-inch pipe	$5\frac{3}{8}$″ centers of sections		5″ centers of sections		$4\frac{5}{8}$″ centers of sections		Actual weights
			52 % of face		Standard 44 % of face		37 % of face		
			Net air space in square feet	Width of Stack in inches	Net air space in square feet	Width of stack in inches	Net air space in square feet	Width of stack in inches	
7	112.0	336	7.62	38	6.45	35	5.47	32	8.20 lbs. per sq. ft. actual 9 lbs. per sq. ft. shipping weight
8	128.0	384	8.70	43	7.37	40	6.25	37	
9	144.0	432	9.77	48	8.29	45	7.03	42	
10	160.0	480	10.85	54	9.21	50	7.81	46	
11	176.0	528	11.93	59	10.13	55	8.59	51	
12	192.0	576	13.00	65	11.05	60	9.37	55	
13	208.0	624	14.08	70	11.97	65	10.15	60	
14	224.0	672	15.15	75	12.89	70	10.93	65	
15	240.0	720	16.23	81	13.81	75	11.71	69	
16	256.0	768	17.31	86	14.73	80	12.49	74	
17	272.0	816	18.39	91	15.65	85	13.27	79	
18	288.0	864	19.46	97	16.57	90	14.05	83	
19	304.0	912	20.54	102	17.50	95	14.83	88	
20	320.0	960	21.62	108	18.42	100	15.61	92	
21	336.0	1,008	22.70	113	19.34	105	16.39	97	
22	352.0	1,056	23.78	118	20.26	110	17.17	102	
23	368.0	1,104	24.85	124	21.18	115	17.95	106	
24	384.0	1,152	25.93	129	22.10	120	18.73	111	

TABLE Xb.—FRICTION OF AIR THROUGH VENTO HOT-BLAST HEATERS

Friction loss—in inches of water—due to air passing through Vento stacks. (Measured at 70°)

Regular section—4⅝, 5 and 5⅜-inches spacing

Velocity feet per mi .	Spacing of sections inches	1 stack	2 stacks	3 stacks	4 stacks	5 stacks	6 stacks	7 stacks	8 stacks
	4⅝	.022	.043	.063	.084	.105	.126	.147	
600	5	.021	.040	.058	.076	.094	.112	.130	.149
	5⅜	.019	.034	.049	.064	.079	.094	.109	.124
	4⅝	.031	.059	.087	.115	.143	.172	.200	
700	5	.028	.054	.079	.105	.130	.155	.180	.205
	5⅜	.025	.046	.066	.087	.108	.128	.149	.170
	4⅝	.040	.077	.114	.150	.187	.224	.259	
800	5	.037	.070	.103	.135	.167	.200	.232	.265
	5⅜	.033	.060	.087	.114	.140	.167	.194	.221
	4⅝	.051	.097	.144	.190	.237	.283	.329	
900	5	.047	.088	.129	.170	.211	.252	.293	.335
	5⅜	.042	.076	.110	.144	.178	.212	.246	.280
	4⅝	.063	.120	.178	.235	.293	.350	.407	
1,000	5	.059	.109	.160	.211	.262	.313	.364	.415
	5⅜	.052	.094	.136	.178	.220	.262	.304	.346
	4⅝	.076	.145	.214	.284	.353	.422	.491	
1,100	5	.071	.132	.193	.255	.316	.377	.438	.501
	5⅜	.062	.113	.164	.215	.265	.316	.367	.418
	4⅝	.090	.172	.255	.337	.420	.502	.584	
1,200	5	.084	.157	.230	.303	.376	.449	.522	.596
	5⅜	.074	.134	.195	.255	.316	.376	.437	.497
	4⅝	.105	.202	.299	.396	.493	.590	.687	
1,300	5	.099	.185	.271	.356	.442	.528	.614	.701
	5⅜	.087	.158	.229	.300	.371	.442	.513	.584
	4⅝	.122	.234	.347	.459	.572	.684	.796	
1,400	5	.115	.214	.314	.414	.513	.612	.712	.813
	5⅜	.101	.183	.266	.348	.430	.512	.595	.677
	4⅝	.140	.269	.398	.527	.656	.785	.914	
1,500	5	.132	.246	.360	.474	.588	.702	.816	.932
	5⅜	.116	.210	.305	.399	.493	.587	.682	.776
	4⅝	.160	.306	.453	.600	.746	.893	1.040	
1,600	5	.150	.280	.410	.540	.670	.800	.930	1.060
	5⅜	.132	.239	.347	.454	.561	.668	.776	.883
	4⅝	.180	.346	.512	.677	.843	1.009	1.174	
1,700	5	.169	.316	.463	.609	.756	.903	1.049	1.197
	5⅜	.149	.270	.391	.512	.634	.755	.876	.997
	4⅝	.202	.387	.573	.759	.944	1.130	1.316	
1,800	5	.190	.354	.518	.683	.848	1.012	1.177	1.342
	5⅜	.167	.303	.439	.575	.710	.846	.982	1.118

TABLE Xc.—FINAL TEMPERATURES AND CONDENSATIONS

Regular section—standard spacing, 5″ centers of sections—steam 227°, 5 lbs. gauge. Air entering above zero

Number of stacks deep	Temperature of entering air	Velocity through heater in feet per minute—measured at 70°																					
		100		120		200		600		800		1,000		1,200		1,400		1,600		1,800		2,000	
		Final temp. air leaving heater	Cond. lbs. per sq. ft. per hour	F. T.	C.	F. T.	C.	F. T.	C.	F. T.	C.	F. T.	C.	F. T.	C.	F. T.	C.	F. T.	C.	F. T.	C.	F. T.	C.
1	20	87	.43	83	.48	75	.70	58	1.46	54	1.75	51	1.99	49	2.23	47	2.42	45	2.56	43	2.65	42	2.82
	30	93	.40	89	.45	81	.65	66	1.39	62	1.64	60	1.92	58	2.17	56	2.33	54	2.46	52	2.54	51	2.69
	40	100	.38	96	.43	88	.62	74	1.31	70	1.54	68	1.80	66	2.00	64	2.16	62	2.26	61	2.42	60	2.56
	60	112	.33	109	.38	102	.54	90	1.15	86	1.34	81	1.54	82	1.69	81	1.89	80	2.05	79	2.19	78	2.31
	70	118	.31	115	.35	109	.50	97	1.04	94	1.23	92	1.41	90	1.54	89	1.71	88	1.85	87	1.96	86	2.05
2	20	129	.35	124	.40	112	.59	87	1.29	81	1.57	76	1.80	72	2.00	69	2.20	66	2.36	64	2.54	62	2.69
	30	133	.33	129	.38	117	.56	93	1.21	87	1.46	83	1.70	79	1.89	76	2.06	73	2.21	71	2.37	69	2.50
	40	137	.31	133	.36	121	.52	100	1.15	94	1.39	90	1.60	86	1.77	83	1.93	81	2.10	79	2.27	77	2.37
	60	145	.27	142	.32	131	.46	112	1.00	107	1.21	103	1.38	100	1.54	98	1.71	96	1.85	94	1.96	92	2.05
	70	149	.25	146	.29	136	.42	118	.92	114	1.13	110	1.28	107	1.42	105	1.57	103	1.69	101	1.79	100	1.92
3	20	157	.29	152	.34	139	.51	110	1.15	103	1.42	97	1.65	92	1.85	88	2.06	85	2.22	82	2.38	79	2.52
	30	160	.28	155	.32	142	.48	115	1.09	108	1.33	103	1.56	98	1.75	94	1.91	91	2.08	88	2.23	85	2.35
	40	163	.26	158	.30	146	.45	121	1.04	114	1.26	109	1.47	104	1.64	100	1.79	97	1.95	94	2.08	92	2.22
	60	168	.23	164	.27	153	.40	131	.91	124	1.09	120	1.28	116	1.44	113	1.58	110	1.71	108	1.85	106	1.97
	70	170	.21	167	.25	156	.37	136	.85	130	1.03	126	1.20	122	1.34	119	1.46	116	1.57	114	1.69	112	1.80
4	20	...	...	...	...	...	...	130	1.06	122	1.31	115	1.52	110	1.73	105	1.91	101	2.08	97	2.22	94	2.37
	30	...	...	...	...	...	...	131	1.00	126	1.23	120	1.44	115	1.63	110	1.80	106	1.95	102	2.08	99	2.21
	40	...	...	...	...	...	...	138	.94	130	1.15	124	1.35	119	1.52	115	1.68	111	1.82	108	1.96	105	2.08
	60	...	...	...	...	...	...	146	.83	139	1.01	134	1.19	129	1.33	125	1.46	122	1.59	119	1.70	117	1.83
	70	...	...	...	...	...	...	150	.77	143	.94	138	1.09	134	1.23	131	1.37	128	1.49	125	1.59	123	1.70

5	20	....	....	.....	.....	......	....	144	.95	136	1.19	130	1.41	124	1.60	119	1.78	114	1.93	110	2.08	107	2.23
	30	....	....	.....	.....	......	....	148	.91	140	1.13	134	1.33	128	1.51	123	1.67	118	1.80	115	1.96	112	2.10
	40	....	....	.....	.....	......	....	151	.85	144	1.07	138	1.26	132	1.42	127	1.56	123	1.70	120	1.85	117	1.98
	60	....	....	.....	.....	......	....	158	.75	151	.93	145	1.09	140	1.23	136	1.36	133	1.50	130	1.62	127	1.72
	70	....	....	.....	.....	......	....	162	.71	155	.87	149	1.01	144	1.14	141	1.27	138	1.40	135	1.30	132	1.59
6	20	....	....	.....	.....	......	....	156	.87	148	1.10	142	1.30	136	1.49	130	1.65	126	1.81	122	1.96	118	2.09
	30	....	....	.....	.....	......	....	159	.83	151	1.04	145	1.23	139	1.40	134	1.56	130	1.71	126	1.85	122	1.97
	40	....	....	.....	.....	......	....	162	.78	154	.97	148	1.15	143	1.32	138	1.47	134	1.60	131	1.75	127	1.86
	60	....	....	.....	.....	......	....	167	.69	160	.85	155	1.02	150	1.15	146	1.29	142	1.40	139	1.52	136	1.62
	70																						
7	20	....	....	.....	.....	......	....	167	.81	159	1.02	152	1.21	146	1.39	141	1.55	136	1.70	132	1.85	128	1.98
	30	....	....	.....	.....	......	....	169	.76	161	.96	155	1.15	149	1.31	144	1.46	139	1.60	135	1.73	132	1.87
	40	....	....	.....	.....	......	....	171	.72	164	.91	158	1.08	153	1.24	148	1.39	143	1.51	139	1.63	136	1.76
	60																						
	70																						
8	20	....	....	.....	.....	......	....	175	.75	167	.94	161	1.13	155	1.30	150	1.46	145	1.60	141	1.74	137	1.87
	30	....	....	.....	.....	......	....	177	.71	169	.89	163	1.07	158	1.23	153	1.38	148	1.51	144	1.64	140	1.76
	40	....	....	.....	.....	......	....	179	.67	171	.84	165	1.00	160	1.15	155	1.29	151	1.42	147	1.54	143	1.65
	60																						
	70																						

TABLE XI.—STEAM AT 227° F. AND 5 LBS. PRESS

Heater size	Capacity C.F.M.	Temperature		B.t.u. per hour supplied	Cond. per hour pounds	Equiv. sq. ft. direct radiation
		Room	Air leaving heater			
6	7,600	50 60 70	97 104 111	386,000 361,000 336,000	407 382 355	1,545 1,440 1,345
7	7,200	50 60 70	115 121 127	505,000 475,000 444,000	536 500 468	2,002 1,400 1,780
8	10,700	50 60 70	97 104 111	544,000 510,000 475,000	542 510 430	2,180 2,040 1,900
9	10,000	50 60 70	116 122 128	714,000 670,000 628,000	753 700 652	2,860 2,680 2,510

TABLE XII.—COEFFICIENT K OF HEAT TRANSMISSION OF AIR OR FLUE GASES THROUGH THIN SHEET IRON SURFACES TO AIR

Velocity of heat emitting air in ft. per sec.	Quantity of heat K per hour passing through a surface of 1 sq. ft. for a temperature difference of 1° F. between the heat emitting and heat absorbing air assuming this difference to be				
	20° F.	40° F.	60° F.	80° F.	100° F. and over
2	0.19	0.26	0.31	0.35	0.37
4	0.35	0.47	0.53	0.57	0.60
6	0.49	0.64	0.72	0.75	0.77
10	0.66	0.80	0.87	0.92	0.97
20	0.84	0.99	1.06	1.11	1.13
30	0.94	1.10	1.17	1.21	1.22

TABLE XIII.—DENSITY OF WATER AT TEMPERATURES FROM 100–212° F.

T°	γ	T°	γ	T°	γ	T°	γ
105	61.92	132	61.51	159	61.01	186	60.44
105.5	61.92	132.5	61.50	159.5	61.00	186.5	60.43
106	61.91	133	60.50	160	60.99	187	60.42
106.5	61.90	133.5	61.49	160.5	60.98	187.5	60.41
107	61.89	134	61.48	161	60.97	188	60.40
107.5	61.89	134.5	61.47	161.5	60.96	188.5	60.38
108	61.88	135	61.46	162	60.95	189	60.37
108.5	61.87	135.5	61.45	162.5	60.94	189.5	60.36
109	61.87	136	61.44	163	60.93	190	60.35
109.5	61.86	136.5	61.44	163.5	60.92	190.5	60.34
110	61.85	137	61.43	164	60.91	191	60.33
110.5	61.85	137.5	61.42	164.5	60.90	191.5	60.32
111	61.84	138	61.41	165	60.89	192	60.30
111.5	61.83	138.5	61.40	165.5	60.88	192.5	60.29
112	61.82	139	61.39	166	60.87	193	60.28
112.5	61.82	139.5	61.38	166.5	60.86	193.5	60.27
113	61.81	140	61.37	167	60.85	194	60.26
113.5	61.80	140.5	61.36	167.5	60.84	194.5	60.25
114	61.80	141	61.36	168	60.83	195	60.23
114.5	61.79	141.5	61.35	168.5	60.82	195.5	60.22
115	61.78	142	61.34	169	60.81	196	60.21
115.5	61.77	142.5	61.33	169.5	60.80	196.5	60.20
116	61.77	143	61.32	170	60.79	197	60.19
116.5	61.76	143.5	61.31	170.5	60.78	197.5	60.18
117	61.75	144	61.30	171	60.77	198	60.16
117.5	61.74	144.5	61.29	171.5	60.76	198.5	60.15
118	61.74	145	61.28	172	60.75	199	60.14
118.5	61.73	145.5	61.27	172.5	60.74	199.5	60.13
119	61.72	146	61.26	173	60.73	200	60.12
119.5	61.71	146.5	61.25	173.5	60.71	200.5	60.11
120	61.71	147	61.25	174	60.70	201	60.09
120.5	61.70	147.5	61.24	174.5	60.69	201.5	60.08
121	61.69	148	61.23	175	60.68	202	60.07
121.5	61.68	148.5	61.22	175.5	60.67	202.5	60.06
122	61.67	149	61.21	176	60.66	203	60.04
122.5	61.67	149.5	61.20	176.5	60.65	203.5	60.03
123	61.66	150	61.19	177	60.64	204	60.02
123.5	61.65	150.5	61.18	177.5	60.63	204.5	60.01
124	61.64	151	61.17	178	60.62	205	60.00
124.5	61.63	151.5	61.16	178.5	60.61	205.5	59.98
125	61.63	152	61.15	179	60.60	206	59.97
125.5	61.62	152.5	61.14	179.5	60.59	206.5	59.96
126	61.61	153	61.13	180	60.58	207	59.95
126.5	61.60	153.5	61.12	180.5	60.56	207.5	59.93
127	61.59	154	61.11	181	60.55	208	59.92
127.5	61.59	154.5	61.10	181.5	60.54	208.5	59.91
128	61.58	155	61.10	182	60.53	209	59.90
128.5	61.57	155.5	61.09	182.5	60.52	209.5	59.88
129	61.56	156	61.08	183	60.51	210	59.87
129.5	61.55	156.5	61.07	183.5	60.50	210.5	59.86
130	61.55	157	61.06	184	60.49	211	59.85
130.5	61.54	157.5	61.05	184.5	60.47	211.5	59.83
131	61.53	158	61.04	185	60.46	212	59.82
131.5	61.52	158.5	61.02	185.5	60.45		

TABLE XIV (FOR ESTIMATION PURPOSES).—ADDITIONAL PRESSURE HEADS AND INCREASE OF HEATING SURFACE FOR A HOT-WATER HEATING SYSTEM WITH OVERHEAD DISTRIBUTION, TAKING THE HEAT LOSS OF THE PIPING INTO CONSIDERATION

NOTE: In case of a one-pipe system the tabular values shown below should be multiplied by 0.5.

A. Additional pressure head in inches w.c.*

The following values are valid for a boiler flow temperature of 195° F.

For a flow temperature of 185° F. they should be decreased 15%

For a flow temperature of 175° F. they should be decreased 30%

I. Return mains uninsulated and installed in front of the wall.†

a. Buildings with 1 or 2 stories

Horizontal length of system, feet	Height of rad. center to boiler center, feet	Distance from riser to return main, feet					
		Up to 30	30–60	60–100	100–150	150–225	225–350
to 75	to 25	0.4 in.	0.4 in.	0.6 in.			
75–150	25	0.4 in.	0.4 in.	0.6 in.	0.8		
150–250	25	0.4 in.	0.4 in.	0.6 in.	0.6	0.8	
250–350	25	0.4 in.	0.4 in.	0.4 in.	0.6	0.8	1.0

b. Buildings with 3 to 5 stories

Horizontal length of system, feet	Height from rad. center to boiler center, feet	Distance from riser to return main, feet					
		Up to 30	30–60	60–100	100–150	150–225	225–350
to 75	to 50	1.0	1.0	1.4			
75–150	50	1.0	1.0	1.2	1.4		
150–250	50	1.0	1.0	1.0	1.2	1.4	
250–350	50	1.0	1.0	1.0	1.2	1.4	1.6

c. Buildings with more than 5 stories

Horizontal length of system, feet	Height from rad. center to boiler center, feet	Distance from riser to return main, feet					
		Up to 30	30–60	60–100	100–150	150–225	225–350
to 75	to 25	1.8	2.0	2.2			
	over 25	1.2	1.4	1.8			
75–150	to 25	2.2	2.4	2.6	3.0		
	over 25	1.6	1.8	2.0	2.2		
150–250	to 25	2.2	2.2	2.4	2.6	3.0	
	over 25	1.6	1.6	1.8	2.0	2.2	
250–350	to 25	2.2	2.2	2.2	2.4	2.6	3.0
	over 25	1.6	1.6	1.6	1.8	2.0	2.6

* This is to be added to the pressure head calculated without taking the cooling effect of the pipes into consideration.

† Based on the following assumptions: No cooling effect in the riser, temp. of attic 32° F., efficiency of insulation 80 %, no cooling effect in common return main. Outside temp. 0° F., room temp. 70° F., temperature-drop in radiators 30° F.

TABLE XIV (FOR ESTIMATION PURPOSES).—ADDITIONAL PRESSURE HEADS AND INCREASE OF HEATING SURFACE FOR A HOT-WATER HEATING SYSTEM WITH OVERHEAD DISTRIBUTION, TAKING THE HEAT LOSS OF THE PIPING INTO CONSIDERATION.—(*Continued*)

II. Return mains installed in chases in the wall.*

a. Buildings with 1 or 2 stories

Horizontal length of system, feet	Height from rad. center to boiler center, feet	Distance from riser to return main, feet					
		Up to 30	30–60	60–100	100–150	150–225	225–350
to 75	up to 25	0.2	0.4	0.4			
75–150	25	0.2	0.2	0.4	0.4		
150–250	25	0.2	0.2	0.2	0.4	0.6	
250–350	25	0.2	0.2	0.2	0.4	0.6	0.8

b. Buildings with 3 to 5 stories

Horizontal length of system, feet	Height from rad. center to boiler center, feet	Distance from riser to return main, feet					
		Up to 30	30–60	60–100	100–150	150–225	225–350
to 75	to 50	0.4	0.6	0.8			
75–150	50	0.4	0.6	0.8	1.0		
150–250	50	0.2	0.4	0.6	0.8	1.0	
250–350	50	0.2	0.2	0.4	0.6	0.8	1.0

c. Buildings with more than 5 stories

Horizontal length of system, feet	Height from rad. center to boiler center, feet	Distance from riser to return main, feet					
		Up to 30	30–60	60–100	100–150	150–225	225–350
to 75	up to 30	0.6	0.8	0.8			
	over 30	0.4	0.6	0.6			
75–150	up to 30	0.6	0.8	0.8	1.2		
	over 30	0.4	0.6	0.6	0.8		
150–250	up to 30	0.6	0.6	0.8	0.8	1.2	
	over 30	0.4	0.4	0.6	0.6	0.8	
250–350	up to 30	0.6	0.6	0.8	0.8	1.2	1.4
	over 30	0.4	0.4	0.6	0.6	0.8	1.0

TABLE XIV (FOR ESTIMATION PURPOSES).—ADDITIONAL PRESSURE HEADS AND INCREASE OF HEATING SURFACE FOR A HOT-WATER HEATING SYSTEM WITH OVERHEAD DISTRIBUTION, TAKING THE HEAT LOSS OF THE PIPING INTO CONSIDERATION.—(*Continued*)

B. Increase in heating surface expressed as a percentage of the values arrived at without considering the cooling effect of the pipes

I. Return mains uninsulated and installed in front of wall.*

Number of stories in building	Increase of the heating surface in %		
	Ground floor	1 and 2 story resp.	3, 4 and 5 story resp.
1 or 2	10	5	
3 to 5	15	10	5
over 5	25	10	5

II. Return mains installed in chases in the wall.†

Number of stories in building	Increase of the heating surface in %		
	Ground floor	1st and 2nd story resp.	3, 4 and 5 story resp.
1 or 2	5	0	
3 to 5	5	3	0
over 5	5	5	3

* See footnote under A.

† Based upon the assumptions under A and including the following: Efficiency of the insulation of the return mains 60 %, air temperature in the wall chases 100° F.

Table XV.—Loft Heating Systems Table for Preliminary Design

A. Effective pressure head in in. w.c.

The following values are valid for a boiler flow temperature of 195° F.

For a flow temperature of 185° F. they should be decreased 15%

For a flow temperature of 175° F. they should be decreased 30%

I. Return mains uninsulated and installed in front of wall.*

Horizontal length of system, feet	Distance from riser to return main in feet						
	To 15	15–30	30–50	50–70	70–100	100–130	130–160
to 30	0.27	0.70					
30– 80	0.27	0.43	0.6	0.8	1.0		
80–160	0.20	0.31	0.43	0.55	0.7	0.95	1.2

II. Return mains installed in chases in the wall.†

Horizontal length of system, feet	Distance from riser to return main in feet						
	To 15	15–30	30–50	50–70	70–100	100–130	130–160
to 30	0.20	0.60					
30– 80	0.20	0.31	0.47	0.63	0.87		
80–160	0.16	0.23	0.31	0.43	0.60	0.8	1.0

* Based upon the following assumptions: No cooling effect in the riser and return mains, distributing pipe bare, outside temperature 0° F., inside temperature 70° F. and temperature drop of the radiators 30° F.

† In addition to the above assumptions, the efficiency of the insulation of the return mains is 60 %, and the air temperature in wall chases is 100° F.

B. Increase of the heating surface in per cent of the values computed without taking the cooling effect of the pipes into consideration

I. Return mains uninsulated and installed in front of wall.*

Horizontal length of system, feet	Distance from riser to return main in feet						
	To 15	15–30	30–50	50–70	70–100	100–130	130–160
to 30	10	15					
30– 80	10	10	15	20	25		
80–160	5	5	10	10	15	20	30

II. Return mains installed in chases in the wall.†

Horizontal length of system, feet	Distance from riser to return main in feet						
	to 15	15–30	30–50	50–70	70–100	100–130	130–160
to 30	5	10					
30– 80	5	5	10	15	20		
80–160	3	3	5	10	15	20	30

* Based upon the following assumptions: No cooling effect in the riser and return mains, distributing pipe bare, outside temperature 0° F., inside temperature 70° F. and temperature drop of the radiators 30° F.

† In addition to the above assumptions, the efficiency of the insulation of the return mains is 60 %, and the air temperature in wall chases is 100° F.

TABLE XVI.—PROPORTION OF THE RESISTANCES OF THE FITTINGS TO THE TOTAL RESISTANCE OF A PIPING SYSTEM

The following table is valid for both one and two pipe systems with either attic or basement mains

	Type of system	Proportion of single resistance
1*	Common heating systems for buildings.	Independent of the horizontal and vertical extent of the building, 50%.
2	Central heating system mains with an average distance between two buildings of 150 ft.	20% of the total resistance of the main.
3	Central heating system mains with an average distance between two buildings of 300 ft.	10% of the total resistance of the main.
4		

* In choosing control devices and valves having a very small resistance the values given may be reduced 10% viz: instead of 50% as tabulated 40% will be used.

TABLE XVII.—DIAMETER OF CONDENSATION PIPE LINES FOR STEAM HEATING SYSTEMS

Diameter, d	Dry return		Wet return		
			Horizontal and vertical		
	Horizontal	Vertical	$l \leq 150$ ft.	$l > 150'$ and $< 300'$	$l > 300$
Ins.	Heat in B.t.u. taken from the steam to form condensation				
1	2	3	4	5	6
½″	16,000	24,000	110,000	72,000	32,000
¾″	60,000	88,000	280,000	180,000	100,000
1″	110,000	170,000	500,000	320,000	160,000
1¼″	272,000	400,000	1,080,000	700,000	340,000
1½″	415,000	620,000	1,500,000	1,000,000	460,000
2″	860,000	1,280,000	2,600,000	1,760,000	860,000

NOTE: Radiator connections not to be smaller than ½″. The diameter of the air lines used in connection with wet returns should be chosen in accordance with column No. 4. In the above table l represents the length of pipe leading to the radiator which is vertically nearest and horizontally most distant from the boiler than any other radiator connected to the same system.

Table XVIII

Temper. F.°	Weight of 1 cu. ft. dry air, lb.	Tension of water vapor in. Hg	Weight of saturated vapor per lb. of dry air in grains	Saturated mixtures of air and water vapor			
				1 cu. ft. contains		1 lb. contains	
				Satur. vapor grains	Air, lb.	Satur. vapor, grains	Air, lb.
0	0.0863	0.0375	5.47	0.472	0.0861	5.6	0.9992
2	0.0860	0.0417	6.08	0.522	0.0858	6.3	0.9991
4	0.0856	0.0462	6.74	0.576	0.0854	7.0	0.9990
6	0.0852	0.0512	7.47	0.636	0.0850	7.7	0.9989
8	0.0849	0.0567	8.28	0.701	0.0847	8.4	0.9988
10	0.0845	0.0628	9.16	0.772	0.0843	9.1	0.9987
12	0.0841	0.0694	10.13	0.850	0.0839	9.8	0.9986
14	0.0838	0.0766	11.19	0.935	0.0835	11.2	0.9984
16	0.0834	0.0846	12.35	1.028	0.0831	12.6	0.9982
18	0.0831	0.0932	13.62	1.128	0.0827	14.0	0.9980
20	0.0827	0.1027	15.01	1.237	0.0824	15.4	0.9978
22	0.0824	0.1130	16.52	1.356	0.0820	16.8	0.9976
24	0.0820	0.1242	18.17	1.485	0.0816	18.2	0.9974
26	0.0817	0.1365	19.98	1.625	0.0813	19.6	0.9972
28	0.0814	0.1499	21.91	1.776	0.0810	21.7	0.9969
30	0.0811	0.1646	24.11	1.943	0.0806	23.8	0.9966
32	0.0807	0.1806	26.47	2.124	0.0802	26.5	0.9962
34	0.0804	0.1957	28.79	2.292	0.0799	29.4	0.9958
36	0.0801	0.2119	31.09	2.471	0.0795	32.2	0.9954
38	0.0798	0.2292	33.66	2.663	0.0791	35.0	0.9950
40	0.0794	0.2478	36.41	2.868	0.0787	37.8	0.9946
42	0.0791	0.2678	39.38	3.087	0.0783	40.6	0.9942
44	0.0788	0.2891	42.55	3.319	0.0780	44.1	0.9937
46	0.0785	0.3120	45.94	3.568	0.0776	47.6	0.9932
48	0.0782	0.3364	49.58	3.832	0.0772	51.1	0.9927
50	0.0779	0.3624	53.47	4.113	0.0769	54.6	0.9922
52	0.0776	0.3903	57.64	4.411	0.0766	58.8	0.9916
54	0.0773	0.4200	62.09	4.729	0.0762	63.0	0.9910
56	0.0770	0.4517	66.85	5.066	0.0758	67.2	0.9904
58	0.0767	0.4855	71.93	5.424	0.0754	71.4	0.9898
60	0.0763	0.5214	77.3	5.804	0.0750	76.3	0.9891
62	0.0760	0.5597	83.2	6.208	0.0746	81.2	0.9884
64	0.0757	0.6005	89.3	6.633	0.0742	86.8	0.9876
66	0.0754	0.6438	95.9	7.084	0.0738	92.4	0.9868
68	0.0751	0.6898	103.0	7.563	0.0734	98.7	0.9859
70	0.0749	0.7386	110.5	8.069	0.0730	105.7	0.9849
72	0.0746	0.7906	118.4	8.603	0.0726	113.4	0.9838
74	0.0743	0.8456	126.9	9.168	0.0722	121.8	0.9826
76	0.0740	0.9040	135.9	9.76	0.0718	130.9	0.9813
78	0.0737	0.9658	145.6	10.39	0.0714	141.4	0.9798
80	0.0735	1.031	155.8	11.06	0.0710	152.6	0.9782
82	0.0732	1.101	166.7	11.76	0.0705	163.8	0.9766
84	0.0729	1.171	178.3	12.50	0.0700	174.3	0.9751
86	0.0726	1.251	190.6	13.28	0.0695	186.2	0.9734
88	0.0724	1.334	203.7	14.10	0.0691	198.5	0.9716
90	0.0722	1.421	217.6	14.96	0.0687	211.5	0.9698
92	0.0720	1.512	232.4	15.87	0.0683	226.8	0.9676
94	0.0717	1.609	247.1	16.82	0.0678	242.2	0.9654
96	0.0714	1.710	264.8	17.82	0.0673	257.6	0.9632
98	0.0711	1.818	282.5	18.88	0.0668	273.0	0.9610
100	0.0709	1.931	301.3	19.98	0.0663	288.4	0.9588
102	0.0706	2.045	321.4	21.15	0.0657	303.1	0.9567
104	0.0703	2.171	342.7	22.36	0.0650	318.5	0.9545
106	0.0700	2.303	365	23.64	0.0646	333.9	0.9523
108	0.0698	2.443	389	24.98	0.0641	349.3	0.9501
110	0.0696	2.589	415	26.38	0.0636	391.7	0.9440
112	0.0693	2.740	442	27.85	0.0630	414.4	0.9408
114	0.0690	2.904	471	29.39	0.0623	439.6	0.9372
116	0.0688	3.073	502	31.00	0.0617	466.2	0.9334
118	0.0686	3.252	534	32.68	0.0611	494.9	0.9293
120	0.0684	3.438	569	34.44	0.0605	526.5	0.9248
122	0.0681	3.635	606	36.20	0.0598	553.7	0.9200
124	0.0679	3.841	647	38.15	0.0592	587.3	0.9161
126	0.0677	4.057	689	40.15	0.0585	627.2	0.9104
128	0.0675	4.282	733	42.25	0.0578	659.4	0.9058
130	0.0673	4.52	780	44.49	0.0571	701.4	0.8998
132	0.0670	4.76	831	46.80	0.0564	739.9	0.8943
134	0.0668	5.02	885	49.2	0.0556	783.3	0.8881
136	0.0666	5.29	942	51.7	0.0548	829.5	0.8815
138	0.0664	5.58	1,080	54.25	0.0540	878.5	0.8745
140	0.0662	5.88	1,072	56.91	0.0532	928.2	0.8674

TABLE XVIII.—(*Continued*)

Temper. F.°	Weight of 1 cu. ft. dry air, lb.	Tension of water vapor in. Hg	Weight of saturated vapor per lb. of dry air in grains	Saturated mixtures of air and water vapor			
				1 cu. ft. contains		1 lb. contains	
				Satur. vapor grains	Air, lb.	Satur. vapor, grains	Air, lb.
142	0.0659	6.18	1,144	59.75	0.0523	983.5	0.8595
144	0.0657	6.51	1,221	62.6	0.0514	1,039.5	0.8515
146	0.0655	6.84	1,300	65.6	0.0505	1,099.7	0.8429
158	0.0653	7.20	1,390	68.75	0.0496	1,162.0	0.8340
150	0.0651	7.57	1,485	72.1	0.0486	1,224.3	0.8254
152	0.0648	7.95	1,580	75.5	0.0476	1,294.3	0.8151
154	0.0646	8.34	1,700	79.0	0.0466	1,365.0	0.8050
156	0.0644	8.76	1,818	82.8	0.0455	1,442.0	0.7940
168	0.0642	9.20	1,950	86.6	0.0445	1,522.5	0.7825
160	0.0640	9.65	2,091	90.6	0.0433	1,611.4	0.7698
162	0.0638	10.12	2,243	94.7	0.0422	1,704.5	0.7565
144	0.0636	10.61	2,412	98.8	0.0410	1,801.1	0.7427
168	0.0634	11.12	2,600	103.4	0.0398	1,897.7	0.7289
166	0.0632	11.65	....	108.0	0.0386	2,002.0	0.7140
170	0.0630	12.20	....	112.8	0.0373	2,111.2	0.6984
172	0.0628	12.77	....	117.7	0.0360	2,216.9	0.6833
174	0.0626	13.37	....	123.0	0.0346	2,332.2	0.6654
176	0.0624	13.98	....	128.0	0.0332	2,476.6	0.6462
178	0.0622	14.62	....	133.6	0.0318	2,616.6	0.6262
180	0.0621	15.29	....	139.4	0.0304	2,769.2	0.6044
182	0.0619	15.98	....	145.3	0.0288	2,923.2	0.5824
184	0.0617	16.70	....	151.3	0.0272	3,092.6	0.5582
186	0.0615	17.45	....	157.3	0.0256	3,270.4	0.5328
188	0.0613	18.22	....	164.0	0.0240	3,465.0	0.5050
190	0.0611	19.02	....	170.9	0.0223	3,658.2	0.4774
192	0.0609	19.85	....	177.6	0.0205	3,878.0	0.4460
194	0.0607	20.71	....	184.5	0.0187	4,119.5	0.4115
196	0.0605	21.60	....	191.5	0.0168	4,375.0	0.3750
198	0.0603	22.52	....	199.5	0.0149	4,625.6	0.3392
200	0.0601	23.47	....	208.0	0.0130	4,869.2	0.3044
202	0.0599	24.45	....	215.0	0.0110	5,201.0	0.2570
204	0.0597	25.48	....	223.7	0.0089	5,540.5	0.2085
206	0.0595	26.53	....	233.1	0.0067	5,876.5	0.1605
208	0.0593	27.63	....	242.5	0.0045	6,216.0	0.1120
210	0.0592	28.76	....	252	0.0023	6,579.3	0.0601
212	0.0590	29.92		262	0.0000	7,060	0.0000
214	0.0588						
216	0.0586						
218	0.0584						
220	0.0583						
222	0.0581						
224	0.0579						
226	0.0577						
228	0.0576						
230	0.0575						
232	0.0573						
234	0.0571						
236	0.0569						
238	0.0568						
240	0.0567						
242	0.0565						
244	0.0563						
246	0.0561						
248	0.0559						
250	0.0558						
252	0.0557						
254	0.0555						
256	0.0554						
258	0.0552						
260	0.0551						
262	0.0549						
264	0.0548						
266	0.0546						
268	0.0545						
270	0.0544						
272	0.0542						
274	0.0541						
276	0.0540						
278	0.0538						
280	0.0537						
282	0.0535						

TABLE XVIII. (*Continued*)

Temper. F.°	Weight of 1 cu. ft. dry air, lb.	Tension of water vapor in. Hg	Weight of saturated vapor per lb. of dry air in grains	Saturated mixtures of air and water vapor			
				1 cu. ft. contains		1 lb. contains	
				Satur. vapor, grains	Air, lb.	Satur. vapor, grains	Air, lb.
284	0.0534						
286	0.0533						
288	0.0531						
290	0.0530						
292	0.0528						
294	0.0527						
296	0.0526						
298	0.0524						
300	0.0523						
302	0.0521						
304	0.0520						
306	0.0518						
308	0.0517						
310	0.0515						
312	0.0514						
314	0.0513						
316	0.0511						
318	0.0510						
320	0.0508						
322	0.0507						
324	0.0506						
326	0.0505						
328	0.0503						
330	0.0502						
332	0.0501						
334	0.0500						
336	0.0498						
338	0.0497						
340	0.0496						
342	0.0495						
344	0.0493						
346	0.0492						
348	0.0491						
350	0.0190						

Table XIX.—Volume of Dry Air Necessary to Supply or Remove 1,000 B.t.u. per Hour

Temp. difference between air entering and leaving the room in °F.	The temperature t of air leaving the room in °F.																				
	60	61	62	63	64	65	66	67	68	69	70	71	72	73	74	75	76	77	78	79	80
	Air change in cu. ft. of t °F.																				
1	54,200	54,300	54,400	54,500	54,600	54,700	54,800	54,900	55,000	55,100	55,200	55,300	55,400	55,500	55,600	55,700	55,800	55,900	56,000	56,100	56,200
2	27,100	27,150	27,200	27,250	27,300	27,350	27,400	27,450	27,500	27,550	27,600	27,650	27,700	27,750	27,800	27,850	27,900	27,950	28,000	28,050	28,100
3	18,060	18,100	18,130	18,160	18,200	18,230	18,260	18,300	18,330	18,360	18,400	18,430	18,460	18,500	18,530	18,560	18,600	18,630	18,660	18,700	18,730
4	13,550	13,570	13,600	13,620	13,650	13,670	13,700	13,720	13,750	13,770	13,800	13,820	13,850	13,870	13,900	13,920	13,950	13,970	14,000	14,020	14,050
5	10,840	10,860	10,880	10,900	10,920	10,940	10,960	10,980	11,000	11,020	11,040	11,060	11,080	11,100	11,120	11,140	11,160	11,180	11,200	11,220	11,240
6	9,030	9,050	9,060	9,080	9,100	9,110	9,130	9,150	9,160	9,180	9,200	9,210	9,230	9,250	9,260	9,280	9,300	9,310	9,330	9,350	9,360
7	7,740	7,760	7,770	7,780	7,800	7,810	7,830	7,840	7,860	7,870	7,880	7,900	7,910	7,930	7,940	7,960	7,970	7,980	8,000	8,010	8,030
8	6,780	6,790	6,800	6,810	6,830	6,840	6,850	6,860	6,880	6,890	6,900	6,910	6,930	6,940	6,950	6,960	6,980	6,990	7,000	7,010	7,030
9	6,020	6,030	6,040	6,060	6,070	6,080	6,090	6,100	6,110	6,120	6,130	6,140	6,160	6,170	6,180	6,190	6,200	6,210	6,220	6,230	6,240
10	5,420	5,430	5,440	5,450	5,460	5,470	5,480	5,490	5,500	5,510	5,520	5,530	5,540	5,550	5,560	5,570	5,580	5,590	5,600	5,610	5,620
11	4,930	4,940	4,950	4,950	4,960	4,970	4,980	4,990	5,000	5,010	5,020	5,030	5,040	5,050	5,050	5,060	5,070	5,080	5,090	5,100	5,110
12	4,520	4,530	4,530	4,540	4,550	4,560	4,570	4,580	4,590	4,590	4,600	4,610	4,620	4,630	4,630	4,640	4,650	4,660	4,670	4,680	4,990
13	4,170	4,180	4,180	4,190	4,200	4,210	4,220	4,220	4,230	4,240	4,250	4,250	4,260	4,270	4,280	4,280	4,290	4,300	4,310	4,320	4,320
14	3,870	3,880	3,890	3,890	3,900	3,910	3,910	3,920	3,930	3,940	3,940	3,950	3,960	3,960	3,970	3,980	3,980	3,990	4,000	4,010	4,010
15	3,610	3,620	3,630	3,630	3,640	3,650	3,650	3,660	3,670	3,670	3,680	3,690	3,690	3,700	3,710	3,710	3,720	3,730	3,730	3,740	3,750
16	3,390	3,390	3,400	3,400	3,410	3,420	3,420	3,430	3,440	3,440	3,450	3,460	3,460	3,470	3,480	3,480	3,490	3,490	3,500	3,500	3,510
17	3,190	3,190	3,200	3,210	3,210	3,220	3,220	3,230	3,240	3,240	3,250	3,250	3,260	3,270	3,270	3,280	3,280	3,290	3,290	3,300	3,310
18	3,010	3,020	3,020	3,030	3,030	3,040	3,040	3,050	3,060	3,060	3,070	3,070	3,080	3,090	3,090	3,100	3,100	3,110	3,110	3,120	3,130
19	2,850	2,860	2,860	2,870	2,870	2,880	2,890	2,890	2,900	2,900	2,910	2,910	2,920	2,920	2,930	2,930	2,940	2,940	2,950	2,950	2,960
20	2,710	2,720	2,720	2,730	2,730	2,740	2,740	2,750	2,750	2,760	2,760	2,770	2,770	2,780	2,780	2,790	2,790	2,800	2,800	2,810	2,810
21	2,580	2,590	2,590	2,600	2,600	2,610	2,610	2,620	2,620	2,630	2,630	2,630	2,640	2,640	2,650	2,650	2,660	2,660	2,670	2,670	2,680
22	2,460	2,470	2,470	2,480	2,480	2,490	2,490	2,500	2,500	2,510	2,510	2,510	2,520	2,520	2,530	2,530	2,540	2,540	2,550	2,550	2,550
23	2,360	2,360	2,370	2,370	2,370	2,380	2,380	2,390	2,390	2,400	2,400	2,400	2,410	2,410	2,420	2,420	2,430	2,430	2,430	2,440	2,440
24	2,260	2,260	2,270	2,270	2,280	2,280	2,280	2,290	2,290	2,300	2,300	2,310	2,310	2,310	2,320	2,320	2,320	2,330	2,330	2,340	2,340
25	2,170	2,170	2,180	2,180	2,180	2,190	2,190	2,200	2,200	2,200	2,210	2,210	2,220	2,220	2,220	2,230	2,230	2,230	2,240	2,240	2,250
26	2,090	2,090	2,090	2,100	2,100	2,100	2,110	2,110	2,120	2,120	2,120	2,130	2,130	2,130	2,140	2,140	2,150	2,150	2,150	2,160	2,160
27	2,010	2,010	2,010	2,020	2,020	2,030	2,030	2,030	2,040	2,040	2,040	2,050	2,050	2,050	2,060	2,060	2,070	2,070	2,070	2,080	2,080
28	1,940	1,940	1,940	1,950	1,950	1,950	1,960	1,960	1,960	1,970	1,970	1,980	1,980	1,980	1,990	1,990	1,990	2,000	2,000	2,000	2,010
29	1,870	1,870	1,880	1,880	1,880	1,890	1,890	1,890	1,900	1,900	1,900	1,910	1,910	1,910	1,920	1,920	1,920	1,930	1,930	1,940	1,940
30	1,810	1,810	1,810	1,820	1,820	1,820	1,830	1,830	1,830	1,840	1,840	1,840	1,850	1,850	1,850	1,860	1,860	1,860	1,870	1,870	1,870
31	1,750	1,750	1,760	1,760	1,760	1,770	1,770	1,770	1,770	1,780	1,780	1,780	1,790	1,790	1,790	1,800	1,800	1,800	1,810	1,810	1,810

32	1,690	1,690	1,700	1,700	1,710	1,710	1,710	1,720	1,720	1,720	1,730	1,730	1,730	1,730	1,740	1,740	1,740	1,750	1,750	1,750	1,760
33	1,640	1,650	1,650	1,650	1,650	1,660	1,660	1,660	1,670	1,670	1,670	1,680	1,680	1,680	1,690	1,690	1,690	1,690	1,700	1,700	1,700
34	1,590	1,600	1,600	1,600	1,610	1,610	1,610	1,610	1,620	1,620	1,620	1,630	1,630	1,630	1,640	1,640	1,640	1,640	1,650	1,650	1,650
35	1,550	1,550	1,550	1,560	1,560	1,560	1,570	1,570	1,570	1,570	1,580	1,580	1,580	1,590	1,590	1,590	1,590	1,600	1,600	1,600	1,610
36	1,510	1,510	1,510	1,510	1,520	1,520	1,520	1,530	1,530	1,530	1,530	1,540	1,540	1,540	1,540	1,550	1,550	1,550	1,560	1,560	1,560
37	1,460	1,470	1,470	1,470	1,480	1,480	1,480	1,480	1,490	1,490	1,490	1,490	1,500	1,500	1,500	1,500	1,510	1,510	1,510	1,530	1,520
38	1,430	1,430	1,430	1,430	1,440	1,440	1,440	1,440	1,450	1,450	1,430	1,450	1,460	1,460	1,460	1,460	1,470	1,470	1,470	1,480	1,480
39	1,390	1,390	1,390	1,400	1,400	1,400	1,400	1,410	1,410	1,410	1,410	1,420	1,420	1,420	1,420	1,430	1,430	1,430	1,430	1,440	1,440
40	1,360	1,360	1,360	1,360	1,370	1,370	1,370	1,370	1,380	1,380	1,380	1,380	1,390	1,390	1,390	1,390	1,400	1,400	1,400	1,400	1,410
41	1,320	1,320	1,330	1,330	1,330	1,330	1,340	1,340	1,340	1,340	1,350	1,350	1,350	1,350	1,360	1,360	1,360	1,360	1,370	1,370	1,370
42	1,290	1,290	1,300	1,300	1,300	1,300	1,310	1,310	1,310	1,310	1,310	1,320	1,320	1,320	1,320	1,330	1,330	1,330	1,330	1,340	1,340
43	1,260	1,260	1,270	1,270	1,270	1,270	1,270	1,280	1,280	1,280	1,280	1,280	1,290	1,290	1,290	1,290	1,300	1,300	1,300	1,300	1,310
44	1,230	1,230	1,240	1,240	1,240	1,240	1,250	1,250	1,250	1,250	1,250	1,260	1,260	1,260	1,260	1,270	1,270	1,270	1,270	1,280	1,280
45	1,200	1,210	1,210	1,210	1,210	1,220	1,220	1,220	1,220	1,230	1,230	1,230	1,230	1,230	1,240	1,240	1,210	1,240	1,240	1,250	1,250
46	1,180	1,180	1,180	1,190	1,190	1,190	1,190	1,190	1,200	1,200	1,200	1,200	1,200	1,210	1,210	1,210	1,210	1,220	1,220	1,220	1,220
47	1,150	1,160	1,160	1,160	1,160	1,160	1,170	1,170	1,170	1,170	1,170	1,180	1,180	1,180	1,180	1,190	1,190	1,190	1,190	1,190	1,200
48	1,130	1,130	1,130	1,140	1,140	1,140	1,140	1,140	1,150	1,150	1,150	1,150	1,150	1,160	1,160	1,160	1,160	1,170	1,170	1,170	1,170
49	1,110	1,110	1,110	1,110	1,110	1,120	1,120	1,120	1,120	1,130	1,130	1,130	1,120	1,130	1,140	1,140	1,140	1,140	1,140	1,150	1,150
50	1,080	1,090	1,090	1,090	1,090	1,090	1,100	1,100	1,100	1,100	1,100	1,110	1,110	1,110	1,110	1,110	1,120	1,120	1,120	1,120	1,120

NOTE For exact calculations the above table will not be sufficiently accurate since an average specific heat for 70° F has been assumed

TABLE XIX.—VOLUME OF DRY AIR NECESSARY TO SUPPLY OR REMOVE 1,000 B.T.U. PER HOUR.—(*Continued*)

Temp. difference between the air entering and leaving the room in °F	The temperature t of the air leaving the room in °F.																				
	60	61	62	63	64	65	66	67	68	69	70	71	72	73	74	75	76	77	78	79	80
	Air change in cu. ft. at t °F.																				
50	1,080	1,090	1,090	1,090	1,090	1,090	1,100	1,100	1,100	1,100	1,100	1,110	1,110	1,110	1,110	1,110	1,120	1,120	1,120	1,120	1,120
51	1,060	1,060	1,070	1,070	1,070	1,070	1,070	1,080	1,080	1,080	1,080	1,080	1,090	1,090	1,090	1,090	1,090	1,100	1,100	1,100	1,100
52	1,040	1,040	1,050	1,050	1,050	1,050	1,050	1,060	1,060	1,060	1,060	1,060	1,070	1,070	1,070	1,070	1,070	1,080	1,080	1,080	1,080
53	1,020	1,020	1,030	1,030	1,030	1,030	1,030	1,040	1,040	1,040	1,040	1,010	1,050	1,050	1,050	1,050	1,050	1,060	1,060	1,060	1,060
54	1,000	1,010	1,010	1,010	1,010	1,010	1,020	1,020	1,020	1,020	1,020	1,020	1,030	1,030	1,030	1,030	1,030	1,040	1,040	1,040	1,040
55	990	990	990	990	990	990	1,000	1,000	1,000	1,000	1,000	1,000	1,010	1,010	1,010	1,010	1,010	1,020	1,020	1,020	1,020
56	970	970	970	970	980	980	980	980	980	980	990	990	990	990	990	1,000	1,000	1,000	1,000	1,000	1,000
57	950	950	950	960	960	960	960	960	960	970	970	970	970	970	970	980	980	980	980	980	980
58	940	940	940	940	940	940	950	950	950	950	950	950	960	960	960	960	960	960	970	970	970
59	920	920	920	920	930	930	930	930	930	930	940	940	940	940	940	940	950	950	950	950	950
60	900	910	910	910	910	910	910	920	920	920	920	920	920	930	930	930	930	930	930	940	940
61	890	890	890	890	900	900	900	900	900	900	910	910	910	910	910	910	920	920	920	920	920
62	870	880	880	880	880	880	880	890	890	890	890	890	890	900	990	900	900	900	900	910	910
63	860	860	860	870	870	870	870	870	870	880	880	880	880	880	880	890	890	890	890	890	890
64	850	850	850	850	850	850	860	860	860	860	860	860	870	870	870	870	870	870	880	880	860
65	830	840	840	840	840	840	840	850	850	850	850	850	850	860	860	860	860	860	860	860	870
66	820	820	820	830	830	830	830	830	830	840	840	840	840	840	840	850	850	850	850	850	850
67	810	810	810	810	810	820	820	820	820	820	820	830	830	830	830	830	830	830	840	840	840
68	800	800	800	800	800	800	810	810	810	810	810	810	810	820	820	820	820	820	830	830	830
69	790	790	790	790	790	790	790	800	800	800	800	800	800	810	810	810	810	810	810	820	820
70	770	780	780	780	780	780	780	780	790	790	790	790	790	790	800	800	800	800	800	800	800
71	760	770	770	770	770	770	770	770	780	780	780	780	780	780	780	780	790	790	790	790	790
72	750	750	760	760	760	760	760	760	760	770	770	770	770	770	770	770	780	780	780	780	780
73	740	740	750	750	750	750	750	750	750	760	760	760	760	760	760	760	760	770	770	770	770
74	730	730	740	740	740	740	740	740	740	750	750	750	750	750	750	750	750	760	760	760	760
75	720	720	730	730	730	730	730	730	730	740	740	740	740	740	740	740	740	750	750	750	750
76	710	710	720	720	720	720	720	720	720	730	730	730	730	730	730	730	730	740	740	740	740
77	700	710	710	710	710	710	710	710	710	720	720	720	720	720	720	720	730	730	730	730	730
78	700	700	700	700	700	700	700	700	710	710	710	710	710	710	710	710	720	720	720	720	720
79	690	690	690	690	690	690	690	690	700	700	700	700	700	700	700	710	710	710	710	710	710
80	680	680	680	680	680	680	690	690	690	690	690	690	690	690	700	700	700	700	700	700	700

81	670	670	670	670	670	680	680	680	680	680	680	680	680	690	690	690	690	690	690	690	690
82	660	660	660	660	670	670	670	670	670	670	670	670	680	680	680	680	680	690	680	680	690
83	650	650	660	660	660	660	660	660	660	660	670	670	670	670	670	670	670	670	680	680	680
84	650	650	650	650	650	650	650	650	660	660	660	660	660	660	660	660	660	670	670	670	670
85	640	640	640	640	640	640	640	650	650	650	650	650	650	650	650	660	660	660	660	660	660
86	630	630	630	630	640	640	640	640	640	640	640	640	640	650	650	650	650	650	650	650	650
87	620	620	630	630	630	630	630	630	630	630	630	640	640	640	640	640	640	640	640	650	650
88	610	620	620	620	620	620	620	620	630	630	630	630	630	630	630	630	630	640	640	640	640
89	610	610	610	610	610	610	620	620	620	620	620	620	620	620	620	630	630	630	630	630	630
90	600	600	600	610	610	610	610	610	610	610	610	610	620	620	620	620	630	630	620	620	620
91	600	600	600	600	600	600	600	600	600	610	610	610	610	610	610	610	610	610	620	620	620
92	590	590	590	590	590	590	600	600	600	600	600	600	600	600	600	610	610	610	610	610	610
93	580	580	580	590	590	590	590	590	590	590	590	590	600	600	600	600	600	600	600	600	600
94	580	580	580	580	580	580	580	580	590	590	590	590	590	590	590	590	590	600	600	600	600
95	570	570	570	570	570	580	580	580	580	580	580	580	580	580	590	590	590	590	590	590	590
96	560	570	570	570	570	570	570	570	570	570	580	580	580	580	580	580	580	580	580	580	590
97	560	560	560	560	560	560	560	570	570	570	570	570	570	570	570	570	580	580	580	580	580
98	550	550	550	560	560	560	560	560	560	560	560	560	570	570	570	570	570	570	570	570	570
99	550	550	550	550	550	550	550	550	560	560	560	560	560	560	560	560	560	560	570	570	570
100	540	540	540	550	550	550	550	550	550	550	550	550	550	560	560	560	560	560	560	560	560

TABLE XX.—HOURLY AIR CHANGE IN CU. FT. AT ROOM TEMPERATURE FOR FULLY OCCUPIED ROOMS BASED ON THE HUMIDITY STANDARD AND THE FOLLOWING CONDITIONS

Outer Air Temp. 50° F. Relative Humidity 80%. Relative Humidity of Room Air 70%

Volume of room in cu. ft. per person	Temp. of the room	Required air change when using the room for z hours z =							
		1	2	3	4	5	6	7	8
150	68	870	1,020	1,070	1,110	1,110	1,120	1,130	1,140
	70	700	850	900	930	940	950	960	960
	72	570	720	770	790	810	820	820	830
	74	460	610	660	680	700	710	720	720
200	68	770	970	1,040	1,070	1,090	1,110	1,120	1,120
	70	600	800	870	900	920	930	940	950
	72	470	670	730	770	790	800	810	820
	74	360	560	630	660	680	600	700	710
250	68	670	920	1,010	1,050	1,070	1,090	1,100	1,110
	70	500	750	830	880	900	920	930	940
	72	370	620	700	740	770	780	800	800
	74	260	510	590	630	660	680	690	700
300	68	570	870	970	1,020	1,050	1,070	1,090	1,100
	70	400	700	800	850	880	900	910	930
	72	270	570	670	720	750	770	780	790
	74	160	460	560	610	640	660	670	680
400	68	370	770	910	970	1,010	1,040	1,060	1,070
	70	200	600	730	800	840	870	890	900
	72	70	470	600	670	700	730	750	770
	74	......	360	490	560	600	630	650	660
500	68	170	670	840	920	970	1,010	1,030	1,050
	70	......	500	670	750	800	830	860	880
	72	......	370	530	620	670	700	720	740
	74	......	260	430	510	560	590	620	630
600	68	......	570	770	870	930	970	1,000	1,020
	70	......	400	600	700	760	800	830	850
	72	......	270	470	570	630	670	700	720
	74	......	160	360	460	520	560	590	610
700	68	......	470	700	820	890	940	970	1,000
	70	......	300	530	650	720	770	800	830
	72	......	170	400	520	590	630	670	690
	74	......	60	290	410	480	520	560	580

For children up to 10 years one-half, and for older children three-quarters of the tabular values should be taken.

TABLE XXI.—MOISTURE IN POUNDS TO BE SUPPLIED TO 1,000 CU. FT. OF (ROOM) AIR, TAKEN FROM OUTSIDE TO RAISE THE HUMIDITY TO 50 PER CENT AFTER HEATING

Outside air		Room temperature, °F							
Temp., °F	Relative Humidity, per cent	50°	52°	54°	56°	58°	60°	62°	64°
0°	70	0.251	0.273	0.296	0.320	0.346	0.372	0.401	0.432
	80	0.245	0.267	0.290	0.314	0.340	0.366	0.395	0.426
	90	0.239	0.261	0.284	0.308	0.334	0.360	0.389	0.420
10°	70	0.222	0.244	0.267	0.292	0.317	0.344	0.373	0.404
	80	0.212	0.234	0.257	0.282	0.307	0.334	0.363	0.394
	90	0.202	0.224	0.247	0.272	0.297	0.324	0.353	0.284
20°	70	0.177	0.199	0.222	0.247	0.273	0.300	0.329	0.359
	80	0.160	0.182	0.206	0.230	0.256	0.283	0.312	0.343
	90	0.144	0.166	0.189	0.214	0.240	0.267	0.296	0.327
32°	70	0.0882	0.110	0.134	0.159	0.185	0.213	0.242	0.273
	80	0.0588	0.0812	0.105	0.130	0.156	0.184	0.213	0.244
	90	0.0295	0.0520	0.0759	0.101	0.127	0.155	0.184	0.216
40°	70	0.0122	0.0347	0.0587	0.0836	0.111	0.139	0.168	0.200
	80	......	......	0.0189	0.0441	0.0708	0.0984	0.128	0.160
	90	......	......	......	0.0044	0.0312	0.0590	0.0889	0.121
		66°	68°	70°	72°	74°	76°	78°	80°
0°	70	0.463	0.497	0.533	0.571	0.611	0.654	0.698	0.745
	80	0.457	0.492	0.527	0.565	0.606	0.648	0.692	0.739
	90	0.451	0.486	0.521	0.560	0.600	0.642	0.687	0.733
10°	70	0.435	0.470	0.506	0.544	0.584	0.627	0.671	0.718
	80	0.426	0.460	0.496	0.534	0.574	0.617	0.661	0.708
	90	0.416	0.450	0.486	0.524	0.565	0.607	0.652	0.699
20°	70	0.391	0.426	0.462	0.500	0.541	0.584	0.628	0.675
	80	0.375	0.410	0.446	0.484	0.525	0.568	0.612	0.659
	90	0.359	0.391	0.430	0.468	0.509	0.552	0.596	0.643
32°	70	0.305	0.340	0.376	0.415	0.456	0.499	0.544	0.591
	80	0.277	0.312	0.348	0.387	0.428	0.471	0.516	0.564
	90	0.248	0.283	0.320	0.359	0.400	0.443	0.488	0.536
40°	70	0.232	0.267	0.303	0.342	0.383	0.427	0.472	0.519
	80	0.193	0.228	0.264	0.304	0.345	0.388	0.434	0.481
	90	0.151	0.189	0.226	0.265	0.306	0.350	0.396	0.443

TABLE XXII

Temper. F.°	Density S lbs. per cu. ft.	$\left(\frac{S}{0.0749}\right)^{0.852}$
0	0.0863	1.129
10	0.0845	1.112
20	0.0827	1.090
30	0.0811	1.072
32	0.0807	1.070
40	0.0794	1.051
50	0.0779	1.032
60	0.0763	1.015
70	0.0749	1.000
80	0.0735	0.984
90	0.0722	0.970
100	0.0709	0.955
120	0.0684	0.926
140	0.0662	0.900
160	0.0640	0.874
180	0.0621	0.851
200	0.0601	0.828
212	0.0590	0.815
250	0.0559	0.780
300	0.0522	0.735
350	0.0490	0.698
400	0.0462	0.663
500	0.0413	0.604
600	0.0375	0.553
700	0.0342	0.514
800	0.0316	0.481
900	0.0292	0.450
1,000	0.0272	0.422
1,200	0.0239	0.379
1,400	0.0214	0.343
1,600	0.0193	0.315
1,800	0.0176	0.287
2,000	0.0161	0.270

TABLE XXIII*a*.—STOVE HEATING SURFACE IN SQ. FT. REQUIRED FOR ROOMS WITH WINDOWS COMPUTED BY MEANS OF THE LENGTH OF THE OUTER WALL SINGLE

	Ordinary living rooms*				Rooms in upper story with unheated attic above			
	Height of room in feet				Height of room in feet			
	8	10	11	13	8	10	11	13
a	*b*	*c*	*d*	*e*	*f*	*g*	*h*	*i*
3	1.18	1.40	1.62	1.83	1.29	1.51	1.83	2.15
6	2.37	2.80	3.23	3.77	2.69	3.12	3.66	4.20
10	3.55	4.20	4.95	5.72	3.98	4.74	5.60	6.46
13	4.63	5.38	6.35	7.32	5.28	6.24	7.32	8.50
16	5.82	6.78	7.96	9.14	6.56	7.75	9.15	10.55
20	6.90	8.08	9.47	10.87	7.96	9.36	10.98	12.70
23	8.30	9.70	11.40	13.15	9.26	10.87	12.80	14.75
26	9.14	10.76	12.70	14.65	10.65	12.50	14.65	16.90
30	10.00	11.75	13.80	15.82	11.95	14.00	16.46	18.95
33	11.40	13.45	15.82	18.30	13.25	15.60	18.30	21.10
36	12.60	14.75	17.45	20.00	14.53	17.10	20.12	23.15
40	13.15	15.50	18.30	21.10	16.05	18.85	22.20	25.50
43	14.96	17.55	20.70	23.80	17.12	20.12	23.70	27.22
46	16.05	18.85	22.20	25.50	18.50	21.50	25.60	29.50
50	17.12	20.01	23.80	27.44	19.90	23.50	27.55	42.50

* Thickness of wall approximately 1¼ ft., window area approx. 35% of the outer wall surface.

For additional factors see Table XXIII*b*.

Table XXIII*b*.—Stove Heating Surface in Sq. Ft. Required for Rooms with Double Windows Computed by Means of the Length of the Outer Wall

Length of wall in feet	Ordinary living rooms*				Rooms in upper story with unheated attic above			
	Height of room in feet				Height of room in feet			
	8	10	11	13	8	10	11	13
a	*b*	*c*	*d*	*e*	*f*	*g*	*h*	*i*
3	0.75	0.86	1.08	1.29	0.97	1.18	1.1	1.62
6	1.51	1.72	2.04	2.37	1.91	2.26	2.69	3.12
10	2.26	2.69	3.12	3.55	2.91	3.44	4.09	4.74
13	2.91	3.45	4.09	4.74	3.98	4.63	5.38	6.24
16	3.77	4.42	5.18	5.92	4.95	5.82	6.78	7.86
20	4.41	5.16	6.14	7.10	5.92	6.89	8.06	9.26
23	5.17	6.02	7.10	8.18	6.88	8.08	9.48	10.87
26	5.92	7.00	8.18	9.36	7.75	9.12	10.76	12.38
30	6.67	7.86	9.26	10.65	8.72	10.23	11.95	13.78
33	7.42	8.72	10.23	12.05	9.70	11.40	13.45	15.50
36	8.18	9.58	11.30	13.03	10.76	12.60	14.85	17.10
40	8.94	10.45	12.27	14.10	11.75	13.78	16.15	18.63
43	9.70	11.30	13.35	15.50	12.60	14.85	17.55	20.10
46	10.23	12.05	14.20	16.35	13.78	16.15	18.85	21.63
50	11.2	13.15	15.50	17.85	14.65	17.23	20.23	23.25

* Thickness of wall approximately 1¼ ft., window area approx. 35% of outer wall surface.

Note: The following additions to the heating surface as computed are necessary.

15–20% for entirely exposed houses and for those exposed to strong winds.

50% for rooms whose outer wall surface is very small in proportion to the room volume.

10% for rooms bordering on unheated stairway halls.

20% for rooms bordering on driveways open to the outside air.

INDEX

A

ccidents, 39
ccelerated circulation systems, 60, 61
lvantages of district heating, 97
exhaust ventilation, 131
high-pressure steam heating, 64
low-pressure steam heating, 72
pressure ventilation, 141
Reck heating system, 60
steam air heating, 93
water air heating, 93
lditional circulating heads for down-feed hot-water systems, 308
factors for heating up and intermittent operation, 171
corner rooms, 297
direction of exposure, 297
high rooms, 172, 297
r changes, calculation of, 135
determination of, 257
empirical data, 129
natural, 133
necessity of, 121
circulation in rooms 7
cooling of 145 154, 281
currents with radiator under window, 20
dehumidification, 145
density, 322
drying, 145
filters 143
heating, by steam, 93
by water, 93
systems, air supply for, 142
dust chamber for, 142
duct work for, 149
humidification of, 145
heaters, tubular, 95
fin-surface, 95
heat output and control, 95
humidity, 145
humidity content influence on room temperature 145, 281
requirements, 128
standard, 258

Air humidifier, 144, 145
layers, insulation effect of, 163
motion in rooms 7 150
pressure room-temperature regulator, 84
purification, 143
putrefaction, 125, 144
supply, in boiler rooms, 38, 136, 142
velocity, 55
through registers, 151
washing, 144
weight of, 313
American stoves, 11
Ammonia, 125
Anthropotoxin, 125
Application of district heating, 97
electric heating, 18
gas heating, 14
high-pressure steam heating, 64
iron stoves, 6
low-pressure steam heating, 73
Reck heating systems, 61
steam hot-water heating, 89
air heating, 93
tile stoves 6
ventilation by means of blowers, 141
Apportioning load among boilers, 34
Arcola heater, 59
Assumption of duct work for ventilation systems, 273
Assumption of piping for high-pressure steam heating, 235
hot-water heating, 197
low-pressure steam heating, 246
Ash handling 7, 97
Automatic regulator for hot-water heating, 58
for low-pressure steam heating 84

B

Basement distribution for hot-water heating, 24
for steam heating, 77
Bolcke storage tanks, 106

Bellows regulator, 33
Bleeder steam heating, 110
turbines, 110
Blowers, 140, 146
Boiler, calculation of high-pressure steam, 230
hot-water heating, 191
low-pressure steam, 243
cast-iron, 25, 74
Catena, 30
connections, 35
efficiency, 36, 37
explosions, 38
gas, 32
heating-up period, 191
high-temperature hot-water heating, 63
high-pressure steam heating, 65
Lollar, 27
low-pressure steam, 74
magazine, 30
oil-burning, 31
safety devices, 35
rating 36
resistance, 197
return connections, 35
room, 37
height, 38
size of, 38
ventilation, 38
steam-heated steam, 75
steel 25, 74
Strebel, 26, 74
soft-coal 30
testing, 36
trimmings, 32
Brabo stove, 11
Brick walls, K values of, 294
Brine, 155
Building materials, K values, 292
porosity of, 134
Bulb heater, 18
Bypass, 35, 68

C

Calculation of boiler sizes for high-pressure steam heating, 230
hot-water heating, 191
low-pressure steam heating, 243
cooling plants, 281
duct sizes, 265, 267
Calculation of electric stoves, 190
gravity hot-water systems, 194
high-pressure steam systems, 230
heating surface, 182
for high-pressure steam, 230
for hot-water heating, 191, 193
for low-pressure steam, 243
iron stoves, 189, 323
K values, 168
low-pressure steam heating systems, 243
necessary air change, 135
pipe sizes for high-pressure steam, 230
hot-water heating, 194
low-pressure steam, 244
radiators, 182
ventilation systems, 265
Cannon stove, 8
Capron acid, 125
Capryl acid, 125
Carbon dioxide content of air, 125
emission by persons and animals, 124
requirements, 128
standard, 260
Catena boiler, 30
Ceiling fan, 138
Cellar temperatures, 173
Centralized control for hot-water heating systems, 85, 104, 153
Central heating, 20
humidification, 146
venting, 78
Centrifugal fans, 147
Characteristic of a fan, 148
Chart 1, method of using, 197
2, 197
3 and 4, 234
5, 246
6 and 7, 272
Checking calculation of piping for high-pressure steam, 235
hot-water systems, 198
low-pressure steam, 247
Chimneys, calculation of, 16, 40, 283
Circulating heads for high-pressure steam systems, 185, 230
hot-water systems, 185
loft heating systems, 185, 311
low-pressure steam systems, 185, 244
ventilation systems, 185 267
Coils low-pressure steam, 248
resistance heater, 18

Coke filters, 143
Combustion efficiency 8
Combination heating systems, 88
Condensation return lines, 72, 85, 236, 248, 312
Conduit heating systems, 6, 188
Conduction, transfer of heat by, 163
Control board 40, 152
Consideration of piping heat loss, 198
Cooling of air, 154, 281
fluids, 155
surfaces, inside the room, 154, 156, 282
outside the room, 155
systems, calculation of, 156, 281
installation, 154, 281
Cookanheat, 28
Cork insulation, 43
Corrosion, 23, 79
Corner rooms, 297
Counter current, 179
apparatus, 89, 179
calculation of, for steam, 180
Crown Jewel stove, 11
Cowls, suction, 138
pressure, 140

D

Damper regulator, 33, 76
for hot-water heating, 33
Dehumidification of air, 281
Diffuser, 270
Disadvantages of high-pressure steam heating, 64
hot-air heating, 254
hot-water heating, 23
low-pressure steam heating, 73
pressure ventilation, 141
Reck heating systems, 61
steam air heating, 93
water air heating 93
District heating steam, advantages, 96
application, 97
design, 254
pipe conduits, 99
supplementary mains, 98
hot water, 100
connections at building, 103
control panels, 104
Double-current apparatus, 178
Drafts, 132
Draining hot-water systems, 44
Drying of air, 145
Dry returns, 77
Ducts, loss of heat from, 265
sizing of, for ventilation systems, 267
Dust accumulation, 126
chambers, 142
collectors, 143

E

Economy of exhaust heat utilization, 117
Eddying-current apparatus, 179
Efficiency of boilers, 37
iron stoves, 8
Einheit stove, 13
Electrical heating pads, 19
stoves, 18
calculation of, 190
testing of, 19
temperature regulators, 84
Emission of carbon dioxide from animals and persons, 124
heat from illumination and persons, 123
moisture from illumination, 122
moisture from persons, 122
Enclosures, 50
Equivalent orifice, 148
Equilibrium (see *Steady state*), 159, 192
Exhaust air discharge, 137, 151
ducts in boiler room 38
heating, 137
gas heating, 116
heat from chimneys, 137
internal-combustion engines, 111, 104 116
steam power plants, 108
heating systems, 106, 112
utilization, 105
steam, air heating, 111
hot-water heating, 110, 114
heating 73
oil separator, 109
regulator, 109
storage 106
utilization 108
ventilation, 131
Expansion tank, 44, 102
indicating pipe, 44
volume of, 44
connections, 68

F

Fans, ceiling, 138
  centrifugal, 146
  characteristic of, 148
  installation of, 149
  location of, 149
  regulation of, 149
  table, 138
  quiet operation of, 149
Factory heating, 93, 96
Factory temperature, 164
Felt, 43
Filter calculation of, 262
  cloth, 143
  coke, 143
  gravel, 143
  peat, 143
  pocket, 143
  stone, 143
  wood, 143
Fin surfaces, 46
Firemen 40
Fireplace, 5, 188
  efficiency, 5, 188
Flue-gas heater 13
Forced-circulation heating system, 61
  advantages, 61
  application, 62
  disadvantages, 61
  general control, 104
  installation, 62
  pump connections, 102
  ventilation systems, 138
Form of radiators, 54
Fresh-air ventilation, 138
Freezing danger in hot-water systems,
    23, 45
Frictional resistance 186
  for high-pressure steam systems, 230
  for hot-water systems, 195
  for low-pressure steam systems, 244
  for ventilation systems, 271
Fuel storage space, 38
Furnace hot-air heating, 92

G

Gas boiler, 32
  heaters, 15, 16, 17
  steam radiators, 15, 48
Gilled pipe, 46
Gravity hot-air heating systems, 93
  hot-water heating systems, 21
  ventilation, 136
Greenhouses, boilers for, 60
  heating, 24, 60

H

Heat, absorption coefficient, 159, 166
  balance, 106
  content of air, 128, 281
  conductivity, coefficient of, 160
    equivalent coefficient of, 163
    for building materials, 289
    for insulating materials, 290
  emission, coefficient of, 160, 167
    of illumination, 122
    of persons, 122
  loss, basing on volume, 177
    calculation, 174
    consideration of in hot-water heating
      systems, 199
    during heating-up period, 171
  standard, 257
  transmission, 159
    coefficient for building materials, 292
      for hot-water heated surfaces, 298
      for steam-heated surfaces, 300
    of air or flue gas through thin iron
      partitions, 306
Heating surface calculation 182
  for high-pressure steam systems, 230
  for hot-water systems, 191, 193
  for low-pressure steam systems, 243
  increase of for loft heating systems, 311
Heating up, 159
  additional factor for, 171
Heating-up period, 159, 172, 191
Height of boiler room, 38
High-pressure steam heating, 64
  advantage of, 64
  application, 64
  boiler piping, 65
  calculation of heating surface for, 230
    of boiler heating surface for, 230
    of piping for, 230
  condensation return, 236
  disadvantages 64
  installation of piping 70, 71
  insulation for, 69, 70
  radiators, 71
    connections, 72

High-pressure steam heating, reducing valve, 70, 74
regulation, 71
steam-water separators for, 66
traps for, 66
High-temperature hot-water heating, 63
district heating, 104
Hot-water air heating systems, 93
application 93
heaters for, 93
regulation of 93
district heating, 100
application, 104
boilers for, 101
conduits, 102
connection of pump 102
expansion tanks for, 102
piping for, 102
heating systems, 21
application, 23
boilers for, 25
damper regulator 33
draining of, 33, 44
expansion tank for, 44
filling of 33
forced circulation, 61
gravity, 21
one-pipe, 201
medium-pressure, 63
high-temperature, 63
life of, 23
pipe sizes for, 194
pipe friction, 195
safety devices for, 35
water content of, 24
water level of, 33
Human comfort, 81
Humidity of air, 281
requirements, 128
standard, 258
Humidifiers, 145, 146

I

Ice, 155
water, 155
Indicating pipe, 44
Influence of exposure, 297
Infusorial earth, 43
Installation of, cooling plants, 154, 281
high-pressure steam heating systems, 70, 71
hot-air heating systems, 251
Installation of hot-water heating systems, 24
steam hot-water heating systems, 89
Inside temperatures, 3
Insulation, 70
Intermittent heating, additional factor for 171
Irish stoves, 10
Iron stoves, 6
advantages, 7
application, 6
calculation of, 189, 323
determination of heating surface, 9, 323
design, 8
efficiency of, 8
testing of, 8

J

Junkers gas air-heating stove, 16

K

Krantz heating systems, 65
K values of building materials, 292
of various forms of heating surface, 298, 299, 300, 301

L

Latent heat of evaporation, 287
Lecture rooms, ventilation of, 150
Local heating, 5, 188
resistances, 186
extent of, in steam systems, 312
in ventilation systems, 273
in water systems, 312
Loft heating systems, 59
advantages, 59
Lollar boiler, 27
Low-pressure steam heating, 72, 243
advantages, 72
application, 73
boiler, 74, 243
cast-iron, 75
equipment, 76
steel, 74
disadvantages, 73
installation, 73
piping, 79, 244
radiators, 80
calculation of, 80, 243

M

Medium-pressure hot-water heating, 63
Methods of heating, 5
Mılddampfheizung, 80
Moisture content of air, 282
  emission by illumination, 122
    by persons, 122

N

Necessity of ventilation, 121
Neutral zone, 131
Noise in forced-circulation systems, 149

O

Oil-burning boiler, 31
Oil contained in steam, 108
  separator, 109
  stoves, 14
One-family houses, pipe systems in, 24, 201
Operating expense and installation cost of tile stoves, 7
  instructions 39
    for safety devices on hot-water heating systems, 39
  of fans, 149
    noiseless, 149
  pressure, 78
Outside temperatures, 3
Overflow, 44
  pipe, 44
Ozonizing the air, 144

P

Parallel current, 179
Performance curves of valves, 57
Petroleum stoves, 14
Pipe, bypass, 35
  coils, 45
  covering, 42
  diameters, 287
  expansion, 67
  external surface, 287
  insulation, 43
  K values, 298, 299, 300, 301
  tunnels (conduits) 99
  weight of, 287
Pocket filters, 143
Poisonous effluvia, 125
Poisons in exhaled air, 126
Porosity of building materials, 134
Pressure cowls 140
  dynamic, 271
  distribution in a room, 131
    in a forced-circulation system, 103
  head in high-pressure steam systems, 185, 230
    in low-pressure steam systems, 244
    in hot-water heating systems, 185
    in ventilation systems, 185, 267
  regulator, 76
  requirements, 129
  standard, 261
  static, 271
Pressure-reducing valves, 70, 74
Pump pressure, for forced-circulation systems, 103

R

Radiator, 45
  air velocity, 55
  average water temperature, 55
  automatic regulation, 58
  brackets, 48, 52
  calculation of surface, 56, 80, 193
  cast-iron, 47
  connections 72
  Corto, 49
  distance from wall, 55
  enclosures, 45, 80
    efficiency of, 50
  forms of, 45, 54
  gas steam 48
  influence of room dimensions on, 55
  location of, 45
  mantels, 50
  rating, 53
  Recesso, 51
  recessed, 52
  regulation automatic for, 58
  steel, 50
  surface, 55, 80
  testing, 53
  trap, 83
  useful heat output, 81
  valves, 57, 82
  water velocity, 55
Radiation exchange coefficient, 165, 291
Radiant heater, 16, 17

Recesso radiator, 51
Recessed radiator, 51
Reck heating system, 60
  advantage, 60
  application, 61
  disadvantage, 60
Recirculated air, 138
Recirculation systems, 138
Reducing valves, 70, 74
Refractory elements, 16
Regulation, ideal, 57
  group, 65
  hand, 57
Regulator, automatic, 84
  Bellow type, 76
Return lines, diameter of, 312
  dry, 77
  wet, 77
Resistance heaters, 19
Room temperatures, 3

S

Safety devices for steam boilers, 77
    for hot-water boilers, 35
  factors, 297
Schlopter fan, 147
School heating, 93, 254
Self-ventilation of a room, 133
Single-current apparatus, 178
Soft-coal boilers, 30
Spring water 155
Standpipe for low-pressure steam systems, 76
Steam air heating, 93, 115
    application, 93
    advantages, 93
    disadvantages, 93
    radiators, 93
    regulation, 93
  boiler, 65
    steam-heated, 74
  containing oil, 108
  coils, 248
  district heating, 100
  heating high-pressure, 64, 230
      application, 64
      installation, 65
    low-pressure, 72, 243
    hot-water, 88
  hot-water heaters, 90
  mains for superheated steam, 233
Steam oil separator, 109
  pipes, 66, 79, 109, 233
    dripping of, 69
    insulation, 70
  pressure at various temperatures, 288
  properties of, 288
  reducing valve 70
  storage tank, 106
  traps, 66
  turbine, 110
  velocity, 235, 247
  vents, 71
  water separators, 66
    heating, 90
Steady state, 159
Stone filter, 143
Stove, Einheit, 13
  iron, 6
    calculation of, 189, 323
  oil, 14
    calculation of 190
  testing of, 8
Storage space for fuel, 38
Strebel boiler, 26, 71
Street steam, 97
Suction cowl, 138
Supplementary mains, 98

T

Table fan, 138
Tees, 79
Temperature, attic 173
  cellar 173
  control for air heater, 95
  factor, 164
  inside, 3
    effect of air change on, 4
  outside, 3
  regulators, 84
  requirements, 127
  unheated rooms, 173
  variation in a room, 173
  vestibules, 173
Testing of boilers, 36
  radiators, 53
  stoves, 8
Tile stoves, 13
Toilets 132
Transoms, 136
Traps 83
Transverse currents, 179
Tubular boiler for hot-water heating 25

U

Useful heat, 56
  output of radiators, 81
Unit heaters, 265

V

Vacuum heating, 85
Valves, 44, 56, 82
Vecto heater, 12
Vento heater, 94, 264, 302
Vents, 16
Venting of forced-circulation systems, 62
Ventilation by gravity circulation, 136
  by blowers 141
  design, 141
Vapor pressure at different temperatures, 288
Viscosity of water, 197

W

Warm-air furnace heating, 91, 251
  advantages, 91
  application, 91
  computation, 91, 251
  design, 91, 251
  disadvantages 91
Water density, 307
  level in hot-water heating systems, 33
  separators 66
  supply 155
  vapor, data on, 288
  velocity 55
Weight of air, 322
  water, 307
Wet return main, 77
Window draft diverter, 136
Wind pressure, 138
Wood filter 143

Z

Z ducts, 151

Rietschel, Hermann Imanuel, 1847-1918.
Heating and ventilation; a handbook for architects and engineers, by C. W. Brabbée, translated for American use from the seventh German edition of Rietschel-Brabbée, "Heizungsund luftungstechnik". 1st ed. New York [etc.] McGraw-Hill book company, inc., 1927.
xi, 332 p. illus. 24 cm.

2 Diame…

Velocity of water in ft. per sec.	Resistance Z in ins. w.c. of fittings for Σζ =														
	1	2	3	4	5	6	7	8	9	10	11	12	13	14	15
0.1	0.0018	0.0036	0.0054	0.0072	0.0091	0.011	0.013	0.015	0.016	0.018	0.020	0.022	0.024	0.026	0.027
0.11	0.0022	0.0044	0.0066	0.0088	0.0110	0.013	0.015	0.018	0.020	0.022	0.024	0.026	0.029	0.031	0.033
0.12	0.0026	0.0052	0.0078	0.0010	0.013	0.016	0.018	0.021	0.024	0.026	0.029	0.031	0.034	0.036	0.039
0.13	0.0031	0.0061	0.0092	0.012	0.015	0.018	0.021	0.025	0.028	0.031	0.034	0.037	0.040	0.043	0.046
0.14	0.0036	0.0071	0.011	0.014	0.018	0.021	0.025	0.028	0.032	0.036	0.039	0.043	0.046	0.050	0.053
0.15	0.0041	0.0081	0.012	0.016	0.020	0.024	0.028	0.033	0.037	0.041	0.045	0.049	0.053	0.057	0.061
0.16	0.0046	0.0093	0.014	0.018	0.023	0.028	0.032	0.037	0.042	0.046	0.051	0.056	0.060	0.065	0.07
0.17	0.0052	0.011	0.016	0.021	0.026	0.031	0.037	0.042	0.047	0.052	0.058	0.063	0.068	0.073	0.079
0.18	0.0059	0.012	0.018	0.024	0.029	0.035	0.041	0.047	0.053	0.059	0.064	0.070	0.076	0.082	0.088
0.19	0.0066	0.013	0.020	0.026	0.033	0.039	0.046	0.052	0.059	0.065	0.072	0.078	0.085	0.091	0.098
0.20	0.0072	0.015	0.022	0.029	0.036	0.043	0.051	0.058	0.065	0.072	0.080	0.087	0.094	0.10	0.11
0.22	0.0088	0.018	0.026	0.035	0.044	0.053	0.061	0.070	0.079	0.088	0.096	0.11	0.12	0.12	0.13
0.24	0.010	0.021	0.03	0.042	0.052	0.063	0.073	0.083	0.094	0.10	0.12	0.13	0.14	0.15	0.16
0.26	0.012	0.025	0.037	0.049	0.061	0.073	0.086	0.098	0.11	0.12	0.14	0.15	0.16	0.17	0.18
0.28	0.014	0.028	0.043	0.057	0.071	0.085	0.099	0.11	0.13	0.14	0.16	0.17	0.19	0.20	0.21
0.30	0.016	0.033	0.049	0.066	0.082	0.098	0.11	0.13	0.15	0.16	0.18	0.20	0.21	0.23	0.24
0.33	0.020	0.038	0.058	0.078	0.089	0.12	0.14	0.16	0.18	0.20	0.22	0.24	0.26	0.28	0.30
0.36	0.024	0.047	0.070	0.094	0.12	0.14	0.16	0.18	0.21	0.24	0.26	0.28	[illegible]	0.35	0.38
0.38	0.026	0.052	0.078	0.105	0.13	0.16	0.18	0.21	0.24	0.26	0.29	0.31	0.34	0.37	0.39
0.40	0.030	0.059	0.087	0.12	0.15	0.17	0.20	0.23	0.26	0.30	0.32	0.35	0.38	0.41	0.44
0.45	0.037	0.073	0.11	0.15	0.18	0.22	0.26	0.29	0.33	0.37	0.40	0.44	0.48	0.51	0.55
0.50	0.045	0.092	0.14	0.18	0.23	0.27	0.32	0.36	0.41	0.45	0.50	0.54	0.59	0.63	0.68
0.55	0.055	0.11	0.16	0.22	0.27	0.33	0.38	0.44	0.49	0.55	0.60	0.66	0.71	0.77	0.82
0.60	0.065	0.13	0.20	0.26	0.33	0.39	0.46	0.52	0.59	0.65	0.72	0.78	0.85	0.91	0.98
0.70	0.089	0.18	0.27	0.36	0.44	0.53	0.62	0.71	0.80	0.89	0.98	1.07	1.15	1.24	1.33
0.80	0.12	0.23	0.35	0.46	0.58	0.70	0.81	0.93	1.04	1.16	1.28	1.39	1.51	1.62	1.74
0.90	0.15	0.29	0.44	0.59	0.73	0.88	1.03	1.17	1.32	1.47	1.61	1.76	1.91	2.05	2.2
1.00	0.18	0.36	0.54	0.73	0.91	1.09	1.27	1.45	1.63	1.81	2.00	2.17	2.36	2.53	2.72
1.10	0.22	0.44	0.66	0.88	1.10	1.31	1.53	1.75	1.97	2.19	2.41	2.63	2.85	3.06	3.28
1.20	0.26	0.52	0.78	1.04	1.30	1.56	1.82	2.08	2.34	2.61	2.87	3.13	3.39	3.65	3.91
1.30	0.31	0.61	0.92	1.22	1.50	1.84	2.14	2.45	2.75	3.06	3.37	3.67	3.98	4.28	4.59
1.40	0.36	0.71	1.06	1.42	1.80	2.13	2.48	2.84	3.19	3.55	3.90	4.26	4.61	4.97	5.32
1.50	0.41	0.81	1.22	1.63	2.04	2.44	2.85	3.26	3.66	4.07	4.48	4.88	5.29	5.70	6.10
1.60	0.46	0.93	1.39	1.85	2.32	2.78	3.24	3.70	4.17	4.63	5.10	5.56	6.02	6.48	6.95

VALUES OF ζ FOR RESISTANCES OF FITTINGS

Radiators	ζ = 3.0	Abrupt velocity changes	ζ = 1.0
Boilers	ζ = 2.5	Crossover	ζ = 0.5
Tees: Straightway	ζ = 1.0	Counter current	ζ = 3.0
Tees: Branch	ζ = 1.5	Double branch long sweep ell	ζ = 1.5

Kind of fitting	Maker	ζ values for inside pipe diameter in inches of						
		⅜"	½"	1"	1¼"	1½"	2"	
Frictional angle valve	Schäffer & Oehlmann	8.0	9.0	9.0	8.0			
Frictional globe valve	Schäffer & Oehlmann	16.0	17.0	19.0	20.0			
Riser valve	Rud. Otto Meyer	10.5	10.0	9.0	9.0	8.0	7.0	4½" inside diameter ζ = 7.0
Flanged check valve	Rud. Otto Meyer							
Angle cock	Korting Bros.	7.0	4.5	4.0	4.0			
Straight cock	Korting Bros.	4.0	2.0	2.0	2.0			
Gate valve (screwed)	Rud. Otto Meyer	1.5	0.5	0.5	0.5	0.5	0.5	4½" inside diameter ζ = 1.0
Gate valve (flanged)	Rud. Otto Meyer							
Throttle valve	Rud. Otto Meyer	3.5	2.0	2.0	1.5	1.5	1.0	
Ell	George Fischer	2.2	2.0	1.5	1.5	1.0	1.0	
Longsweep ell 90°	George Fischer	1.5	1.5	1.0	1.0	0.5	0.5	
Bends: Short radius	George Fischer	2.0	2.0	2.0	2.0	2.0	2.0	
Bends: Long radius	George Fischer	1.0	1.0	1.0	1.0	1.0	1.0	
Standard pipe coupling	George Fischer	0.5	0.0	0.0	0.0	0.0	0.0	

Frictional resistance of one ft. of pipe in 1/1,000" w.c.	½"	¾"	1"	1¼"	1½"	2"	2½"
0.5	.....	.....	.....	.....	8,100 0.1	17,700 0.12	32,600 0.15
0.6	.....	.....	.....	.....	9,000 0.11	19,600 0.14	35,600 0.16
0.7	.....	.....	.....	6,000 0.11	9,700 0.12	21,100 0.15	39,000 0.17
0.8	.....	.....	.....	6,500 0.12	10,100 0.13	22,700 0.16	41,700 0.18
0.9	.....	.....	3,770 0.11	6,900 0.12	11,800 0.14	24,200 0.17	44,700 0.20
1.0	.....	.....	4,000 0.11	7,300 0.13	11,800 0.16	25,300 0.18	47,300 0.22
1.2	.....	.....	4,400 0.12	8,100 0.14	13,000 0.16	28,200 0.20	52,200 0.24
1.4	.....	2,180 0.11	4,800 0.13	8,800 0.16	14,200 0.18	31,000 0.22	56,700 0.26
1.6	.....	2,390 0.12	5,130 0.14	9,500 0.17	15,300 0.19	33,200 0.24	61,000 0.28
1.8	.....	2,540 0.13	5,500 0.15	10,000 0.18	16,300 0.20	35,900 0.26	65,000 0.30
2.0	.....	2,600 0.13	5,800 0.16	10,700 0.19	17,200 0.22	37,000 0.26	69,000 0.30
2.3	.....	3,000 0.14	6,300 0.18	11,500 0.20	18,500 0.24	40,900 0.28	74,300 0.33
2.6	.....	3,130 0.15	6,700 0.18	12,300 0.22	20,000 0.26	43,200 0.30	79,300 0.35
3.0	.....	3,370 0.17	7,300 0.20	13,300 0.24	21,700 0.28	47,000 0.33	86,000 0.38
3.5	1,040 0.12	3,680 0.18	7,900 0.22	14,500 0.26	23,300 0.30	51,000 0.36	93,000 0.40
4.0	1,200 0.14	3,920 0.18	8,500 0.24	15,800 0.28	25,100 0.30	55,000 0.38	100,000 0.45
4.5	1,350 0.15	4,190 0.20	9,100 0.26	16,600 0.30	26,800 0.33	58,500 0.40	106,000 0.50
5.0	1,470 0.16	4,440 0.22	9,800 0.28	17,500 0.33	28,300 0.36	62,000 0.45	112,500 0.50
6.0	1,610 0.17	4,900 0.24	10,500 0.30	19,400 0.38	31,000 0.38	68,200 0.50	123,000 0.55
7.0	1,760 0.20	5,300 0.26	11,300 0.33	21,100 0.38	34,000 0.40	74,300 0.50	135,000 0.60
8.0	1,800 0.22	5,710 0.28	12,500 0.36	22,600 0.40	37,000 0.45	80,000 0.56	145,000 0.70
9.0	2,000 0.24	6,100 0.30	13,100 0.38	24,200 0.45	39,000 0.50	85,000 0.60	155,000 0.70
10.0	2,140 0.24	6,470 0.33	14,000 0.40	25,500 0.50	41,500 0.50	90,000 0.60	160,000 0.70
12.0	2,350 0.30	7,150 0.30	15,400 0.45	28,200 0.50	46,000 0.55	99,300 0.70	180,000 0.80
14.0	2,550 0.30	7,800 0.38	16,700 0.50	30,500 0.55	49,300 0.60	107,000 0.80	198,000 0.90
16.0	2,780 0.30	8,300 0.4	18,000 0.50	33,000 0.60	53,300 0.70	110,500 0.80	210,000 0.90
18.0	2,920 0.33	8,900 0.45	19,100 0.55	35,000 0.60	57,000 0.70	122,500 0.90	228,000 1.00
20.0	3,100 0.36	9,490 0.50	20,200 0.60	37,300 0.70	60,000 0.80	130,000 0.90	240,000 1.10
22.0	3,380 0.38	10,100 0.50	21,900 0.60	40,000 0.70	65,000 0.80	140,000 1.00	256,000 1.10
25.0	3,600 0.40	10,730 0.55	23,400 0.70	43,000 0.80	69,000 0.90	150,000 1.10	272,000 1.20
30.0	3,880 0.45	11,700 0.60	25,300 0.70	46,800 0.80	78,000 0.90	162,800 1.20	296,000 1.30
35.0	4,200 0.50	12,700 0.60	27,500 0.80	50,300 0.90	80,200 1.00	177,000 1.20	322,000 1.40
40.0	4,510 0.50	13,600 0.70	29,600 0.80	54,200 1.00	88,000 1.10	190,000 1.30	347,000 1.50
45.0	4,820 0.58	14,500 0.70	31,300 0.90	58,000 1.00	93,000 1.20	200,000 1.40	370,000 1.60
50.0	5,100 0.55	15,400 0.80	33,500 0.90	61,000 1.10	98,000 1.20	212,000 1.50	395,000 1.7

Velocity of water in ft. per sec.	Resistance Z of fittings in ins. w.c. for Σζ =														
	1	2	3	4	5	6	7	8	9	10	11	12	13	14	15
1.6	0.46	0.93	1.4	1.8	2.3	2.8	3.2	3.7	4.2	4.6	5.1	5.5	6.0	6.5	6.9
1.7	0.52	1.1	1.6	2.1	2.6	3.1	3.6	4.2	4.7	5.2	5.7	6.2	6.8	7.3	7.8
1.8	0.58	1.2	1.8	2.3	3.0	3.5	4.1	4.7	5.3	5.9	6.5	7.1	7.7	8.3	8.8
1.9	0.65	1.3	2.0	2.6	3.3	3.9	4.6	5.2	5.9	6.6	7.2	7.8	8.5	9.1	9.8
2.0	0.72	1.5	2.2	2.9	3.6	4.3	5.0	5.8	6.5	7.2	7.9	8.6	9.4	10.0	10.8
2.2	0.88	1.8	2.6	3.5	4.4	5.3	6.2	7.0	7.9	8.8	9.7	10.6	11.5	12.3	13.2
2.4	1.1	2.1	3.1	4.2	5.2	6.2	7.3	8.3	9.4	10.4	11.4	12.5	13.5	14.6	15.6
2.6	1.2	2.4	3.7	4.9	6.2	7.4	8.6	9.8	11.1	12.3	13.5	14.8	16.0	17.2	18.5
2.8	1.4	2.8	4.3	5.7	7.1	8.6	9.9	11.4	12.8	14.2	15.6	17.0	18.5	19.9	21.3
3.0	1.6	3.3	4.9	6.5	8.2	9.8	11.4	13.0	14.7	16.3	17.9	19.6	21.2	22.8	24.4
3.3	2.0	3.9	5.9	7.9	9.9	11.8	13.8	15.8	17.7	19.7	21.7	23.6	25.6	27.6	29.6
3.6	2.4	4.7	7.1	9.4	11.8	14.1	16.5	18.8	21.2	23.5	25.9	28.2	30.6	32.9	35.3
3.8	2.6	5.2	7.8	10.5	13.1	15.7	18.3	20.9	23.5	26.1	28.7	31.3	33.9	36.5	39.2
4.0	2.9	5.8	8.7	11.6	14.5	17.4	20.3	23.1	26.0	28.9	31.8	34.7	37.6	40.5	43.4
4.5	3.7	7.3	11.0	14.7	18.3	22.0	25.6	29.3	33.0	36.6	40.3	44.0	47.6	51.3	55.0
5.0	4.5	9.0	13.6	18.1	22.6	27.1	31.6	36.2	40.7	45.2	49.7	54.2	58.8	63.3	67.8
5.5	5.5	11.0	16.4	21.9	27.4	32.9	38.4	43.8	49.3	54.8	60.3	65.8	71.2	76.7	82.2
6.0	6.5	13.0	19.6	26.1	32.6	39.1	45.6	52.2	58.7	65.2	71.7	78.3	84.8	91.3	97.8
6.5	7.7	15.3	23.0	30.6	38.3	45.9	53.6	61.2	68.9	76.5	84.2	91.8	99.5	107	115
7.0	8.9	17.7	26.6	35.4	44.3	53.2	62.0	70.9	79.7	88.6	97.5	106	115	124	133
8.0	11.6	23.2	34.7	46.3	57.9	69.5	81.1	92.6	104	116	127	139	150	162	174
9.0	14.7	29.3	44.0	58.6	73.3	88.0	103	117	132	147	161	176	190	205	220
10.0	18.0	36.2	54.3	72.4	90.5	109	127	145	163	181	199	217	235	253	272

VALUES OF ζ FOR RESISTANCES OF FITTINGS

Radiators	ζ = 3.0	Abrupt velocity changes	ζ = 1.0
Boilers	ζ = 2.5	Crossover	ζ = 0.5
Tee { Straightway	ζ = 1.0	Counter current	ζ = 3.0
Tee { Branch	ζ = 1.5	Double branch long sweep el.	ζ = 1.5

Kind of fitting	Maker	ζ values for inside pipe diameter in inches of						
		½"	¾"	1"	1¼"	1½"	2"	
Fractional angle valve	Schäffer & Oehlmann	9.0	9.0	9.0	9.0			
Fractional globe valve	Schäffer & Oehlmann	15.0	17.0	19.0	30.0			
Riser valve	Rud. Otto Meyer	16.0	10.0	6.0	9.0	8.0	7.0	{ 4½" inside diameter ζ = 7.0
Flanged check valve	Rud. Otto Meyer	...		...	...	...	...	
Angle cock	Korting Bros.	7.0	4.0	5.0	4.0			
Straight cock	Korting Bros.	4.0	2.0	2.0	2.0			
Gate valve (screwed)	Rud. Otto Meyer	1.5	0.8	0.5	0.5	0.5	0.5	{ 4½" inside diameter ζ = 1.0
Gate valve (flanged)	Rud. Otto Meyer	...		...	...	...	...	
Throttle valve	Rud. Otto Meyer	3.5	2.0	2.0	1.5	1.5	1.0	
Ell	George Fischer	2.0	2.0	1.5	1.5	1.0	1.0	
Longsweep ell 90°	George Fischer	1.5	1.5	1.0	1.0	0.5	0.5	
Bends { Short radius	George Fischer	2.0	2.0	2.0	2.0	3.0	3.0	
Bends { Long radius	George Fischer	1.0	1.0	1.0	1.0	1.0	1.0	
Standard pipe coupling	George Fischer	0.5	0.0	0.0	0.0	0.0	0.0	

Frictional resistance per running ft. of pipe in $\frac{1}{1,000}$" w.c.	He… Ve…								
	½"	¾"	1"	1¼"	1½"	2"	2½"	3"	…
30	3,880 0.45	11,700 0.6	25,300 0.7	46,500 0.8	73,000 0.9	162,500 1.2	298,000 1.3	400,000 1.5	7…
35	4,300 0.5	12,700 0.6	27,500 0.8	50,300 0.9	80,200 1.0	175,000 1.2	322,000 1.4	520,000 1.6	7…
40	4,510 0.5	13,800 0.7	29,000 0.8	54,200 1.0	88,000 1.1	190,000 1.3	347,000 1.5	590,000 1.7	8…
45	4,820 0.55	14,500 0.7	31,300 0.9	58,000 1.0	93,000 1.2	200,000 1.4	370,000 1.6	600,000 1.9	9…
50	5,100 0.55	15,400 0.8	33,300 0.9	61,000 1.1	98,000 1.2	215,000 1.5	380,000 1.7	630,000 2.0	9…
60	5,620 0.6	17,000 0.8	37,000 1.0	66,300 1.2	108,000 1.3	233,000 1.6	425,000 1.9	700,000 2.2	1,0…
70	6,100 0.7	18,500 0.9	40,000 1.1	72,200 1.3	117,500 1.5	255,000 1.8	466,000 2.0	760,000 2.4	1,1…
80	6,500 0.7	20,000 1.0	43,000 1.2	78,000 1.4	126,000 1.6	272,000 1.9	500,000 2.2	815,000 2.6	1,2…
90	7,000 0.8	21,300 1.0	46,000 1.3	83,500 1.5	135,000 1.7	291,000 2.0	530,000 2.4	870,000 2.8	1,3…
100	7,400 0.8	22,200 1.1	49,000 1.4	88,000 1.6	142,500 1.8	309,000 2.2	555,000 2.6	920,000 2.8	1,3…
120	8,200 0.9	24,800 1.2	53,000 1.5	97,000 1.7	157,500 1.9	340,000 2.4	622,000 2.8	1,000,000 3.0	1,5…
140	9,000 1.0	26,800 1.3	57,800 1.6	105,000 1.8	171,000 2.0	370,000 2.6	675,000 3.0	1,100,000 3.3	1,6…
160	9,700 1.1	29,000 1.4	62,500 1.7	113,000 2.0	185,000 2.2	398,000 2.8	730,000 3.3	1,180,000 3.6	1,8…
180	10,300 1.2	30,600 1.5	66,000 1.8	120,000 2.2	197,000 2.4	423,000 3.0	778,000 3.5	1,260,000 3.8	1,9…
200	10,900 1.2	32,300 1.6	70,000 1.9	127,500 2.4	210,000 2.6	450,000 3.0	820,000 3.5	1,330,000 4.0	2,0…
230	11,800 1.3	35,000 1.7	76,000 2.0	137,000 2.4	225,000 2.6	482,000 3.5	890,000 4.0	1,440,000 4.5	2,2…
260	12,600 1.4	37,500 1.8	81,000 2.2	147,000 2.6	240,000 3.0	520,000 3.5	950,000 4.0	1,575,400 5.0	2,3…
300	13,600 1.5	40,300 2.0	87,300 2.4	158,000 2.8	260,000 3.3	560,000 3.8	1,020,000 4.5	1,670,000 5.0	2,5…
350	14,700 1.6	44,000 2.2	95,000 2.6	172,500 3.0	280,000 3.6	610,000 4.0	1,110,000 5.0	1,810,000 5.5	2,7…
400	15,800 1.7	47,200 2.4	102,000 2.8	185,000 3.3	300,000 3.8	653,000 4.5	1,200,000 5.0	1,950,000 6.0	2,9…
450	16,800 1.8	50,300 2.6	108,000 3.0	198,000 3.6	320,000 4.0	695,000 5.0	1,270,000 5.5	2,070,000 6.5	3,1…
500	18,000 2.0	53,300 2.6	115,000 3.3	210,000 3.8	340,000 4.0	735,000 5.0	1,350,000 6.0	2,200,000 7.0	3,5…
600	19,700 2.2	59,000 2.8	126,000 3.6	230,000 4.0	377,000 4.5	810,000 5.5	1,480,000 6.5	2,420,000 8.0	3,8…
700	21,400 2.4	64,000 3.0	137,000 3.8	250,000 4.5	410,000 5.0	880,000 6.0	1,620,000 7.0	2,620,000 8.0	[illegible]
800	23,000 2.6	68,300 3.3	147,000 4.0	270,000 5.0	440,000 5.5	950,000 6.5	1,750,000 8.0	2,800,000 9.0	4,… I
900	24,400 2.8	73,900 3.6	157,000 4.5	286,000 5.0	470,000 6.0	1,000,000 7.0	1,860,000 8.0	3,000,000 9.0	I II
1,000	26,100 3.0	78,000 3.8	167,000 5.0	301,000 5.5	500,000 6.0	1,075,000 8.0	1,960,000 9.0	3,200,000 10.0	I II
1,200	29,000 3.3	86,000 4.0	183,000 5.0	333,000 6.0	550,000 6.5	1,180,000 8.0	2,180,000 10.0	I II	
1,400	31,400 3.5	93,300 4.5	200,000 5.5	367,000 6.5	601,000 7.0	1,250,000 9.0	2,360,000 10.0	I II	
1,600	33,000 3.8	100,000 5.0	216,000 6.0	385,000 7.0	642,000 8.0	1,380,000 10.0	I II		
1,800	35,300 4.0	107,000 5.5	230,000 6.5	423,000 8.0	685,000 9.0	1,475,000 10.0	I II		
2,000	38,000 4.0	113,000 6.0	244,000 7.0	445,000 8.5	725,000 9.0	I II			
2,300	41,000 4.5	122,000 6.0	262,000 7.0	484,000 9.0	790,000 10.0	I II			
2,600	44,700 6.0	130,000 6.5	280,000 8.0	515,000 9.0	I II				
3,000	47,300 6.5	140,000 7.0	302,000 8.0	560,000 10.0	I II				
3,500	51,500 6.0	150,000 8.0	330,000 9.0	I II					
4,000	55,300 8.0	162,000 8.0	355,000 10.0	I II					

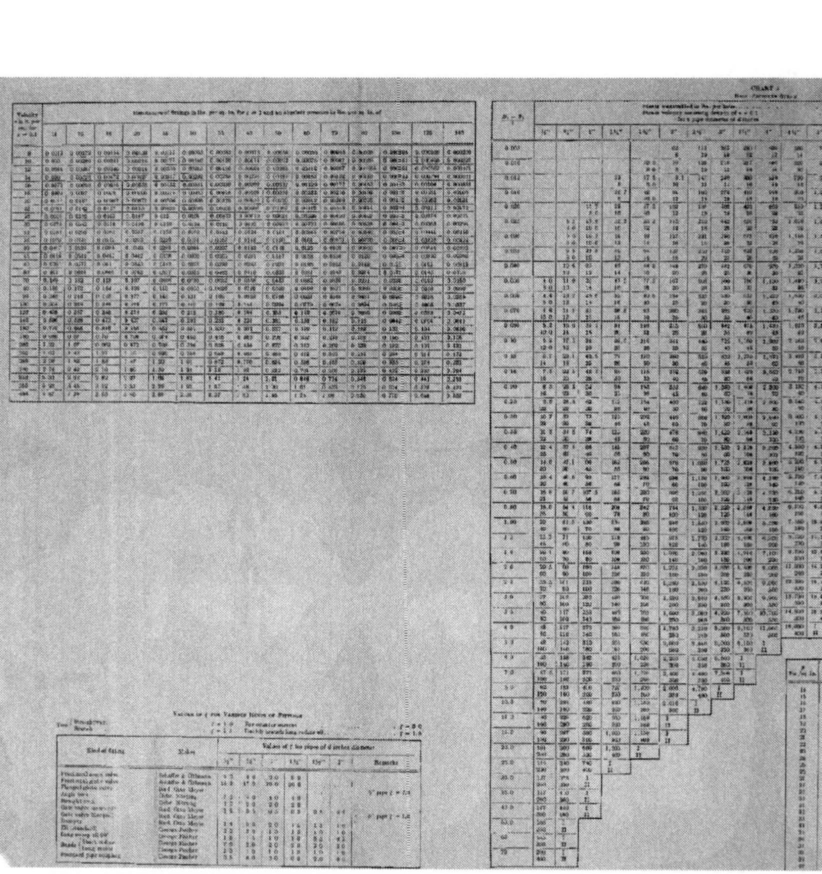

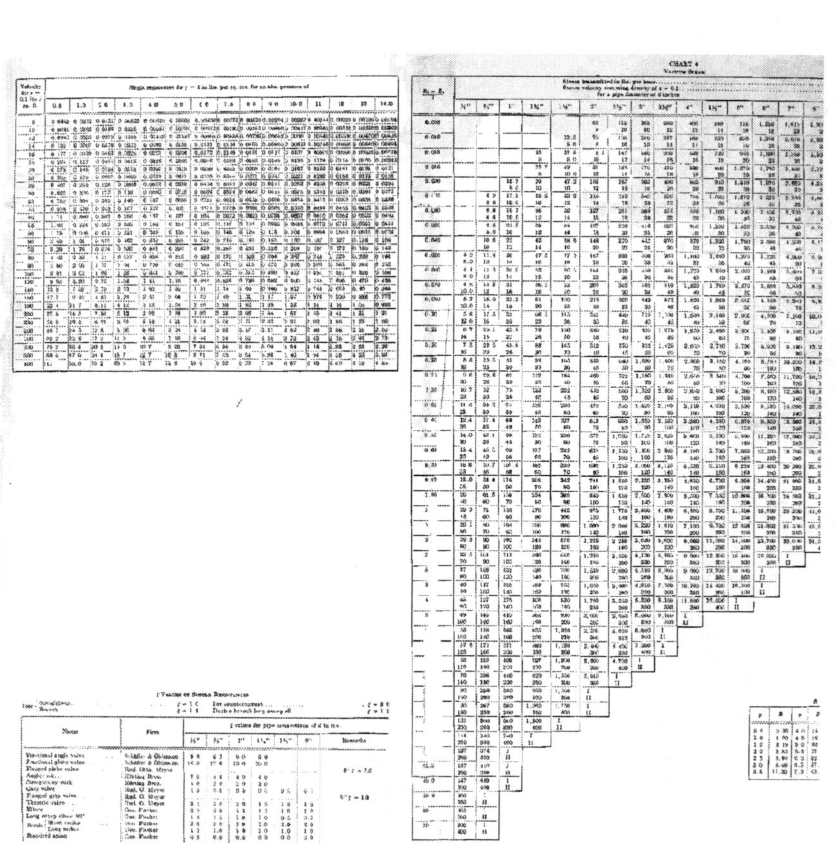

CHART 4

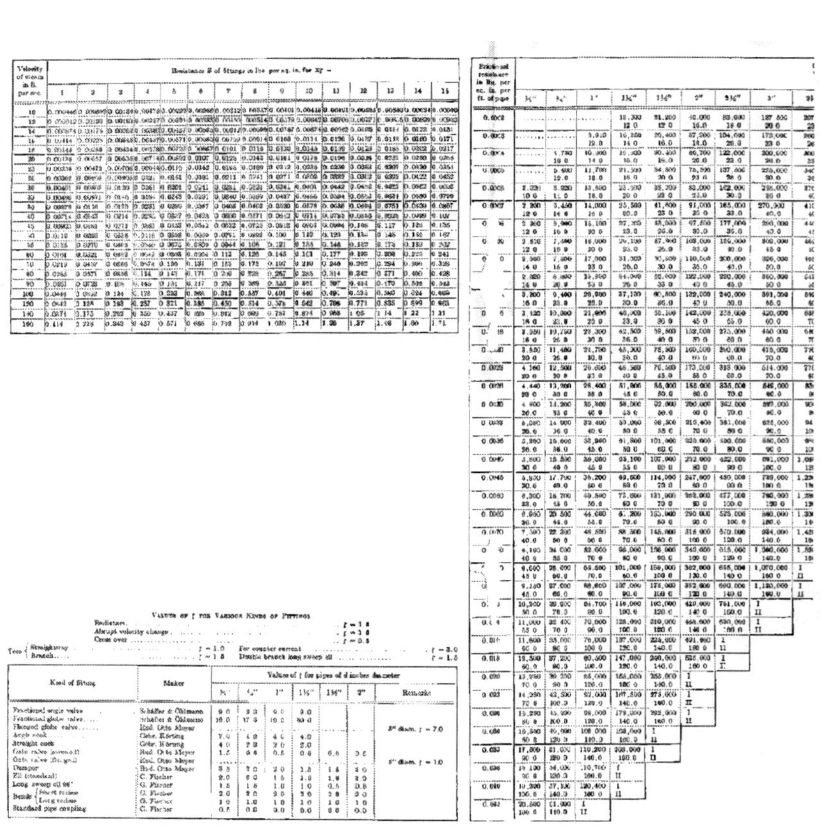

Velocity of steam in ft. per sec.	Resistance Z of fittings in lbs. per sq. in. for Σζ =														
	1	2	3	4	5	6	7	8	9	10	11	12	13	14	15
10	0.000446	0.000892	0.00134	0.00179	0.00223	0.00268	0.00312	0.00357	0.00401	0.00446	0.00491	0.00535	0.00580	0.00624	0.00669
12	0.000642	0.00128	0.00193	0.00257	0.00321	0.00385	0.00450	0.00514	0.00579	0.00642	0.00706	0.00771	0.00835	0.00899	0.00963
14	0.000874	0.00175	0.00262	0.00350	0.00437	0.00524	0.00612	0.00699	0.00787	0.00874	0.00962	0.0105	0.0114	0.0122	0.0131
16	0.00114	0.00228	0.00343	0.00457	0.00571	0.00685	0.00799	0.00914	0.0103	0.0114	0.0126	0.0137	0.0148	0.0160	0.0171
18	0.00144	0.00288	0.00434	0.00578	0.00723	0.00867	0.0101	0.0116	0.0130	0.0145	0.0159	0.0173	0.0188	0.0202	0.0217
20	0.00178	0.00357	0.00535	0.00714	0.00892	0.0107	0.0125	0.0143	0.0161	0.0178	0.0196	0.0214	0.0232	0.0250	0.0268
22	0.00216	0.00432	0.00648	0.00864	0.0110	0.0130	0.0151	0.0173	0.0194	0.0216	0.0238	0.0259	0.0281	0.0303	0.0324
26	0.00302	0.00604	0.00906	0.0121	0.0151	0.0181	0.0211	0.0241	0.0271	0.0302	0.0332	0.0362	0.0392	0.0422	0.0452
30	0.00401	0.00803	0.0120	0.0161	0.0201	0.0241	0.0281	0.0321	0.0361	0.0401	0.0442	0.0482	0.0522	0.0562	0.0602
33	0.00486	0.00971	0.0146	0.0194	0.0243	0.0291	0.0340	0.0389	0.0437	0.0486	0.0534	0.0583	0.0631	0.0680	0.0729
36	0.00578	0.0116	0.0173	0.0231	0.0289	0.0347	0.0405	0.0462	0.0520	0.0578	0.0636	0.0694	0.0751	0.0809	0.0867
40	0.00714	0.0143	0.0214	0.0285	0.0357	0.0428	0.0500	0.0571	0.0642	0.0714	0.0785	0.0856	0.0928	0.0999	0.107
45	0.00903	0.0181	0.0271	0.0361	0.0452	0.0542	0.0632	0.0723	0.0813	0.0903	0.0994	0.108	0.117	0.126	0.135
50	0.0112	0.0223	0.0335	0.0446	0.0558	0.0669	0.0781	0.0892	0.100	0.112	0.123	0.134	0.145	0.156	0.167
55	0.0135	0.0270	0.0405	0.0540	0.0675	0.0809	0.0944	0.108	0.121	0.135	0.148	0.162	0.175	0.189	0.202
60	0.0161	0.0321	0.0482	0.0642	0.0803	0.0964	0.112	0.128	0.145	0.161	0.177	0.193	0.209	0.225	0.241
70	0.0219	0.0437	0.0656	0.0874	0.109	0.131	0.153	0.175	0.197	0.219	0.240	0.262	0.284	0.306	0.328
80	0.0285	0.0571	0.0856	0.114	0.143	0.171	0.200	0.228	0.257	0.285	0.314	0.342	0.371	0.400	0.428
90	0.0361	0.0723	0.108	0.145	0.181	0.217	0.253	0.289	0.325	0.361	0.397	0.434	0.470	0.506	0.542
100	0.0446	0.0892	0.134	0.178	0.223	0.268	0.312	0.357	0.401	0.446	0.491	0.535	0.580	0.624	0.669
120	0.0642	0.128	0.193	0.257	0.321	0.385	0.450	0.514	0.578	0.642	0.706	0.771	0.835	0.899	0.963
140	0.0874	0.175	0.262	0.350	0.437	0.525	0.612	0.699	0.787	0.874	0.962	1.05	1.14	1.22	1.31
160	0.114	0.228	0.343	0.457	0.571	0.686	0.799	0.914	1.030	1.14	1.26	1.37	1.48	1.60	1.71

VALUES OF ζ FOR VARIOUS KINDS OF FITTINGS

Radiators ζ = 1.5
Abrupt velocity change ζ = 1.0
Cross over ζ = 0.5

Tee { Straightway ζ = 1.0 For counter current ζ = 3.0
Tee { Branch ζ = 1.5 Double branch long sweep ell ζ = 1.5

Kind of fitting	Maker	Values of ζ for pipes of d inches diameter						
		½"	¾"	1"	1¼"	1½"	2"	Remarks
Fractional angle valve	Schäffer & Ohlmann	9.0	9.0	6.0	3.0			
Fractional globe valve	Schäffer & Ohlmann	10.0	17.5	19.0	30.0			
Flanged globe valve	Rud. Otto Meyer							3" diam. ζ = 7.0
Angle cock	Gebr. Körting	7.0	4.0	4.0	4.0			
Straight cock	Gebr. Körting	4.0	2.0	2.0	2.0			
Gate valve (screwed)	Rud. Otto Meyer	1.5	0.5	0.5	0.5	0.5	0.5	
Gate valve (flanged)	Rud. Otto Meyer							5" diam. ζ = 1.0
Damper	Rud. Otto Meyer	3.5	2.0	2.0	1.5	1.5	1.0	
Ell (standard)	C. Fischer	2.0	2.0	1.5	1.0	1.0	1.0	
Long sweep ell 90°	G. Fischer	1.5	1.5	1.0	1.0	0.5	0.5	
Bends { Short radius	G. Fischer	2.0	2.0	2.0	2.0	2.0	2.0	
Bends { Long radius	G. Fischer	1.0	1.0	1.0	1.0	1.0	1.0	
Standard pipe coupling	G. Fischer	0.5	0.5	0.0	0.0	0.0	0.0	

Frictional resistance in lbs. per sq. in. per ft. of pipe	½"	¾"	1"	1¼"	1½"	2"	2½"	3"	3½"
0.0002	......	......	......	18,300 12.0	21,200 12.0	40,000 16.0	83,600 18.0	137,800 20.0	[illegible]
0.0003	......	......	3,910 12.0	16,250 14.0	26,400 16.0	57,200 18.0	104,000 20.0	172,000 23.0	[illegible]
0.0004	......	4,730 10.0	10,300 14.0	19,300 16.0	30,400 16.0	66,700 20.0	122,000 22.0	200,600 26.0	[illegible]
0.0005	......	5,400 12.0	11,700 18.0	21,500 18.0	34,600 20.0	75,800 23.0	137,600 28.0	235,000 30.0	[illegible]
0.0006	2,030 10.0	5,320 12.0	13,300 18.0	23,500 20.0	38,300 23.0	83,000 25.0	152,000 30.0	238,000 30.0	[illegible]
0.0007	2,200 12.0	3,450 14.0	14,000 16.0	25,580 20.0	41,600 23.0	91,000 25.0	165,000 33.0	270,000 40.0	[illegible]
[illegible]	2,360 12.0	5,900 16.0	15,100 20.0	27,300 23.0	45,000 26.0	97,800 30.0	177,000 35.0	280,000 40.0	[illegible]
[illegible]	2,510 12.0	7,400 16.0	16,000 20.0	29,100 25.0	47,900 26.0	103,000 33.0	189,000 30.0	300,000 45.0	[illegible]
[illegible]	2,380 14.0	7,800 16.0	17,000 23.0	31,000 26.0	50,600 30.0	110,000 35.0	200,000 40.0	326,000 50.0	[illegible]
[illegible]	2,820 14.0	6,800 20.0	18,300 23.0	34,000 26.0	55,000 33.0	122,000 40.0	220,000 45.0	360,000 50.0	[illegible]
[illegible]	3,200 16.0	9,400 23.0	20,200 25.0	37,100 30.0	60,800 35.0	132,000 40.0	240,000 50.0	391,000 55.0	[illegible]
[illegible]	3,420 18.0	10,000 23.0	21,600 25.0	40,000 33.0	55,100 30.0	142,000 45.0	238,000 55.0	420,000 60.0	[illegible]
[illegible]	3,580 18.0	10,750 26.0	23,300 30.0	42,500 36.0	59,800 40.0	152,000 50.0	275,000 50.0	460,000 60.0	[illegible]
[illegible]	3,850 20.0	11,480 26.0	24,700 30.0	45,300 36.0	78,500 40.0	160,000 50.0	290,000 60.0	415,000 70.0	[illegible]
0.0025	4,160 20.0	12,500 30.0	26,600 33.0	48,500 40.0	79,500 45.0	173,000 55.0	313,000 60.0	514,000 70.0	[illegible]
0.0028	4,440 20.0	13,200 30.0	28,400 38.0	51,800 45.0	85,000 50.0	185,000 60.0	335,000 70.0	546,000 80.0	[illegible]
0.0030	4,600 20.0	14,200 33.0	30,300 40.0	58,000 45.0	92,000 50.0	200,000 60.0	362,000 70.0	597,000 90.0	[illegible]
0.0033	5,060 26.0	14,900 36.0	32,400 40.0	59,000 50.0	98,500 55.0	210,400 70.0	381,000 80.0	625,000 90.0	[illegible]
0.0036	5,260 26.0	15,600 36.0	33,900 45.0	61,800 50.0	101,000 60.0	220,000 70.0	400,000 80.0	650,000 90.0	[illegible]
0.0040	5,800 30.0	16,500 40.0	38,000 45.0	65,100 55.0	107,000 60.0	232,000 80.0	422,000 90.0	691,000 100.0	[illegible]
0.0045	5,830 30.0	17,700 40.0	36,200 50.0	69,500 60.0	114,000 70.0	247,000 80.0	450,000 90.0	730,000 100.0	[illegible]
0.0050	6,300 33.0	18,700 45.0	40,500 50.0	73,000 60.0	121,000 70.0	262,000 80.0	477,000 100.0	780,000 120.0	[illegible]
0.0060	6,960 36.0	20,500 44.0	44,000 54.0	[illegible] 70.0	135,000 80.0	290,000 90.0	525,000 100.0	860,000 120.0	[illegible]
0.0070	7,300 40.0	22,300 50.0	48,500 60.0	88,500 70.0	145,000 80.0	316,000 100.0	570,000 120.0	934,000 140.0	[illegible]
[illegible]	8,100 40.0	24,000 55.0	52,600 70.0	95,000 80.0	156,000 90.0	340,000 100.0	615,000 120.0	1,000,000 140.0	[illegible]
[illegible]	8,600 45.0	25,000 60.0	56,500 70.0	101,000 80.0	166,000 100.0	362,000 120.0	655,000 140.0	1,070,000 160.0	I II
[illegible]	9,150 45.0	27,000 60.0	58,600 80.0	107,000 90.0	176,000 100.0	382,000 120.0	690,000 140.0	1,120,000 160.0	I II
[illegible]	10,300 50.0	29,900 70.0	64,700 80.0	116,000 100.0	190,000 120.0	420,000 140.0	761,000 160.0	I II	
[illegible]	11,000 55.0	32,400 70.0	70,000 90.0	128,000 100.0	210,000 120.0	466,000 140.0	830,000 160.0	I II	
0.016	11,800 60.0	35,000 80.0	76,000 100.0	137,000 120.0	224,000 140.0	491,000 160.0	I II		
0.018	12,500 60.0	37,200 80.0	80,500 100.0	147,000 120.0	240,000 140.0	525,000 160.0	I II		
0.020	13,250 70.0	39,300 90.0	85,000 120.0	156,000 120.0	258,000 140.0	I II			
0.022	14,250 70.0	42,500 100.0	92,000 120.0	167,500 140.0	275,000 160.0	I II			
0.026	15,250 80.0	45,900 100.0	98,000 120.0	178,000 140.0	292,000 160.0	I II			
0.028	16,500 80.0	49,000 120.0	106,000 140.0	193,000 160.0	I II				
0.032	17,800 90.0	51,000 120.0	110,200 140.0	203,000 160.0	I II				
0.036	18,100 90.0	54,000 120.0	110,700 160.0	I II					
0.040	19,200 100.0	57,100 140.0	120,400 160.0	I II					
0.042	20,500 100.0	61,000 110.0	I II						

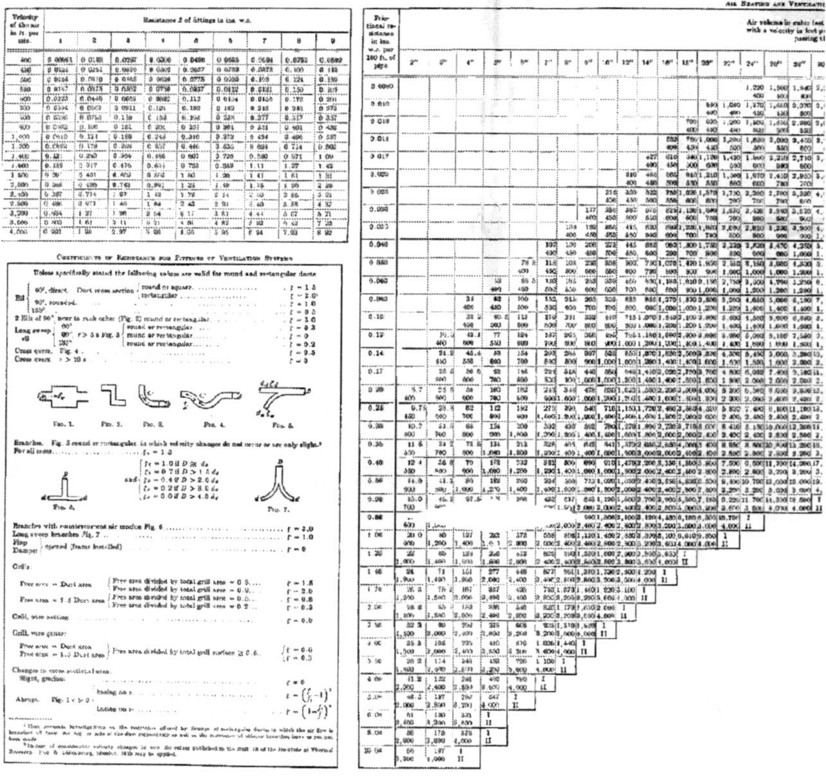

Velocity of the air in ft. per min.	Resistance Z of fittings in ins. w.c.								
	1	2	3	4	5	6	7	8	9
400	0.00661	0.0188	0.0297	0.0396	0.0496	0.0595	0.0694	0.0793	0.0892
450	0.0126	0.0251	0.0376	0.0502	0.0627	0.0753	0.0878	0.100	0.113
500	0.0155	0.0310	0.0465	0.0620	0.0775	0.0930	0.108	0.124	0.139
550	0.0187	0.0375	0.0562	0.0750	0.0937	0.112	0.131	0.150	0.169
600	0.0223	0.0446	0.0669	0.0892	0.112	0.134	0.156	0.178	0.201
700	0.0304	0.0608	0.0911	0.121	0.152	0.182	0.213	0.243	0.273
800	0.0396	0.0793	0.119	0.158	0.198	0.238	0.277	0.317	0.357
900	0.0502	0.100	0.151	0.201	0.251	0.301	0.351	0.401	0.452
1,000	0.0619	0.124	0.186	0.248	0.310	0.372	0.434	0.496	0.557
1,200	0.0892	0.178	0.268	0.357	0.446	0.535	0.624	0.714	0.803
1,400	0.121	0.243	0.364	0.486	0.607	0.728	0.850	0.971	1.09
1,600	0.159	0.317	0.476	0.634	0.793	0.952	1.11	1.27	1.43
1,800	0.201	0.401	0.602	0.803	1.00	1.20	1.41	1.61	1.81
2,000	0.248	0.496	0.743	0.991	1.24	1.49	1.73	1.98	2.23
2,400	0.357	0.714	1.07	1.43	1.78	2.14	2.50	2.85	3.21
2,800	0.486	0.971	1.46	1.94	2.43	2.91	3.40	3.88	4.37
3,200	0.634	1.27	1.90	2.54	3.17	3.81	4.44	5.07	5.71
3,600	0.803	1.61	2.41	3.21	4.01	4.82	5.62	6.42	7.22
4,000	0.991	1.98	2.97	3.96	4.96	5.95	6.94	7.93	8.92

COEFFICIENTS OF RESISTANCE FOR FITTINGS OF VENTILATION SYSTEMS

Unless specifically stated the following values are valid for round and rectangular ducts

Ell	90°, direct. Duct cross section	round or square	$\zeta = 1.5$
		rectangular	$\zeta = 2.0$
	90°, rounded		$\zeta = 1.0$
	135°		$\zeta = 0.5$
2 Ells of 90° near to each other (Fig. 2) round or rectangular			$\zeta = 3.0$
Long sweep ell	90°	round or rectangular	$\zeta = 0.3$
	90°, r > 5 s Fig. 3	round or rectangular	$\zeta = 0$
	135°	round or rectangular	$\zeta = 0.2$
Cross overs.	Fig. 4		$\zeta = 0.5$
Cross overs	r > 10 s		$\zeta = 0$

Fig. 1. Fig. 2. Fig. 3. Fig. 4. Fig. 5.

Branches. Fig. 5 round or rectangular, in which velocity changes do not occur or are only slight.[2]

For all cases .... $\zeta_c = 1.5$

and
$\zeta_b = 1.0$ if $D \cong d_b$
$\zeta_b = 0.7$ if $D > 1.5\ d_b$
$\zeta_b = 0.4$ if $D > 2.0\ d_b$
$\zeta_b = 0.2$ if $D > 3.0\ d_b$
$\zeta_b = 0.0$ if $D > 4.0\ d_b$

Fig. 6. Fig. 7.

Branches with countercurrent air motion Fig. 6	$\zeta = 3.0$
Long sweep branches Fig. 7	$\zeta = 1.0$
Flap Damper, opened (frame installed)	$\zeta = 0$

Grills:

Free area = Duct area	Free area divided by total grill area = 0.8	$\zeta = 1.5$
	Free area divided by total grill area = 0.6	$\zeta = 2.0$
Free area = 1.5 Duct area	Free area divided by total grill area = 0.8	$\zeta = 0.8$
	Free area divided by total grill area = 0.2	$\zeta = 0.3$
Grill, wire netting		$\zeta = 0.5$

Grill, wire gauze:

Free area = Duct area	Free area divided by total grill surface ≥ 0.6	$\zeta = 0.6$
Free area = 1.5 Duct area		$\zeta = 0.3$

Changes in cross sectional areas.

Slight, gradual		$\zeta = 0$
Abrupt. Fig. 1 < 5 · 0	leaving ...	$\zeta = \left(\frac{f}{f_1} - 1\right)^2$
	leaving ...	$\zeta = \left(1 - \frac{f}{f_1}\right)^2$

[1] [illegible]

[2] In case of considerable velocity changes in [illegible] the values published in the [illegible] may be applied.

CHART 5

AIR HEATING AND VENTILATI[ON]

Air volume in cubic feet [...] with a velocity in feet p[er ...] passing th[...]

Frictional resistance in ins. w.c. per 100 ft. of pipe	2"	3"	4"	5"	6"	7"	8"	9"	10"	12"	14"	16"	18"	20"	22"	24"	26"	28"	30"
0.0080																1,220 400	1,500 400	1,940 [illegible]	[illegible]
0.010														690 400	1,080 400	1,370 450	1,660 450	2,070 500	[illegible]
0.012													700 400	930 450	1,200 450	1,500 500	[illegible] 500	2,380 550	[illegible]
0.015												582 400	760 450	1,000 450	1,200 500	1,630 500	2,000 550	2,450 600	[illegible]
0.017											427 400	610 450	840 500	1,120 600	1,430 650	1,800 600	2,210 600	2,710 600	[illegible]
0.020										310 400	466 450	665 500	910 550	1,210 550	1,560 600	1,970 600	2,410 700	2,950 700	[illegible]
0.025									216 400	330 450	522 500	750 550	1,030 600	1,370 600	1,750 700	2,200 700	2,700 700	3,320 800	[illegible]
0.030								177 400	236 450	382 500	575 550	826 600	1,130 600	1,500 700	1,930 700	2,420 800	2,990 800	3,670 900	[illegible]
0.035							135 400	192 450	256 450	415 550	630 600	896 600	1,220 700	1,600 700	2,090 800	2,620 800	3,230 900	3,950 900	[illegible]
0.040						107 400	150 450	206 450	273 500	445 550	685 600	960 700	1,300 700	1,750 800	2,220 800	2,820 900	3,470 900	4,250 1,000	[illegible]
0.050	[illegible]																		
0.060	[illegible]																		
0.080	[illegible]																		
0.10	[illegible]																		
0.12	[illegible]																		
0.14	[illegible]																		
0.17	[illegible]																		
0.20	[illegible]																		
0.25	[illegible]																		
0.30	[illegible]																		
0.35	[illegible]																		
0.40	[illegible]																		
0.50	[illegible]																		
0.60	[illegible]																		
0.80	[illegible]																		
1.00	[illegible]																		
1.20	[illegible]																		
1.40	[illegible]																		
1.70	[illegible]																		
2.00	[illegible]																		
2.50	[illegible]																		
3.00	[illegible]																		
3.50	[illegible]																		
4.00	[illegible]																		
5.00	[illegible]																		
6.00	[illegible]																		
8.00	[illegible]																		
10.00	[illegible]																		

CHART 7

EQUIVALENT DIAMETERS d_e AND CROSSSECTIONAL AREAS FOR RECTANGULAR DUCTS*

I. Values of the equivalent diameter corresponding to the values of a and b in ins.

II. Values of $a \times b$ in sq. ft.

a →, b ↓

	2″	4″	6″	8″	10″	12″	14″	16″	20″	25″	30″	35″	40″	45″	50″	60″	70″	80″	90″	100″	
2″	2″	3	3	3	3	3	3	4	4	4	4	4	4	4	4	4	4	4	4	4	I
	0.0278	0.0556	0.0834	0.111	0.139	0.167	0.195	0.222	0.278	0.347	0.417	0.486	0.556	0.625	0.694	0.834	0.972	1.11	1.25	1.39	II
3″	2″	3	4	4	5	5	5	5	5	5	5	6	6	6	6	6	6	6	6	6	I
	0.0417	0.0834	0.125	0.167	0.209	0.250	0.292	0.334	0.417	0.521	0.626	0.729	0.834	0.938	1.04	1.25	1.46	1.67	1.88	2.09	II
4″	3	4	5	5	6	6	6	6	7	7	7	7	7	7	7	8	8	8	8	8	I
	0.0556	0.111	0.167	0.222	0.278	0.334	0.389	0.445	0.556	0.694	0.834	0.972	1.11	1.25	1.39	1.67	1.94	2.22	2.50	2.78	II
5″	3	4	5	6	7	7	7	8	8	8	9	9	9	9	9	9	9	9	9	10	I
	0.0694	0.139	0.209	0.278	0.348	0.417	0.486	0.556	0.694	0.868	1.04	1.22	1.39	1.56	1.74	2.09	2.43	2.78	3.12	3.48	II
6″	3	5	6	7	7	8	8	9	9	10	10	10	10	10	10	10	12	12	12	12	I
	0.0834	0.167	0.250	0.334	0.417	0.500	0.583	0.667	0.834	1.04	1.25	1.46	1.67	1.875	2.08	2.50	2.91	3.34	3.75	4.17	II
7″	3	5	6	7	8	9	9	10	10	10	12	12	12	12	12	12	12	12	12	14	I
	0.0972	0.194	0.292	0.389	0.486	0.583	0.680	0.778	0.972	1.22	1.46	1.70	1.94	2.19	2.43	2.92	3.40	3.89	4.38	4.86	II
8″	3	5	7	8	9	10	10	10	12	12	12	14	14	14	14	14	14	14	14	14	I
	0.111	0.222	0.333	0.444	0.555	0.666	0.778	0.888	1.11	1.39	1.67	1.95	2.22	2.50	2.78	3.33	3.89	4.44	5.00	5.55	II
9″	3	6	7	8	9	10	10	12	12	14	14	14	14	16	16	16	16	16	16	16	I
	0.125	0.250	0.375	0.500	0.625	0.750	0.875	1.00	1.25	1.56	1.88	2.19	2.50	2.81	3.12	3.75	4.37	5.00	5.62	6.25	II
10″	3	6	8	9	10	10	12	12	14	14	14	16	16	16	16	18	18	18	18	18	I
	0.139	0.278	0.417	0.556	0.694	0.834	0.972	1.11	1.39	1.74	2.09	2.43	2.78	3.13	3.47	4.17	4.86	5.50	6.25	6.94	II
12″	3	6	8	10	10	12	12	14	16	16	18	18	18	18	20	20	20	20	22	22	I
	0.167	0.334	0.500	0.668	0.834	1.00	1.17	1.31	1.67	2.08	2.51	2.92	3.34	3.75	4.16	5.00	5.83	6.68	7.5	8.34	II
14″	4	6	8	10	12	12	14	14	14	18	20	20	20	22	22	22	24	24	24	24	I
	0.194	0.388	0.582	0.778	0.972	1.16	1.36	1.55	1.94	2.43	2.91	3.40	3.88	4.38	4.86	5.83	6.80	7.78	8.75	9.72	II
16″	4	6	9	10	12	14	14	16	18	20	20	22	22	24	24	26	26	26	28	28	I
	0.222	0.444	0.666	0.888	1.11	1.33	1.56	1.78	2.22	2.78	3.33	3.89	4.44	5.00	5.56	6.66	7.78	8.88	10.00	11.1	II
18″	4	7	9	12	12	14	16	16	18	20	22	24	24	26	26	28	28	30	30	30	I
	0.250	0.500	0.750	1.00	1.25	1.50	1.75	2.00	2.50	3.13	3.75	4.38	5.00	5.62	6.25	7.50	8.75	10.0	11.2	12.5	II
20″	4	7	9	12	14	16	16	18	20	22	24	26	26	28	28	30	32	32	32	34	I
	0.278	0.556	0.834	1.11	1.39	1.67	1.95	2.22	2.78	3.47	4.17	4.86	5.56	6.25	6.94	8.34	9.72	11.1	12.5	13.9	II
22″	4	7	9	12	14	16	18	18	20	24	26	28	28	30	30	32	34	34	36	36	I
	0.306	0.612	0.918	1.22	1.53	1.84	2.14	2.45	3.06	3.82	4.59	5.35	6.12	6.88	7.64	9.18	10.7	12.2	13.8	15.3	II
24″	4	7	10	12	14	16	18	20	22	24	26	28	30	32	32	34	36	36	38	38	I
	0.333	0.666	1.00	1.33	1.67	2.00	2.33	2.66	3.33	4.17	5.00	5.84	6.66	7.50	8.34	10.0	11.7	13.3	15.0	16.7	II
26″	4	7	10	12	14	16	18	20	22	26	28	30	32	32	34	36	38	40	40	40	I
	0.361	0.722	1.08	1.44	1.81	2.17	2.53	2.89	3.61	4.51	5.42	6.32	7.22	8.12	9.02	10.8	12.6	14.4	16.2	18.1	II
28″	4	7	10	12	14	16	18	20	24	26	28	32	32	34	36	38	40	40	45	45	I
	0.389	0.778	1.17	1.56	1.95	2.33	2.72	3.11	3.89	4.86	5.84	6.80	7.78	8.75	9.72	11.7	13.6	15.6	17.5	19.5	II
30″	4	7	10	12	16	18	20	20	24	28	30	32	34	36	38	40	40	45	45	45	I
	0.417	0.834	1.25	1.67	2.09	2.50	2.92	3.34	4.17	5.21	6.26	7.29	8.34	9.38	10.4	12.5	14.6	16.7	18.7	20.9	II
32″	4	7	10	12	16	18	20	22	24	28	30	34	36	38	40	40	45	45	45	50	I
	0.445	0.890	1.34	1.78	2.22	2.67	3.11	3.56	4.45	5.56	6.67	7.78	8.90	10.0	11.1	13.4	15.6	17.8	20.0	22.2	II
34″	4	7	10	12	16	18	20	22	26	28	32	34	36	38	40	45	45	50	50	50	I
	0.472	0.944	1.42	1.89	2.36	2.83	3.30	3.78	4.72	5.90	7.08	8.26	9.44	10.6	11.8	14.2	16.5	18.9	21.3	23.6	II
36″	4	7	10	14	16	18	20	22	26	30	32	36	38	40	40	45	50	50	50	55	I
	0.500	1.00	1.50	2.00	2.50	3.00	3.50	4.00	5.00	6.25	7.50	8.75	10.0	11.3	12.5	15.0	17.5	20.00	22.5	25.0	II
38″	4	7	10	14	16	18	20	22	26	30	34	36	38	40	45	45	50	50	55	55	I
	0.528	1.06	1.58	2.11	2.64	3.17	3.69	4.22	5.28	6.60	7.92	9.23	10.6	11.9	13.2	15.8	18.5	21.1	23.8	26.4	II
40″	4	7	10	14	16	18	20	22	26	30	34	38	40	40	45	50	50	55	55	55	I
	0.556	1.11	1.67	2.22	2.78	3.34	3.89	4.45	5.56	6.94	8.34	9.72	11.1	12.5	13.9	16.7	19.4	22.2	25.0	27.8	II
45″	4	7	10	14	16	18	22	24	28	32	36	40	40	45	45	50	55	60	60	60	I
	0.625	1.25	1.88	2.50	3.13	3.75	4.37	5.00	6.25	7.82	9.38	10.9	12.5	14.1	15.6	18.8	21.9	25.0	28.1	31.3	II
50″	4	7	10	14	16	20	22	24	28	34	38	40	45	45	50	55	60	60	65	65	I
	0.694	1.39	2.09	2.78	3.48	4.17	4.86	5.56	6.94	8.68	10.4	12.2	13.9	15.6	17.4	20.9	24.3	27.8	31.2	34.8	II
55″	4	7	10	14	16	20	22	24	30	34	38	45	45	50	50	55	60	65	70	70	I
	0.764	1.53	2.29	3.06	3.82	4.58	5.34	6.11	7.64	9.55	11.5	13.4	15.3	17.2	19.1	22.9	26.7	30.6	34.4	38.2	II
60″	4	8	10	14	18	20	22	26	30	36	40	45	50	50	55	60	65	70	70	75	I
	0.834	1.67	2.50	3.34	4.17	5.00	5.83	6.67	8.34	10.4	12.5	14.6	16.7	18.8	20.8	25.0	29.1	33.4	37.5	41.7	II
65″	4	8	10	14	18	20	24	26	30	36	40	45	50	55	55	60	65	70	75	80	I
	0.903	1.81	2.71	3.61	4.52	5.42	6.32	7.22	9.03	11.3	13.6	15.8	18.1	20.3	22.6	27.1	31.6	36.1	40.6	45.2	II
70″	4	8	12	14	18	20	24	26	32	36	40	45	50	55	60	65	70	75	80	80	I
	0.972	1.94	2.92	3.89	4.86	5.83	6.80	7.78	9.72	12.2	14.6	17.0	19.4	21.0	24.3	29.2	34.0	38.9	43.8	48.6	II
75″	4	8	12	14	18	20	24	26	32	38	45	50	50	55	60	65	70	75	80	90	I
	1.04	2.08	3.12	4.16	5.20	6.24	7.29	8.32	10.4	13.0	15.6	18.2	20.8	23.4	26.0	31.2	36.4	41.6	46.8	52.0	II
80″	4	8	12	14	18	20	24	26	32	38	45	50	55	60	60	70	75	80	80	90	I
	1.11	2.22	3.33	4.44	5.55	6.66	7.78	8.88	11.1	13.9	16.7	19.4	22.2	25.0	27.8	33.3	38.9	44.4	50.0	55.5	II
90″	4	8	12	14	18	22	24	28	32	40	45	50	55	60	65	70	80	80	90	90	I
	1.25	2.50	3.75	5.00	6.25	7.50	8.75	10.0	12.5	15.6	18.8	21.9	25.0	28.1	31.3	37.5	43.8	50.0	56.2	62.5	II
100″	4	8	12	14	18	22	24	28	34	40	45	50	55	60	65	75	80	90	90	100	I
	1.39	2.78	4.17	5.56	6.94	8.34	9.72	11.1	13.9	17.4	20.9	24.3	27.8	31.3	34.7	41.7	48.6	55.6	62.5	69.4	II

* This table is based on the equivalent velocity and frictional resistance for the rectangular and equivalent round ducts. For square ducts with side of length a the equivalent diameter is a.

CPSIA information can be obtained
at www.ICGtesting.com
Printed in the USA
LVHW081457080323
741196LV00003B/24

9 781017 038835